Carcinogens, DNA Damage and Cancer Risk

Mechanisms of Chemical Carcinogenesis

Carcinogens, DNA Damage and Cancer Risk

Mechanisms of Chemical Carcinogenesis

Editor

Miriam C. Poirier

Bethesda, MD, USA

World Scientific

NEW JERSEY • LONDON • SINGAPORE • BEIJING • SHANGHAI • HONG KONG • TAIPEI • CHENNAI • TOKYO

Published by

World Scientific Publishing Co. Pte. Ltd.
5 Toh Tuck Link, Singapore 596224
USA office: 27 Warren Street, Suite 401-402, Hackensack, NJ 07601
UK office: 57 Shelton Street, Covent Garden, London WC2H 9HE

Library of Congress Cataloging-in-Publication Data
Names: Poirier, M. C. (Miriam C.), editor.
Title: Carcinogens, DNA damage and cancer risk : mechanisms of chemical carcinogenesis / editor, Miriam C. Poirier.
Description: New Jersey : World Scientific, 2018. | Includes bibliographical references.
Identifiers: LCCN 2018023920 | ISBN 9789813237193 (hardcover : alk. paper)
Subjects: | MESH: Carcinogenesis | Carcinogens--chemical synthesis | DNA Damage--drug effects | Neoplasms--chemically induced
Classification: LCC RC268.5 | NLM QZ 203 | DDC 616.99/4071--dc23
LC record available at https://lccn.loc.gov/2018023920

British Library Cataloguing-in-Publication Data
A catalogue record for this book is available from the British Library.

For any available supplementary material, please visit
https://www.worldscientific.com/worldscibooks/10.1142/10908#t=suppl

Typeset by Stallion Press
Email: enquiries@stallionpress.com

Printed in Singapore

Preface

Human exposures to chemicals that cause cancer (carcinogens) are typically chronic, and occur *via* occupation, environmental pollution, diet, lifestyle choices, pharmaceutical use, or combinations thereof. Currently there are approximately 100 chemicals that have been designated carcinogenic to humans. Carcinogen-DNA damage, either in the form of stable covalent DNA-bound products (DNA adducts), or DNA fragmentation with strand breaks and aneuploidy, is a major mechanism contributing to the mutagenesis essential for cancer initiation. This book explores DNA damage in humans as an essential link between chemical carcinogen exposure and cancer, and a critical biomarker that necessary although not sufficient for tumor formation. In addition, chapters focused on cancer chemoprevention, epigenetics and adductomics point to important new strategies that may help reduce the effects of carcinogen exposures.

In the future one can reasonably expect that new chemicals and/or sources of exposure will be constantly on the horizon. A recent example is the World Trade Center disaster of 2001, where, in a very short time (< 15 years), increased rates of thyroid and bladder cancers have been reported among the rescue and recovery workers. Furthermore, as chemicals become more numerous in our environment, combinations of carcinogens, or carcinogens combined with infection/inflammation, may significantly enhance cancer risk, compared with single exposures.

The purpose of this volume is to share knowledge that has accumulated for the last 40 years documenting widespread DNA adduct formation in human tissues and ascertaining associations between DNA adduct formation and human cancer risk. It is our anticipation that this information will assist in addressing the effects of spills, environmental disasters, and chemicals newly-discovered to be carcinogenic, all of which will surely be encountered by mankind in the future. In addition to its use as a teaching tool for students, we envision this book as providing guidance whereby newly-emerging exposures may be understood in the context of past experience to inform experimental approaches and decision-making in the years to come.

Miriam C. Poirier
August 2018

Disclaimer

Contributions provided by Miriam C. Poirier to the Work were written in a personal capacity and do not necessarily represent the opinions or endorsement of the National Institutes of Health, Department of Health and Human Services, or the Federal Government. Notwithstanding any other provisions of the Agreement, Miriam C. Poirier represents and warrants that rights of her employer, the National Institutes of Health (NIH), shall not be abrogated by any commitment or obligation she has incurred under the Agreement. The Publisher recognizes that as a Federal employee, she is bound by federal laws, regulations and policies, including those governing conflict of interest, standards of conduct, and intellectual property, including 45 CFR Part 7, and she agrees that her performance of services hereunder shall not violate any of the forgoing and that she shall take all appropriate action to ensure her compliance.

Contents

Chapter 1

The Role of DNA Damage in Cancers Caused by Chemicals

Miriam C. Poirier

Abstract

Many cancers caused by chemical carcinogens do so through a mutagenic mechanism initiated by the formation of carcinogen–DNA damage. Some chemical carcinogens can damage DNA with no metabolic intervention, and others must be biotransformed. DNA adduct formation is considered necessary but not sufficient for tumorigenesis, as dosimetry for DNA adducts and tumorigenesis has been amply demonstrated in experimental models. Once methods became available to measure DNA adducts in humans, studies showed widespread distribution of DNA adducts in human tissues, dosimetry for DNA adduct formation in circumstances where exposure was accurately known, and consistent positive associations between DNA adduct formation and human cancer risk. Case–control and nested case–control epidemiological studies, examining DNA adducts in individuals exposed to a single carcinogen or carcinogenic mixtures, have shown, for many cancers, that individuals with the highest DNA adduct levels are also those with the highest cancer risk. The modest nature of these associations (odds ratio or relative risk between 1.25 and 9.10) indicates that, although DNA damage is essential, other events typically contribute to human cancer etiology.

1. Introduction

Most human exposures to chemicals that cause cancer (*carcinogens* — see Glossary for definitions of italicized words) occur as a result of occupation, pollution in the ambient environment, lifestyle choices, pharmaceutical use, or combinations thereof. For more than 200 years human cancer induction has been associated with exposures to chemicals or chemical mixtures. The International Agency for Research on Cancer (IARC, World Health Organization, Lyon, France) has identified approximately 100 Group 1 chemicals, defined as chemicals which, after evaluation of all the available data, are designated as carcinogenic to humans (https://www.cancer.org/cancer/cancer-causes/general-info/known-and-probable-human-carcinogens.html). In the future, one can reasonably expect that new chemicals and/or sources of exposure will be constantly on the horizon. Furthermore, combinations of carcinogenic exposures may be more dangerous than individual single exposures. A recent example is the World Trade Center disaster of 2001, where, less than 15 years later, increased rates of thyroid and bladder cancers were reported in World Trade Center rescue and recovery workers, compared to matched residents of different states [1]. The purpose of this volume is to share knowledge that has been accumulated for many years concerning the role played by DNA damage and mutagenic events in human cancer induction. It is our anticipation that this information may assist in addressing the effects of spills, environmental disasters, and chemicals newly discovered to be carcinogenic, all of which will surely be encountered by mankind in the future.

In 1775, Sir Percival Pott, a London surgeon, published an article linking soot exposure with scrotal cancers occurring in high incidence in chimney sweeps [2]. His diagnosis that the etiology of these cancers was related to chimney soot exposure, initiated the field of occupational cancer epidemiology. He also observed that young men who spent their early years as chimney sweeps were susceptible to scrotal cancers appearing much later in life. Overall, his studies were among the first to show that human tumors arise as a result of long-term chronic carcinogen exposure, that tumors may be induced by irreversible events occurring early in life, and that reduction in exposure decreases risk of tumor formation.

Since those early years, scientific investigations have revealed that most cancer-causing chemicals, or chemical carcinogens, act through

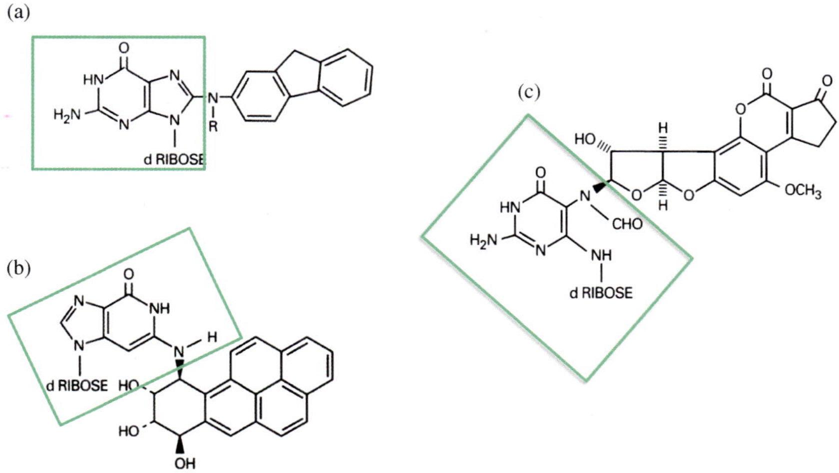

Figure 1. Molecular structures of stable carcinogen–DNA adducts formed by covalent binding with deoxyguanosine (shown in boxes) in DNA. (a) *N*-(deoxyguanosin-8-yl)-2-(amino)fluorene (dG-C8-AF, R=H or dG-C8- AAF, R=acetyl), formed when *N*-hydroxyl-aminofluorene reacts with the C8 position of deoxyguanine. (b) (7R)-N^2-(10-[7β,8α,9α-trihydroxy-7,8,9,10-tetrahydro-benzo[*a*]pyrene]-yl)-deoxyguanosine (BPdG) is the major carcinogenic adduct formed between metabolites of benzo[*a*]pyrene and the exocyclic amino group of deoxyguanosine. (c) 8,9-dihydro-8-(2,6-diamino-6-formamido-4-oxo-3,4–dihydropyrimid-5-yl-amino-guanine) (AFB_1-N^7-G) is the major ring-opened adduct of aflatoxin B_1 formed with the C8 position of the deoxyguanine imidazole ring.

DNA damaging or *genotoxic* mechanisms [3]. For example, of the hundreds of polycyclic aromatic hydrocarbons (PAHs), a major class of chemical found in soot, a small fraction are currently considered to be carcinogenic [4, 5], and one characteristic that differentiates between carcinogenic and noncarcinogenic PAHs is *biotransformation* to intermediates that bind to DNA. Examples of binding of a chemical to DNA, or the production of covalent addition products known as DNA adducts, are shown in Figure 1 [6]. Most cells can remove DNA adducts using *DNA repair* mechanisms, which both arrest DNA replication and remove the DNA damage [7]. However, unrepaired DNA adducts which remain during replication, may direct insertion of a wrong base, with subsequent formation of a *mutation*, or heritable change in the DNA sequence [8, 9]. Some mutations are of no consequence to the cell, but others, including

加合物

those that increase proliferation (oncogenes), reduce DNA repair capacity, and/or reduce cancer protective mechanisms (tumor suppressor genes), may play critical roles in processes leading to tumorigenesis [10, 11], the essence of which is the heritable loss of growth control.

2. Studies in Experimental Models

In the earliest days of cancer research, it was often the discovery of human cancers [2] that led the researchers to model the same exposure and disease in animals. Tumor studies in mice and rats [12] were used to predict which chemicals might be human carcinogens, and to understand mechanisms of metabolism and other events underlying *tumor incidence*. These studies demonstrated dosimetry between carcinogen exposure level and tumor incidence [13], often required lifetime exposures, and revealed that cancers were specific for certain organs termed the *target organs*. The species examined (rat, mouse, and guinea pig), the route of exposure (oral, topical, intraperitoneal, and transplacental), and the course of exposure (weeks, months, and lifetime) were all shown to impact the tumor incidence and the target organ. However, a chemical that induced tumors in any experimental model was deemed cause for concern.

Mechanism studies showed that many chemical carcinogens, characterized as such in experimental tumor models, not only formed DNA adducts in target organs for tumorigenesis, but also formed mutations in those same target organs [14, 15]. Mechanism studies lead to the hypothesis that chemical carcinogens act by binding to DNA, inducing incorporation of a wrong base during DNA replication, and fixing that mutation so it becomes a permanent heritable trait. Early validation was found in cell culture studies, which revealed strong correlations between dose of carcinogen, formation of DNA adducts, rate of mutagenesis, and *neoplastic transformation* [16].

Studies in experimental models have revealed that: chronic administration of chemical carcinogens at different dose levels show dose-related increases in DNA adduct formation and tumor formation in target tissues [17]; DNA adduct formation may directly correlate with mutagenesis [18]; blocking the ability to remove/repair DNA adducts increases cancer risk [19]; and, concomitant events (inflammation, proliferation, and infection) may enhance the rate at which tumors appear [20]. Furthermore, in mouse skin studies, the mutating or *initiating* agent could be given at a dose insufficient to produce tumors alone, but frequent epidermal application of an inflammatory *promoting* agent over the lifetime of the animal would result in skin tumor growth [12].

3. DNA Damage in Humans: Methodologies

While experimental models revealed the importance of DNA damage in chemical carcinogenesis, there was little solid evidence that human cancer etiology was driven by similar mechanistic events. A mutagenic mode of action was hypothesized for most chemical carcinogens, but before about 1980 there were no methods able to measure carcinogen-induced DNA damage in human tissues. The administration and tracking of radiolabeled compounds, the classic approach used in rodents, was clearly not appropriate for use in human subjects. Without the development of new methodologies there could be no direct evidence that the paradigm for genotoxic mechanisms of chemical carcinogenesis, so elegantly elucidated in mice, existed in men.

To answer this need, multiple different methods were developed. Using an antibody-based approach for the determination of carcinogen–DNA adducts in human tissues [21], early studies revealed the presence of DNA adducts of carcinogenic PAHs in humans [22]. Over time, many different antisera were raised against either DNA modified to at least 1% by a carcinogen, or a carcinogen–DNA adduct covalently bound to a carrier protein. These antisera typically cross-reacted with a single DNA adduct, or a family of chemically similar DNA adducts [6], but not with unmodified DNA, the carcinogen alone, or DNA modified with chemically different carcinogens. Because antibody recognition is three dimensional, for some classes of chemicals multiple structurally-similar DNA adducts are recognized by the same antiserum. For example, for the PAHs mentioned earlier, antiserum raised against DNA modified with benzo[a]pyrene (BP) has been shown to cross-react with DNA modified by at least eight other carcinogenic PAHs [23, 24]. Therefore, the evaluation of human samples with this antiserum yields evidence of a family of carcinogenic PAH–DNA adducts, but does not characterize the individual adducts. To quantify DNA adducts in biological samples, carcinogen–DNA antisera are most often used in competitive enzyme-linked immunosorbent assays (ELISAs) [6]. However, they can also be used with immunohistochemistry (IHC) to localize chemical carcinogens within the nuclei of specific cells [23, 25].

^{32}P-postlabeling [26], a frequently used method, is highly sensitive for chemically stable DNA damage, and is particularly appealing for molecular epidemiology studies because it may require ≤1 μg of DNA. The ^{32}P-postlabeling method involves hydrolysis of DNA to 3′-nucleotides, phosphorylation at the 5′ position with ^{32}P, and separation of normal nucleotides from adducted nucleotides by thin-layer or some other form of chromatography. This method detects DNA adducts induced by a broad range of high-molecular-weight chemical carcinogens including PAHs, aromatic amines, nitroaromatics, heterocyclic amines, tamoxifen, and hormones. In an animal model where exposure is to a single carcinogen, the adducts formed are easily identifiable with standards. However, in human samples where the exposures are unknown, chromatography reveals multiple components and the chemical nature of each adduct is unknown without additional analyses. Because the structures of DNA adducts detected in human samples by ^{32}P-postlabeling are largely obscure, they are termed *bulky* and/or hydrophobic DNA adducts.

Given the limitations of the antibody-based methods and ^{32}P-postlabeling, early efforts to identify DNA adducts in human tissues combined techniques to prepurify biological sample DNA, with subsequent characterization of specific chemical structures by mass spectrometry (MS) [27]. In the 1980s, early MS-based methods were not sufficiently sensitive for most human studies; however, with recent developments in the instrumentation, direct characterization of human DNA adduct structures has become feasible using 1–5 μg of DNA [28]. MS-based methodologies are considered the gold standard for DNA adduct characterization, and are now sufficiently sensitive and available [29] to feature prominently in molecular dosimetry and molecular epidemiology studies. Furthermore, a new promising MS-based application is the use of DNA extracted from formalin-fixed paraffin-embedded tissues. It is possible to extract good quality–adducted DNA from formalin-fixed tissues, making available DNA adduct analyses in many human samples from archived banks of paraffin blocks [30].

Accelerator mass spectrometry (AMS) [31], the most sensitive approach for the determination of human DNA adducts, requires exposure to very low levels of radiolabeled chemical. The technology is similar to that used for carbon dating, and is able to measure DNA adducts in the range of $1/10^{12}$ nucleotides. The levels of radioactivity administered to subjects are considered too low to cause biological harm, and human protocols are typically approved by Institutional Review Boards. However, the instrumentation is complex, expensive, and not widely available. Furthermore, because it relies on radiolabeled tags AMS cannot detail DNA adduct structure. However, it is an elegant approach to the determination of interindividual variability in disposition of a known chemical carcinogen.

4. DNA Damage in Humans: Dosimetry

Studies in animal models have indicated that both DNA adduct formation and tumorigenesis show dosimetry with level of chronic carcinogen exposure [13]. Furthermore, if the extent of DNA adduct formation can be reduced in a target tissue, in the presence of unchanging carcinogen exposure, there is typically a reduction in tumor incidence [32]. Experimental model investigations have revealed that DNA adducts are necessary but not sufficient for tumorigenesis. However, in the absence of methods able to evaluate DNA adducts in humans, there was a considerable gap in our ability to extrapolate cancer risk from animal models.

Currently, based on data generated by the methods discussed earlier, there is a substantial body of work documenting DNA adduct formation in human tissues. One of the earliest challenges in the endeavor to evaluate the role played by DNA damage in human carcinogenesis was to establish dosimetry and thereby validate the relationship between dose and DNA adduct formation. Exposure to ubiquitous environmental carcinogens, such as PAHs, is difficult to quantify because accurate dosimetry requires personal monitoring with cumbersome instrumentation. However, one environmental carcinogen, aflatoxin B_1 (AFB_1), a secondary metabolite of Aspergillus mold growth in corn and peanuts, forms an adduct that is released chemically from DNA and can be monitored because it is excreted in the urine [32]. Dosimetry for AFB_1–DNA adducts was shown to be linear in a study performed in China, in which subjects

Table 1. Representative molecular dosimetry studies: comparison of chemical dosimetry with DNA adduct formation in human tissues

Chemical exposure	Cohort	Linear dosimetry[a]	Assay	Tissue	Reference
Dietary aflatoxin	Qidong China	r^2 0.6 for [a]AFB_1 intake versus AFB-N^7-Gua excretion	[b]Immunoaffinity/ HPLC	Urine	33
Cisplatin/ carboplatin chemotherapy	Ovarian cancer patients	r^2 0.8 for dose versus[a] Pt-GG & Pt-AG	[b]ELISA	Blood	34
Procarbazine/ dacarbazine chemotherapy	Leukemia/ melanoma patients	Dose versus[a] N7-me-dG and O^6-me-dG linearity	ELISA	Blood	36, 37

[a]AFB_1, aflatoxin B_1; AFB_1-N^7-Gua, 8,9-dihydro-8-(2,6-diamine-6-formamido-4-oxo-3,4-dihydropyramid-5-yl-amino-guanine; Pt-GG, *cis*-diammineplatinum-N^7-d[GpG]; Pt-AG, *cis*-diammineplatinum-N^7-d[ApG]; N7-me-dG, N7-methyl-deoxyguanine; O^6-me-dG, O^6-methyl-deoxyguanine.
[b]Immunoaffinity chromatography combined with high performance liquid chromatography (HPLC). ELISA, enzyme-linked immunosorbent assay.

spent several weeks eating weighed amounts of food with known AFB_1 content, and their urine was analyzed daily for AFB_1–DNA [33] (Table 1). This was one of the earliest studies that validated the implied relationship between carcinogen dose and DNA adduct formation in humans.

Unlike environmental exposures, during administration of chemotherapy the cumulative dose is precisely known, the subjects are generally available to provide blood samples, and there is the added advantage that it is possible to evaluate correlations with therapy success or failure (see Chapter by S. Kyrtopoulos.). One such example is platinum drug–based therapy given for ovarian cancer. In various studies, blood cell DNA adducts of cisplatin or carboplatin (Table 1) showed a linear increase with cumulative dose of chemotherapy [34]. Furthermore, the most favorable clinical responses occurred in patients having the highest platinum–DNA adduct levels [35]. In other studies (Table 1), dose linearity for the N7-methyl-deoxyguanosine (N7-me-dG) and O^6-methyl-deoxyguanosine

(O^6-me-dG) adducts was observed in blood cell DNA of patients given procarbazine [36] and dacarbazine [37] for leukemia and melanoma. Taken together, these and other important studies demonstrated that humans, like mice, form DNA adducts in a dose-responsive fashion [38].

5. DNA Damage in Humans: Relationship to Cancer Risk

The presence of an adduct in human DNA clearly indicates that exposure to the parent compound has occurred. However, to evaluate the impact of DNA adducts on human cancer risk, it is necessary to invoke epidemiology. Two of the most common types of epidemiologic study designs used to validate biomarkers are the case–control study and the prospective nested case–control study. In the case–control study a biological sample is obtained on one occasion from cancer cases, and matched controls who do not have cancer, and all are assayed simultaneously for the biomarker DNA adduct formation. It is possible, then, to compare DNA adduct levels in the cases and the controls.

The prospective nested case–control study design is more informative, but also more difficult. The first requirement is collection and storage of samples (blood, urine, or any other human organ) from thousands of individuals who are at risk for a particular type of cancer. Then, after waiting some years for a sufficient number of cancer cases to occur, matched controls are chosen from the tissue bank, samples from cases and controls are retrieved, and all are assayed for the biomarker of choice, in this case DNA adduct formation. Ultimately, it is possible to compare DNA adducts in cancers cases with those in the matched controls.

With both study designs, the cancer rates in the individuals positive for the DNA adduct biomarker are compared to cancer rates in individuals negative for the biomarker, and the results are presented as relative risk (RR) or odds ratio (OR) for cancer in the biomarker positive group. Additionally, the 25% (quartile) of individuals with the highest DNA adduct levels can be compared to the 25% with the lowest DNA adduct levels, and the cases and controls analyzed separately.

Table 2 shows studies where DNA adducts have been evaluated in peripheral blood mononuclear cells using the case–control and nested

Table 2. Associations ($p \leq 0.05$) between cancer risk (RR or OR), and DNA adduct value in blood cell DNA for the quartile (~25%) of individuals with the highest DNA adducts compared with the quartile having the lowest DNA adducts

Carcinogen: DNA adduct source	Organ: cancer site	Method of DNA adduct determination	[a]RR or OR (confidence interval)	Reference
PAHs	Colon	[b]ELISA	2.8 (1.1–6.5)	39
PAHs	Lung	ELISA	7.7 (1.7–34.0)	40
[c]Bulky	Stomach	^{32}P-postlabeling	2.2 (1.0–202)	42
Bulky	Colon	^{32}P-postlabeling	2.3 (1.1–1.9)	42
Bulky	Bladder	^{32}P-postlabeling	4.1 (1.9–9.0)	43
Bulky (current smokers)	Lung	^{32}P-postlabeling	3.0 (1.1–8.4)	41
Bulky (current smokers)	Lung	^{32}P-postlabeling	2.0 (1.1–15.4)	44
Bulky (current smokers)	Lung	^{32}P-postlabeling	1.6 (1.0–2.6)	45

[a]RR, relative risk, OR, odds ratio in the 25% of individuals having the highest DNA adduct levels compared with those having the lowest DNA adduct levels. All values shown are statistically significant to $p \leq 0.05$.

[b]ELISA, enzyme-linked immunosorbent assay.

[c]Bulky DNA adducts, determined by ^{32}P-postlabeling, comprise a group of stable covalently bound DNA adducts that are typically formed from high-molecular-weight hydrophobic carcinogens that include polycyclic aromatic hydrocarbons (PAHs), aromatic amines, nitroaromatics, heterocyclic amines, hormones, and/or other hydrophobic DNA-binding agents.

case–control study designs described earlier. These studies have addressed the question: "Is the risk of cancer greater in individuals with the highest blood cell DNA adduct levels compared to those with the lowest blood cell DNA adduct levels?" In the first two studies shown in Table 2 [39, 40], the antisera used were specific for the group of carcinogenic PAHs bound to DNA, not just a single DNA adduct. In the remaining studies bulky DNA adducts were measured by ^{32}P-postlabeling [41–45]. Interestingly, ^{32}P-postlabeling has been criticized for a lack of specificity, but the generalized DNA damage that is detected by this method gives RR and OR values very much in the same range as in the studies where single adducts of known structure have been measured.

Table 3 is similar to Table 2, but shows studies in which the target tissue for cancer induction was evaluated for the presence of DNA adducts [46–51]. In the first four studies shown in Table 3, the DNA adducts

Table 3. Associations ($p \leq 0.05$) between cancer risk (OR or RR), and DNA adduct value in target tissue for cancer for the quartile of individuals with the highest DNA adducts compared with the quartile having the lowest DNA adducts

Carcinogen: DNA adduct source	Organ: DNA adduct measurement	Organ: cancer site	Method of DNA adduct measurement	[a]RR or OR (confidence interval)	Reference
AFB_1	Urine	Liver	HPLC	9.1 (2.9–29.2)	46
4-ABP	Liver	Liver	IHC[b]	6.5 (1.6–26.5)	47
AA	Kidney	Kidney	LC-EI/MS[c]	1.25 (1.004–1.5)	48
PhIP	Breast	Breast	IHC	4.0 (1.4–11.5)	49
PAHs	Liver	Liver	IHC	3.9 (1.0–14.9)	50
Bulky (meta-analysis, $n = 22$)	Bronchus and BAL[d]	Lung	^{32}P-postlabeling[d]	MR 2.64 (2.0–3.5)[d]	51[d]

4-ABP, 4-aminobiphenyl; AFB_1, aflatoxin B_1; HPLC, high pressure liquid chromatography; PhIP, 2-amino-1-methyl-6-phenylimidazo[4,5-b]-pyridine; PAHs, polycyclic aromatic hydrocarbons.

[a]RR and OR defined in the footnote to Table 2.

[b]EI/MS, ultraperformance liquid chromatography-electrospray ionization/multistage scan mass spectrometry; IHC, immunohistochemistry.

[c]In this study, 76% (39 of 51) kidney cancer patients examined were positive for aristolochic acid (AA)–DNA adducts in kidney. A subset of 10 AA-exposed kidney cancer patients, subjected to whole-exome sequencing, had evidence of AA-specific mutational signatures. The OR of 1.25 (1.004–1.547) was found in the quartile of patients with the highest AA exposure.

[d]BAL, bronchoalveolar lavage. Bulky DNA adducts are defined in the footnote to Table 2. In this study, 2.64 represents the MR or mean ratio of bulky DNA adducts in lung cancer cases/background bulky DNA adducts in healthy individuals.

measured were formed by a single carcinogen, and these included AFB_1 [46], 2-amino-1-methyl-6-phenylimidazo[4,5-b]-pyridine (PhIP) [49], 4-aminobiphenyl (4-ABP) [47], and aristolochic acid (AA) [48]. In the study examining PAHs by IHC [50], the OR was similar to the other studies even though the antiserum measured a family of carcinogenic PAH–DNA adducts. Finally, the paper on lung cancer by Ceppi *et al.* [51] involved a meta-analysis of 22 studies measuring bulky DNA adducts by ^{32}P-postlabeling in DNA from either bronchus or bronchoalveolar lavage (BAL). In this study, the authors presented a different metric, mean ratio (MR), defined as the ratio of bulky DNA adducts in lung cancer patients/bulky DNA adducts in healthy individuals. Compared to other studies, this turns the question around, but still shows that individuals with lung cancer had >2.5-fold higher DNA adducts in the target bronchus and BAL, than those without lung cancer. Furthermore, these studies suggest that even though blood is a surrogate tissue and DNA adduct data obtained with blood may not reflect events in the target tissue, in some instances there is concordance between the two.

Several studies have reported a negative or equivocal relationship between DNA adducts and cancer risk. Increases in breast cancer risk were not associated with high levels of PAH–DNA adducts in breast tissue [52, 53] or in peripheral blood [54]. In a large European study [55], there was no association between formation of bulky DNA adducts, measured by ^{32}P-postlabeling, and breast cancer. Another study showed no significant association between consumption of heavily cooked or grilled meat known to contain PAHs and prostate cancer [56].

Taken together, these studies demonstrate predominantly positive associations between increased cancer risk and elevated DNA adduct levels, whether the adducts are measured in blood cell DNA or in target tissue DNA. The modest magnitude of these numbers suggests that other factors, many of which we already know about, may also contribute significantly to human cancer risk. The lack of associations between some exposures and some cancers (e.g., PAH exposure and breast cancer) suggests that events other than the typical carcinogen/DNA damage/mutagenesis scenario may be driving cancer etiology in these organs. In the real world, a person is likely to be exposed to multiple classes of carcinogens and therefore likely to carry many different DNA adducts. Although the area of *adductomics* (see chapter by D. Phillips) is in its infancy, there is preliminary evidence that sustaining DNA damage from different sources or the combination of DNA adducts plus proliferative risk factors, such as infection or inflammation, may be associated with substantial increases in cancer risk. One such example is the combination of hepatitis B infection and exposure to carcinogenic aflatoxins. Qian *et al.* [46] reported an RR of 9.1 for liver cancer in subjects excreting 8,9-dihydro-8-(2,6-diamine-6-formamido-4-oxo-3,4-dihydropyramid-5-yl-amino-guanine) (AFB_1-N7-G) who had no evidence of hepatitis B infection, and an RR of 7.3 in individuals carrying the hepatitis B antigen who had no evidence of AFB_1-N7-G. However, individuals with both exposures had a liver cancer RR of

59.4, demonstrating a multiplicative increase in risk. Clearly more research is required, both with individual carcinogens to confirm published studies, and with complex mixtures to expand our knowledge of the type, extent, and risk of DNA damage sustained by modern-day human populations.

Abbreviations

AA	aristolochic acid
AFB_1	aflatoxin B_1
AFB_1-N^7-G	8,9-dihydro-8-(2,6-diamine-6-formamido-4-oxo-3,4-dihydropyramid-5-yl-amino-guanine)
4ABP	4-aminobiphenyl
AMS	accelerator mass spectrometry
BP	benzo[*a*]pyrene
BPDE	7β,8α-dihydroxy-9α,10α-epoxy-7,8,9,10-tetrahydrobenzo[*a*]-pyrene
BPdG	r7,t8,t9-trihydroxy-c-10-(N^2-deoxyguanosyl)-7,8,9,10-tetrahydrobenzo[*a*]pyrene
BAL	bronchoalveolar lavage
Bulky DNA adducts	determined by ^{32}P-postlabeling in biological samples, comprise a group of stable covalently bound DNA adducts considered to be largely formed from carcinogens that may include PAHs, aromatic amines, nitroaromatics, heterocyclic amines, aflatoxins, and/or hormones
CI	confidence interval
ELISA	enzyme-linked immunosorbent assay
dG-C8-ABP	*N*-(deoxyguanosin-8-yl)-2-aminobiphenyl
dG-C8-AAF	*N*-(deoxyguanosin-8-yl)-2-acetylaminofluorene
HPLC	high pressure liquid chromatography
IARC	International Agency for Research on Cancer
IHC	immunohistochemistry
MR	mean ratio of bulky DNA adducts in lung cancer cases/background bulky DNA adducts in healthy individuals

MS	mass spectrometry
N7-me-dG	N7-methyl-deoxyguanosine
O^6-me-dG	O^6-methyl-deoxyguanosine
OR	odds ratio
PAHs	polycyclic aromatic hydrocarbons
PhIP	2-amino-1-methyl-6-phenylimidazo[4,5-b]-pyridine
Pt-AG	*cis*-diammineplatinum-N^7-d[ApG]
Pt-GG	*cis*-diammineplatinum-N^7-d[GpG]
RR	relative risk

References

1. Li J, Brackbill RM, Liao TS, Qiao B, Cone JE, Farfel MR, *et al.* Ten-year cancer incidence in rescue/recovery workers and civilians exposed to the September 11, 2001 terrorist attacks on the World Trade Center. *Am J Ind Med* 2016; 59: 709–721.
2. Potter M. Percivall Pott's contribution to cancer research. *Natl Cancer Inst Monogr* 1963; 10: 1–13.
3. Miller JA. The metabolism of xenobiotics to reactive electrophiles in chemical carcinogenesis and mutagenesis: a collaboration with Elizabeth Cavert Miller and our associates. *Drug Metab Rev* 1998; 30: 645–674.
4. Dipple A. Formation, metabolism, and mechanism of action of polycyclic aromatic hydrocarbons. *Cancer Res* 1983; 43: 2422s–2425s.
5. IARC. Some non-heterocyclic polycyclic aromatic hydrocarbons and some related exposures. *IARC Monogr Series* 2010; 92: 1–853.
6. Poirier MC, Santella RM, Weston A. Carcinogen macromolecular adducts and their measurement. *Carcinogenesis* 2000; 21: 353–359.
7. Cleaver JE. Cancer in xeroderma pigmentosum and related disorders of DNA repair. *Nat Rev Cancer* 2005; 5: 564–573.
8. Hollstein M, Alexandrov LB, Wild CP, Ardin M, Zavadil J. Base changes in tumour DNA have the power to reveal the causes and evolution of cancer. *Oncogene* 2017; 36: 158–167.
9. Vogelstein B, Papadopoulos N, Velculescu VE, Zhou S, Diaz LA, Jr., Kinzler KW. Cancer genome landscapes. *Science* 2013; 339: 1546–1558.
10. Lengauer C, Kinzler KW, Vogelstein B. Genetic instability in colorectal cancers. *Nature* 1997; 386: 623–627.
11. Vogelstein B, Kinzler KW. Cancer genes and the pathways they control. *Nat Med* 2004; 10: 789–799.

12. Boutwell RK. Model systems for defining initiation, promotion, and progression of skin neoplasms. *Prog Clin Biol Res* 1989; 298: 3–15.
13. Beland FA, Poirier MC. Significance of DNA adduct studies in animal models for cancer molecular dosimetry and risk assessment. *Environ Health Perspect* 1993; 99: 5–10.
14. Hemminki K. Nucleic acid adducts of chemical carcinogens and mutagens. *Arch Toxicol* 1983; 52: 249–285.
15. Liu B, Xue Q, Tang Y, Cao J, Guengerich FP, Zhang H. Mechanisms of mutagenesis: DNA replication in the presence of DNA damage. *Mutat Res Rev Mutat Res* 2016; 768: 53–67.
16. Maher VM, McCormick JJ. Role of DNA lesions and repair in the transformation of human cells. *PharmacolTher* 1984; 25: 395–408.
17. Beland FA, Poirier MC. Significance of DNA adduct studies in animal models for cancer molecular dosimetry and risk assessment. *Environ Health Perspect* 1993; 99: 5–10.
18. Maher VM, McCormick JJ. Role of DNA lesions and DNA repair in mutagenesis by carcinogens in diploid human fibroblasts. *Prog Clin Biol Res* 1986; 209A: 245–253.
19. Maher VM, McCormick JJ, Grover PL, Sims P. Effect of DNA repair on the cytotoxicity and mutagenicity of polycyclic hydrocarbon derivatives in normal and xeroderma pigmentosum human fibroblasts. *Mutat Res* 1977; 43: 117–138.
20. Moss SF, Blaser MJ. Mechanisms of disease: Inflammation and the origins of cancer. *Nat Clin Pract Oncol* 2005; 2: 90–97; quiz 1 p following 113.
21. Poirier MC, Yuspa SH, Weinstein IB, Blobstein S. Detection of carcinogen-DNA adducts by radiommunoassay. *Nature* 1977; 270: 186–188.
22. Perera FP, Poirier MC, Yuspa SH, Nakayama J, Jaretzki A, Curnen MM, *et al.* A pilot project in molecular cancer epidemiology: determination of benzo[a]pyrene — DNA adducts in animal and human tissues by immunoassays. *Carcinogenesis* 1982; 3: 1405–1410.
23. Pratt MM, John K, MacLean AB, Afework S, Phillips DH, Poirier MC. Polycyclic aromatic hydrocarbon (PAH) exposure and DNA adduct semi-quantitation in archived human tissues. *Int J Environ Res Public Health* 2011; 8: 2675–2691.
24. Weston A, Manchester DK, Poirier MC, Choi JS, Trivers GE, Mann DL, *et al.* Derivative fluorescence spectral analysis of polycyclic aromatic hydrocarbon-DNA adducts in human placenta. *Chem Res Toxicol* 1989; 2: 104–108.
25. Santella RM. Immunological methods for detection of carcinogen-DNA damage in humans. *Cancer Epidemiol Biomarkers Prev* 1999; 8: 733–739.

26. Randerath K, Randerath E. 32P-postlabeling methods for DNA adduct detection: overview and critical evaluation. *Drug Metab Rev* 1994; 26: 67–85.
27. Manchester DK, Wilson VL, Hsu IC, Choi JS, Parker NB, Mann DL, *et al.* Synchronous fluorescence spectroscopic, immunoaffinity chromatographic and 32P-postlabeling analysis of human placental DNA known to contain benzo[a]pyrene diol epoxide adducts. *Carcinogenesis* 1990; 11: 553–559.
28. Farmer PB, Singh R. Use of DNA adducts to identify human health risk from exposure to hazardous environmental pollutants: the increasing role of mass spectrometry in assessing biologically effective doses of genotoxic carcinogens. *Mutat Res* 2008; 659: 68–76.
29. Boysen G, Hecht SS. Analysis of DNA and protein adducts of benzo[a]pyrene in human tissues using structure-specific methods. *Mutat Res* 2003; 543: 17–30.
30. Yun BH, Rosenquist TA, Nikolic J, Dragicevic D, Tomic K, Jelakovic B, *et al.* Human formalin-fixed paraffin-embedded tissues: an untapped specimen for biomonitoring of carcinogen DNA adducts by mass spectrometry. *Anal Chem* 2013; 85: 4251–4258.
31. Dingley KH, Roberts ML, Velsko CA, Turteltaub KW. Attomole detection of 3H in biological samples using accelerator mass spectrometry: application in low-dose, dual-isotope tracer studies in conjunction with 14C accelerator mass spectrometry. *Chem Res Toxicol* 1998; 11: 1217–1222.
32. Kensler TW, Egner PA, Davidson NE, Roebuck BD, Pikul A, Groopman JD. Modulation of aflatoxin metabolism, aflatoxin-N7-guanine formation, and hepatic tumorigenesis in rats fed ethoxyquin: role of induction of glutathione S-transferases. *Cancer Res* 1986; 46: 3924–3931.
33. Groopman JD, Sabbioni G, Wild CP. Molecular dosimetry of human aflatoxin exposures. In *Molecular Dosimetry and Human Cancer,* Groopman JD, ed., CRC Press, Boca Raton, 1991, pp. 303–324.
34. Reed E, Ozols RF, Fasy T, Yuspa SH, Poirier MC. Biomonitoring of cisplatin-DNA adducts in cancer patients receiving cisplatin chemotherapy. *Prog Clin Biol Res* 1986; 209B: 247–252.
35. Gupta-Burt S, Shamkhani H, Reed E, Tarone RE, Allegra CJ, Pai LH, *et al.* Relationship between patient response in ovarian and breast cancer and platinum drug — DNA adduct formation. *Cancer EpidemiolBiolPrev* 1993; 2: 229–234.
36. Souliotis VL, Valavanis C, Boussiotis VA, Pangalis GA, Kyrtopoulos SA. Comparative dosimetry of O6-methylguanine in humans and rodents treated with procarbazine. *Carcinogenesis* 1994; 15: 1675–1680.

37. van Delft JH, van den Ende AM, Keizer HJ, Ouwerkerk J, Baan RA. Determination of N7-methylguanine in DNA of white blood cells from cancer patients treated with dacarbazine. *Carcinogenesis* 1992; 13: 1257–1259.
38. Beland FA, Poirier MC. DNA adducts and carcinogenesis. In *Pathobiology of Neoplasia*, Sirica AE, ed., Plenum Press, New York, 1989, pp. 57–80.
39. Gunter MJ, Divi RL, Kulldorff M, Vermeulen R, Haverkos KJ, Kuo MM, *et al.* Leukocyte polycyclic aromatic hydrocarbon-DNA adduct formation and colorectal adenoma. *Carcinogenesis* 2007; 28: 1426–1429.
40. Tang D, Santella RM, Blackwood AM, Young TL, Mayer J, Jaretzki A, *et al.* A molecular epidemiological case-control study of lung cancer. *Cancer EpidemiolBiomarkers Prev* 1995; 4: 341–346.
41. Tang D, Phillips DH, Stampfer M, Mooney LA, Hsu Y, Cho S, *et al.* Association between carcinogen-DNA adducts in white blood cells and lung cancer risk in the physicians health study. *Cancer Res* 2001; 61: 6708–6712.
42. Agudo A, Peluso M, Munnia A, Lujan-Barroso L, Sanchez MJ, Molina-Montes E, *et al.* Aromatic DNA adducts and risk of gastrointestinal cancers: a case-cohort study within the EPIC-Spain. *Cancer Epidemiol Biomarkers Prev* 2012; 21: 685–692.
43. Peluso M, Airoldi L, Armelle M, Martone T, Coda R, Malaveille C, *et al.* White blood cell DNA adducts, smoking, and NAT2 and GSTM1 genotypes in bladder cancer: a case-control study. *Cancer Epidemiol Biomarkers Prev* 1998; 7: 341–346.
44. Peluso M, Munnia A, Hoek G, Krzyzanowski M, Veglia F, Airoldi L, *et al.* DNA adducts and lung cancer risk: a prospective study. *Cancer Res* 2005; 65: 8042–8048.
45. Veglia F, Loft S, Matullo G, Peluso M, Munnia A, Perera F, *et al.* DNA adducts and cancer risk in prospective studies: a pooled analysis and a meta-analysis. *Carcinogenesis* 2008; 29: 932–936.
46. Qian GS, Ross RK, Yu MC, Yuan JM, Gao YT, Henderson BE, *et al.* A follow-up study of urinary markers of aflatoxin exposure and liver cancer risk in Shanghai, People's Republic of China. *Cancer Epidemiol Biomarkers Prev* 1994; 3: 3–10.
47. Wang LY, Chen CJ, Zhang YJ, Tsai WY, Lee PH, Feitelson MA, *et al.* 4-Aminobiphenyl DNA damage in liver tissue of hepatocellular carcinoma patients and controls. *Am J Epidemiol* 1998; 147: 315–323.
48. Hoang ML, Chen CH, Chen PC, Roberts NJ, Dickman KG, Yun BH, *et al.* Aristolochic acid in the etiology of renal cell carcinoma. *Cancer Epidemiol Biomarkers Prev* 2016; 25: 1600–1608.

49. Zhu J, Chang P, Bondy ML, Sahin AA, Singletary SE, Takahashi S, *et al.* Detection of 2-amino-1-methyl-6-phenylimidazo[4,5-b]-pyridine-DNA adducts in normal breast tissues and risk of breast cancer. *Cancer Epidemiol Biomarkers Prev* 2003; 12: 830–837.
50. Chen SY. Polycyclic aromatic hydrocarbon-DNA adducts in liver tissues of hepatocellular carcinoma patients and controls. *Int J Cancer* 2002; 99: 14–21.
51. Ceppi M, Munnia A, Cellai F, Bruzzone M, Peluso MEM. Linking the generation of DNA adducts to lung cancer. *Toxicology* 2017; 390: 160–166.
52. Rundle A, Tang D, Hibshoosh H, Estabrook A, Schnabel F, Cao W, *et al.* The relationship between genetic damage from polycyclic aromatic hydrocarbons in breast tissue and breast cancer. *Carcinogenesis* 2000; 21: 1281–1289.
53. Rundle A, Tang D, Zhou J, Cho S, Perera F. The association between glutathione S-transferase M1 genotype and polycyclic aromatic hydrocarbon-DNA adducts in breast tissue. *Cancer EpidemiolBiomarkers Prev* 2000; 9: 1079–1085.
54. Sagiv SK, Gaudet MM, Eng SM, Abrahamson PE, Shantakumar S, Teitelbaum SL, *et al.* Polycyclic aromatic hydrocarbon-DNA adducts and survival among women with breast cancer. *Environ Res* 2009; 109: 287–291.
55. Saieva C, Peluso M, Masala G, Munnia A, Ceroti M, Piro S, *et al.* Bulky DNA adducts and breast cancer risk in the prospective EPIC-Italy study. *Breast Cancer Res Treat* 2011; 129: 477–484.
56. Tang D, Liu JJ, Rundle A, Neslund-Dudas C, Savera AT, Bock CH, *et al.* Grilled meat consumption and PhIP-DNA adducts in prostate carcinogenesis. *Cancer Epidemiol Biomarkers Prev* 2007; 16: 803–808.

Glossary

Biotransformation — Biochemical processing of a xenobiotic compound through metabolic enzymatic pathways, often predominantly cytochrome P450s, producing hydrophilic intermediates disposed of by circulation, excretion, or binding to DNA.

Bulky DNA adducts — DNA adducts determined by ^{32}P-postlabeling, a method highly sensitive for stable DNA damage induced by a broad range of chemical carcinogens (polycyclic aromatic hydrocarbons [PAHs], aromatic amines, aflatoxins, nitroaromatics, heterocyclic amines, tamoxifen, hormones, and others). In human DNA, ^{32}P-postlabeling cannot provide chemical characterization of individual DNA adducts, and the DNA adducts observed are typically grouped together and termed "bulky."

Carcinogen — A substance, exposure to which can cause cancer in living tissue.

DNA adduct — DNA damage, addition product, formed as a result of covalent binding of a chemical carcinogen to DNA.

DNA repair — A collection of processes by which damaged DNA bases are identified, removed, and replaced with normal bases.

Genotoxic — A chemical or process that is damaging to DNA.

Mutation — Alteration in DNA sequence resulting in a stable variant form that is transmitted to future generations.

Neoplastic transformation — Conversion of normal cells into tumor cells.

Target organ — Organ in which cancer appears.

Tumor initiation — DNA damage resulting in mutagenesis of a critical gene, initiation is considered to be an irreversible early step in the pathway to cancer induction.

Tumor promotion — A process by which a single initiated cell is stimulated to expand and undergo clonal growth via the influence of a noncarcinogen.

Tumor incidence — Percentage of animals with tumors in a single experimental group. For example, 10 animals with tumors in a group of 20 is a 50% tumor incidence.

Chapter 2

The Role of Cytochrome P450 Enzymes in Carcinogen Metabolism: Lessons Learned From Studies With Benzo[*a*]pyrene and Aristolochic Acid

Marie Stiborová*, Laura E. Wohak[†] and Volker M. Arlt[†,‡]

**Department of Biochemistry, Faculty of Science, Charles University, Albertov 2030, 128 40 Prague 2, Czech Republic*

[†]Department of Analytical, Environmental and Forensic Sciences, MRC-PHE Centre for Environment and Health, King's College London, 150 Stamford Street, London SE1 9NH, United Kingdom

[‡]NIHR Health Protection Research Unit in Health Impact of Environmental Hazards at King's College London in partnership with Public Health England and Imperial College London, 150 Stamford Street, London SE1 9NH, United Kingdom

Abstract

This chapter summarizes the latest data increasing our knowledge on a function of cytochrome P450 (P450) enzymes in metabolism of several carcinogens and on the mechanisms of the reaction cycle catalyzed by the P450-monooxygenase system. Particularly it describes the P450-dependent metabolism, toxicity, and carcinogenicity of two human environmental carcinogens, benzo[*a*]pyrene (BP) and aristolochic acid I (AAI). This chapter focuses on studies that have specifically investigated the role of P450 enzymes in the bioactivation and detoxification of BP and AAI, and the lessons learned from these studies in examining the potential role of P450-mediated metabolism for other known or putative human environmental carcinogens. P450 enzymes play a key role both in metabolic activation of these human carcinogens to the reactive intermediates forming DNA adducts, and in their detoxification to metabolites that are excreted. While BaP is oxidatively metabolized (activated or detoxified) by P450s, AAI can be metabolized both oxidatively (to its detoxification metabolite aristolochic acid Ia [AAIa]) and reductively (to be activated to DNA-binding metabolites). Concerning the mechanisms of reactions catalyzed by the microsomal P450 monooxygenase system reviewed in this chapter, it is demonstrated that besides NADPH:cytochrome P450 oxidoreductase (POR), the system consisting of microsomal NADH:cytochrome b_5 reductase together with its substrate cytochrome b_5 and cofactor NADH can function as a donor of electrons both for the first and the second reduction of at least some of P450 enzymes in their reaction cycle. These findings indicate that the old myth, the generally accepted mechanism on the exclusivity of the POR/NADPH system for reduction of several P450s, should be revised.

1. Introduction

Chemical carcinogens can exert their effect either through genotoxic (involving damage to DNA) or nongenotoxic mechanisms [1]. Some highly reactive genotoxic carcinogens are capable of directly interacting with DNA (e.g., alkylating agents) but most (e.g., many polycyclic aromatic hydrocarbons [PAHs] [2], heterocyclic aromatic amines [HAAs] [3]) are chemically inert and require metabolic conversion before exerting their genotoxic properties. These inert chemicals are termed procarcinogens or indirect-acting carcinogens. They are recognized as foreign to the organism (xenobiotic) and aimed to be detoxified through biotransformation processes (xenobiotic metabolism) which often instead leads to the formation of highly reactive metabolites that are electrophilic and react with nucleophilic sites in cellular macromolecules such as proteins or nucleic acids (Figure 1) [4–7].

The metabolism of xenobiotics (environmental chemicals or drugs) can be divided into two phases [9]. Phases I and II aim to render the originally

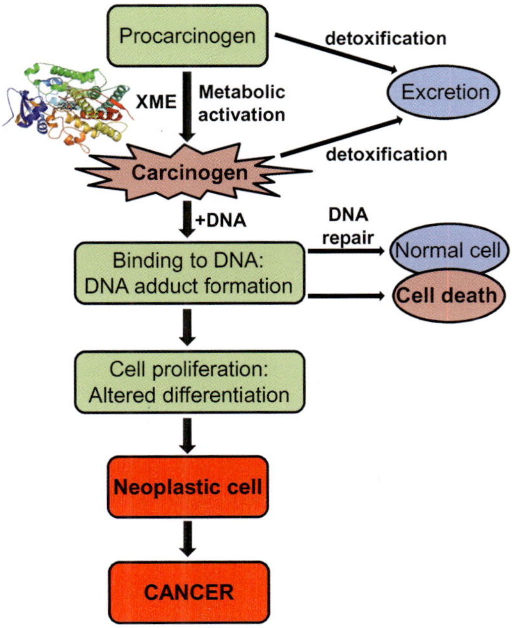

Figure 1. Key events in chemical carcinogenesis. See text for details. XME, xenobiotic-metabolizing enzyme. *Insert*: Structure of human CYP1A1 protein in a complex with α-naphthoflavone [8].

hydrophobic (not water-soluble) compounds more hydrophilic (water-soluble), thus making them readily excretable *via* urine, feces, or sweat. Phase I (functionalization) reactions include oxidation, reduction, and hydroxylation catalyzed by enzymes such as cytochrome P450s (CYPs), cyclooxygenase (COX), aldo-keto reductases (AKRs), and microsomal flavin-containing monooxygenases (FMOs). In the second phase (conjugation), the functional groups that were attached in phase I are used to conjugate small polar molecules to further increase the polarity. Examples of phase II enzymes include sulfotransferases (SULTs), *N,O*-acetyltransferases (NATs), methyltransferases such as catechol-*O*-methyltransferase (COMT), glutathione *S*-transferases (GSTs), and uridine diphosphate glucuronosyltransferases (UGTs). The classification of enzymes in phase I or II is, however, not rigid and some enzymes can arguably be in either category. The contribution of these xenobiotic-metabolizing enzymes to the activation or detoxification of carcinogens in humans is shown in Figure 2 [10].

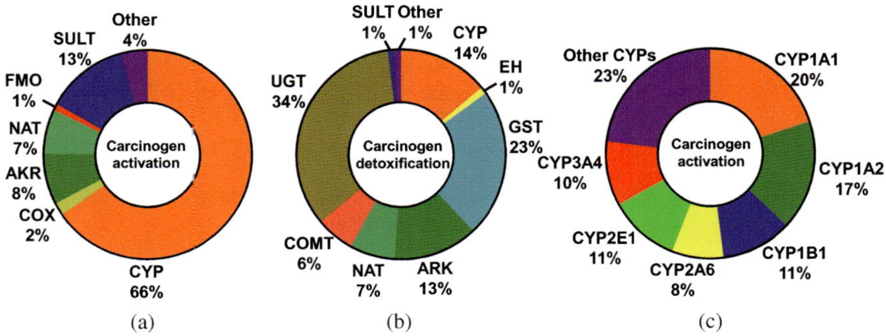

Figure 2. Contribution of human enzymes to carcinogen metabolism. Fractions of activation (a) and detoxification (b) reactions as well as fractions of P450 activation reactions catalyzed by individual human CYP isoenzymes (c) according to [10]. CYP, cytochrome P450; COX, cyclooxygenase; AKR, aldo-keto reductase; NAT, *N*-acetyltransferase; FMO, Flavin-containing monooxygenase; SULT, sulfotransferase; EH, epoxide hydrolase; GST, Glutathione S-transferase; COMT, Catechol-O-methyltransferase; UGT, UDP-glucuronosyltransferase.

It is noteworthy, however, that although these calculations represent the knowledge to date, they might not accurately represent the contribution of certain xenobiotic-metabolizing enzymes in the biotransformation of carcinogens. One challenge is that some of the standard assays used to study the genotoxicity of known or putative carcinogens do not allow for expression of certain xenobiotic-metabolizing enzymes. For example, the classic Ames mutagenicity assay does typically not assess mutagens activated by SULTs [11]. Hence, the impact of this class of enzymes might be vastly underrepresented.

To excrete the converted, now hydrophilic, molecules they are shuttled through cellular lipid bilayer membranes, a process which is facilitated by transporter proteins of the adenosine 5′-triphosphate (ATP)-binding cassette (ABC) or solute carrier (SLC) families [12]. However, since the primary focus of this chapter is on the role of P450 enzymes in carcinogen metabolism, and how their activity leads to DNA damage, this aspect will now be discussed in more detail.

2. Cytochromes P450 — Nomenclature, Localization, and Catalysis

P450 enzymes are generally designated with the abbreviation CYP (Cytochrome P450) for the superfamily, followed by a number showing the

Table 1. Allele nomenclature for human P450 enzymes.

CYP family	CYP isoenzymes
CYPÍ	CYP1A1; CYP1A2; CYP1B1
CYP2	CYP2A6; CYP2A13; CYP2B6; CYP2C8; CYP2C9; CYP2C19; CYP2D6; CYP2E1; CYP2F1; CYP2J2; CYP2R1; CYP2S1; CYP2W1
CYP3	CYP3A4; CYP3A5; CYP3A7; CYP3A43
CYP4	CYP4A11; CYP4A22; CYP4B1; CYP4F2
CYP > 4	CYP8A1; CYP8A1; CYP19A1; CYP21A2; CYP26A1

Further information on CYP enzymes including genetic information on CYP variants can be found on the Human Cytochrome P450 Allele Nomenclature (CYP-allele) webpage (http://www.cypalleles.ki.se/) [13].

gene family, a capital letter assigning the subfamily, and further numeral for the individual gene. The convention is to italicize the name of the gene. Hence, for example, the abbreviation *CYP1A1* is the gene that encodes the enzyme CYP1A1. The current nomenclature guidelines suggest that members of new CYP families share at least 40% amino acid identity, while members of subfamilies have to share at least 55% amino acid identity. The allele information for human P450 enzymes is shown in Table 1 [13].

The most common reaction catalyzed by P450 enzymes is a monooxygenase reaction, namely, the insertion of one atom of molecular oxygen into a variety of xenobiotic substrates, while reducing the other atom to water, a reaction that requires two electrons [14, 15]:

$$R{-}H + O_2 + NADPH + H^+ \rightarrow R{-}OH + H_2O + NADP^+$$

P450 enzymes located in the membrane of endoplasmic reticulum of the mammalian cells (microsomal P450 systems) are components of a mixed-function oxidase (MFO) system. This enzymatic system also contains other enzymes, the multidomain flavoprotein nicotinamide adenine dinucleotide phosphate reduced (NADPH): cytochrome P450 oxidoreductase (POR) and cytochrome b_5 accompanied by its nicotinamide adenine dinucleotide reduced (NADH):cytochrome b_5 reductase. A generally accepted theory hypothesizes that the two electrons needed for the reduction of P450 during its reaction cycle are transferred from NADPH *via* POR. Nevertheless, cytochrome b_5 can also contribute reducing power to this system (as a donor of the second electron) after being reduced by NADH:cytochrome b_5 reductase (Figure 3).

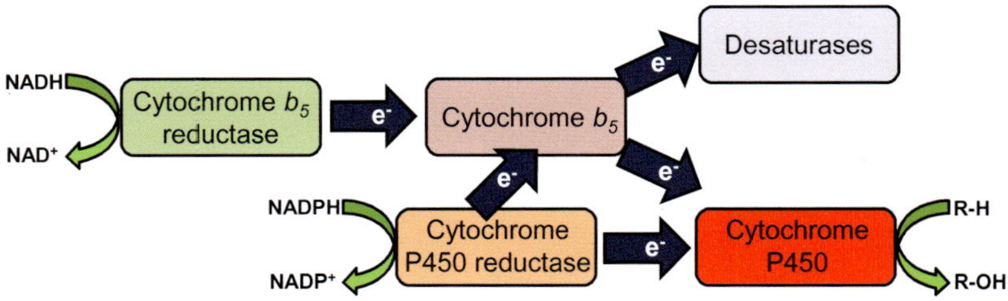

Figure 3. Electon transport pathways to P450s in the endoplasmatic reticulum (adapted from [17]).

Mammals appear to use a set of P450 enzymes (e.g., families CYP5, 8, 11, 17, 19, 21, 24, 26, and 27) for the synthesis of important steroids and eicosanoids, aside from their function in the catabolism of natural products [14, 15]. Of the remainder of the mammalian P450s, a relatively small set of these enzymes accounts for most of metabolism of drugs (i.e., human CYP1A2, 2C9, 2C19, 2D6, and 3A4). It should be emphasized that ~75% of enzymatic reactions converting drugs are catalyzed by P450 enzymes and that ~90% of these P450 reactions can be accounted for by the set of human P450s mentioned earlier (i.e., CYP1A2, 2C9, 2C19, 2D6, and 3A4) [10, 16]. The largest proportion of P450 reactions is catalyzed by P450 enzymes of the 3A subfamily, particularly CYP3A4. Another small set of human P450s (i.e., human CYP1A1, 1A2, 1B1, 2A6, 2E1, and 3A4) is involved in the metabolism of most procarcinogens and protoxicants that are P450 substrates (Figure 2C) [10, 14].

3. Cytochromes P450 — The Major Enzymes Metabolizing Xenobiotics

P450 enzymes (EC 1.14.14.1), a superfamily of hemoproteins distributed widely throughout nature, are involved in the metabolism of a broad variety of substrates and catalyze a variety of chemical reactions [14, 15]. P450 enzymes are the major catalysts for the metabolism of xenobiotic chemicals, and comprise a significant focus of scientists working in the areas of toxicology and pharmacology [18]. The oxidation reactions catalyzed by P450s can result in poor drug bioavailability, leading to acute and chronic toxicities with adverse drug interactions, and subsequently cancer susceptibility [14]. Phase I metabolism of environmental carcinogens is predominantly catalyzed by P450 enzymes [10]. Specifically, P450 enzymes contribute to ~66% of carcinogen bioactivation, whereas they only make a minor contribution to carcinogen detoxification (~14% overall contribution) (Figure 2) [10]. Often the P450-mediated metabolism of the parent compound, which produces excreted hydrophilic conjugated products, leads to the formation of reactive oxygenated intermediates that can cause genotoxicity and mutation by covalently binding to DNA and proteins (Figure 1) [7, 19, 20]. Damage to the structural integrity of DNA through covalent binding of chemicals is referred to as DNA adduct formation. Such DNA damage is generally considered to be causative and directly related to tumor formation [6, 19, 21, 22]. Thus, information on the mechanism by which P450 enzymes are involved in the metabolism of known or putative human carcinogens is critical for hazard identification and risk assessment.

4. Cytochromes P450 Involved in Metabolism of Carcinogens: Activation

The human P450 enzymes involved in carcinogen metabolism are most abundant in the liver (CYP1A2, 2A6, 2E1, and 3A4), except for CYP1A1 and 1B1 which are mainly extrahepatic enzymes (see Table 2 for an overview) [10, 14, 15, 23, 24]. However, many of these P450 enzymes are also found in some extrahepatic tissues, and P450-mediated bioactivation of carcinogens within a target tissue may be very important, particularly if a generated reactive product is not stable enough to migrate out of the cell in which it is formed [23]. Many P450 enzymes are inducible by chemicals. The inducer may be a substrate, but this is not necessarily the case [24].

The participation of individual P450 enzymes in carcinogen metabolism has been widely investigated, and the roles of numerous human P450s in carcinogen activation have been characterized [15, 18, 25]. A recent review of the literature showed that of the human P450 enzymes involved in the metabolism of carcinogens, a major role for P450s might be attributable predominantly to the three P450 enzymes of Family 1 (CYP1A1, 1A2, 1B1) and CYP2A6, 2E1, and 3A4 accounting for ~80% of the reported carcinogen activations (Figure 2C) [10]. As shown in Table 2, it can be exemplified that of these P450s the enzymes of Family 1 are most important in the activation of PAHs [2], aromatic amines [26], HAAs [3], and a variety of other compounds. CYP2A6 and 2E1 play a crucial role in

Table 2. Examples of individual P450 enzymes involved in the metabolism of xenobiotics, particularly environmental carcinogens.

CYP isoenzyme (selected)	Organ*	Important substrates for carcinogen metabolism (selected)
CYP1A1	Many, particularly after induction	• Polycyclic aromatic hydrocarbons (e.g. benzo[*a*]pyrene {oxidation, 3-hydroxylation}) • Heterocyclic amines (e.g. 2-amino-3-methylimidazo[4,5-*f*]quinolone {*N*-hydroxylation, oxidation}) • Aromatic amines (e.g. 6-aminochrysene {oxidation}) • Nitroarene (e.g. 1,8-dinitropyrene {nitroreduction}) • Aristolochic acid I {nitroreduction, *O*-demethylation on C8 methoxy group} • Ellipticine {hydroxylation}
CYP1A2	Liver	• Aromatic amine (e.g. 4-aminobiphenyl {*N*-hydroxylation, oxidation}, 2-aminofluorene {*N*-hydroxylation, oxidation}, 6-aminochrysene {oxidation}) • Heterocyclic amines (e.g. 2-amino-1-methyl-6-phenylimidazo[4,5-*b*]pyridine {*N*-hydroxylation, oxidation}, 2-amino-3-methylimidazo[4,5-*f*]quinolone {*N*-hydroxylation, oxidation}) • Polycyclic aromatic hydrocarbons (e.g. benzo[*a*]pyrene {oxidation, 3-hydroxylation}) • Aristolochic acid I {nitroreduction, *O*-demethylation on C8 methoxy group} • Estragole {C1'-hydroxylation} • Aflatoxin B_1 {epoxidation, oxidation}
CYP1B1	Many extra-hepatic organs	• Polycyclic aromatic hydrocarbons (e.g. benzo[*a*]pyrene {oxidation, 3-hydroxylation}) • Nitroarene (e.g. 1,8-dinitropyrene {nitroreduction}, 1-nitropyrene {nitroreduction, epoxidation}) • Aromatic amines (e.g. 2-aminofluorene {*N*-hydroxylation, oxidation}, 6-aminochrysene {oxidation})

(*Continued*)

Table 2. (*Continued*)

CYP isoenzyme (selected)	Organ*	Important substrates for carcinogen metabolism (selected)
CYP2A6	Liver	• Nitrosamines (e.g. 4-(methylnitrosoamino)-1-(3-pyridyl)-1-butanol {Cα-hydroxylation-methylene, *N*-oxidation}, 4-(methylnitrosoamino)-1-(3-pyridyl)-1-butanone {Cα-hydroxylation-methylene, *N*-oxidation}, *N*-nitrosodiethylamine {oxidation}) • Estragole {C1'-hydroxylation} • Aflatoxin B_1 {hydroxylation, O-demethylation}
CYP2E1	Liver, gastrointestinal tract	• 1,3-butadiene {butadiene monoxide} • Carbon tetrachloride {dechlorination reductive} • Chloroform (dehalogenation reductive, oxidation} • Ethanol {oxidation}; vinyl chloride {oxidation} • Acrylamide {epoxidation to glycidermide}
CYP3A4	Liver, small intestine	• Aflatoxin B_1 {epoxidation, oxidation} • Aromatic amines (e.g. 6-aminochrysene {oxidation}) • Cyclophosphamide {C4-hydroxylation, oxidation} • Ellipticine {hydroxylation, oxidation N2} • Tamoxifen {Cα-hydroxylation, catechol formation, oxidation}

*Cytochrome P450 enzyme detected in this organ but is potentially also expressed in other organs. { }, indicates the activation and/or detoxification reactions catalysed; adapted from [10].

the metabolism of *N*-nitrosamines [27] and CYP3A4 activates the mycotoxin aflatoxin B_1 (AFB_1) [28].

5. Cytochromes P450 Involved in Metabolism of Carcinogens: Detoxification

Beside the P450 enzymes contributing to carcinogen activation, many of them also catalyze detoxification of these compounds. Moreover, a single enzyme can generate different products from the same substrate; among them several are the detoxification metabolites but others are (geno)toxic. For instance, CYP3A4-catalyzed oxidation of AFB_1 to aflatoxin Q1 (i.e., 3-hydroxylation) is a detoxification reaction, while the oxidation of this carcinogen by the same P450 to AFB_1-8,9-epoxide represents the activation pathway [28]. A similar situation can be found in the metabolism of the human carcinogen benzo[*a*]pyrene (BP), where CYP1A1 and 1B1 are the most important P450 enzymes in its activation [29–31] in combination with microsomal epoxide hydrolase (*m*EH). Likewise, the human carcinogen aristolochic acid I (AAI) is also both activated and detoxified by P450s (reviewed in [32–34]). Therefore, because of the crucial importance of P450s in catalyzing the opposing processes in the metabolism of BP and AAI (activation and detoxification), the metabolic pathways of both these human carcinogens and P450 enzymes involved in these processes are described in this chapter in detail (see below).

6. P450 Enzymes Contributing to the Oxidation of BP and AAI

In this chapter, we focus on the metabolic mechanisms catalyzed by the P450-dependent monooxygenase enzyme system, and the metabolism, toxicity, and carcinogenicity of two human environmental carcinogens, BP and aristolochic acid (AA). We focus on studies that have specifically investigated the role of P450 enzymes in the bioactivation and detoxification of BP and AA, and the lessons learned from these studies in examining the potential role of P450-mediated metabolism for other known or putative human carcinogens. This chapter will not discuss the role of P450 enzymes in drug discovery and drug safety assessment.

6.1. *Benzo[a]pyrene*

BP, a PAH, has been classified as a human carcinogen (Group 1) by the International Agency for Research on Cancer (IARC) [35]. BP and other PAHs are produced mainly by incomplete combustion of organic matter, with major human exposures occurring through inhalation (tobacco smoke) and the diet (charbroiled meats) [36]. Many studies, using mammalian cellular and subcellular systems, have demonstrated that PAHs, including BP, are metabolically activated by P450s [31, 37]. BP requires metabolic activation catalyzed by P450 enzymes prior to reaction with DNA [2, 31], and there is a dose-dependent relationship between the concentration of BP in target DNA and the induction of malignant transformation. (For details of BP-induced DNA adducts in human tissues see chapters by R. M. Santella and A. Weston.) In addition to the total amount of BP exposure, metabolism and other factors dictate the effective concentration of BP that becomes bound to DNA causing genotoxicity. For example, in addition to metabolic rates, confounding factors may include additional xenobiotic exposures that alter metabolism, polymorphisms that change metabolic rates, human health status, and diet [35, 38]. Nevertheless, the identification of enzymes principally involved in BP metabolism in humans and detailed knowledge of their catalytic specificities is of major importance.

The most efficient P450 enzymes metabolizing BP, CYP1A1 (see insert in Figure 1), and 1B1, oxidize BP to an epoxide (i.e., BP-7,8-epoxide) that is then converted to BP-7,8-dihydrodiol by *m*EH. Further, bioactivation by CYP1A1 and 1B1 leads to the ultimate reactive species 7β,8α-dihydroxy-9α,10α-epoxy-7,8,9,10-tetrahydrobenzo[*a*]pyrene (BPDE) that can react with DNA to form adducts preferentially at guanine residues [39–43] (Figure 4). However, BP is also oxidized to other metabolites such as the other BP-dihydrodiols (i.e., BP-9,10-dihydrodiol, BP-4,5-dihydrodiol), BP-diones (i.e., BP-1,6-dione, BP-3,6-dione), and hydroxylated BP metabolites (i.e., BP-3-ol, BP-7-ol, BP-9-ol), which are considered as detoxification products [2, 44–49]. One of these metabolites, 9-hydroxy-BP (BP-9-ol) is formed predominantly by CYP1A1 and is a precursor of 9-hydroxy-BP-4,5-epoxide, which can also form an adduct with deoxyguanosine in DNA (Figure 4) [39, 46, 50–54]. Many other P450 enzymes, human CYP1B1, 2B6, 2C8, 2C9, 2C19, and 3A4 have all been reported to oxidize BP, but their efficiencies are about one order of magnitude lower than CYP1A1 (Figure 5a) [2, 44, 45, 47].

Of the human P450s, CYP1A2, 2A6, 2C8, and 2E1 do not catalyze BP oxidation [2, 44, 45] (Figure 5a). However, it should be noted that contrasting results have been found with regards to the role of human CYP1A2 in

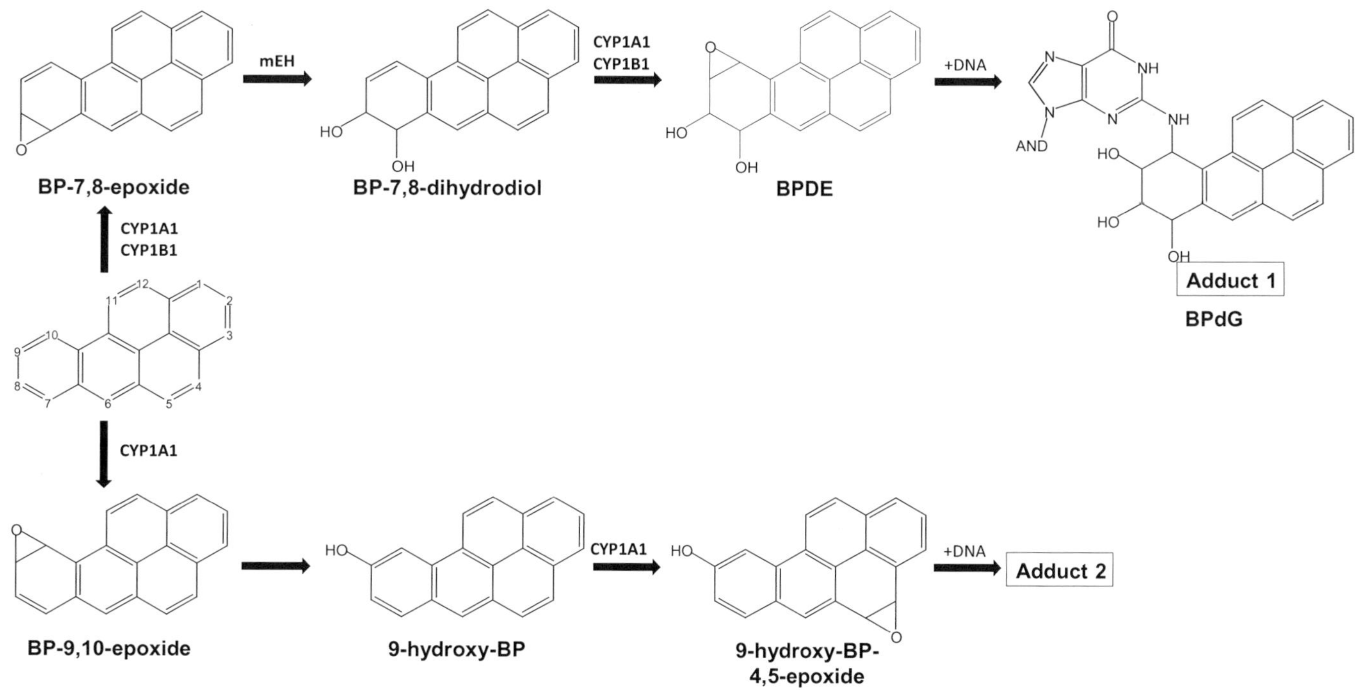

Figure 4. Proposed pathways of biotransformation and DNA adduct formation of BP catalyzed by CYP1A1 and *m*EH. *Upper part*, the typical three-step activation process by CYP1A1 followed by hydrolysis by *m*EH leads to 7α,8β-dihydroxy-9α,10α-7,8,9,10-tetrahydrobenzo[*a*]pyrene (BPDE) which forms BPdG (*r*7,*t*8,*t*9-trihydroxy-*c*-10-(N^2-deoxyguanosyl)-7,8,9,10-tetrahydrobenzo[*a*]pyrene; adduct 1). *Lower part*, the two-step activation process by CYP1A1 leads to the formation of 9-hydroxy-BP-4,5-epoxide that can react with deoxyguanosine in DNA (adduct 2).

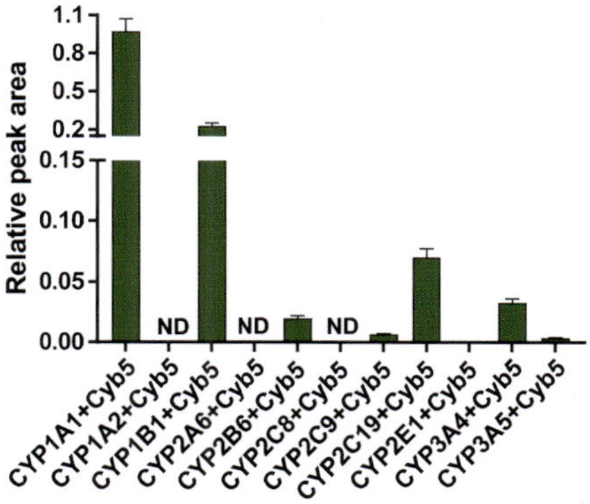

(a)

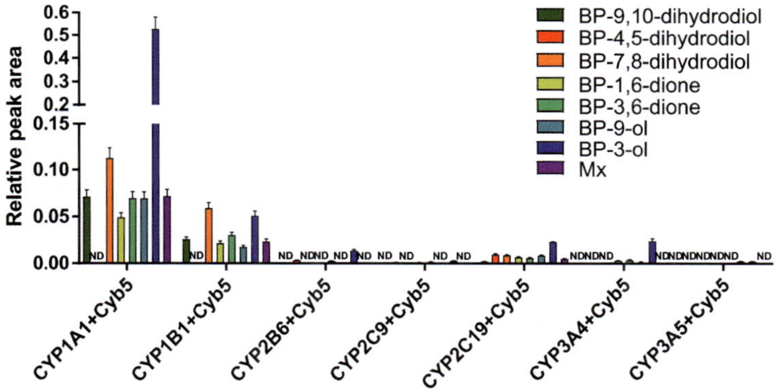

(b)

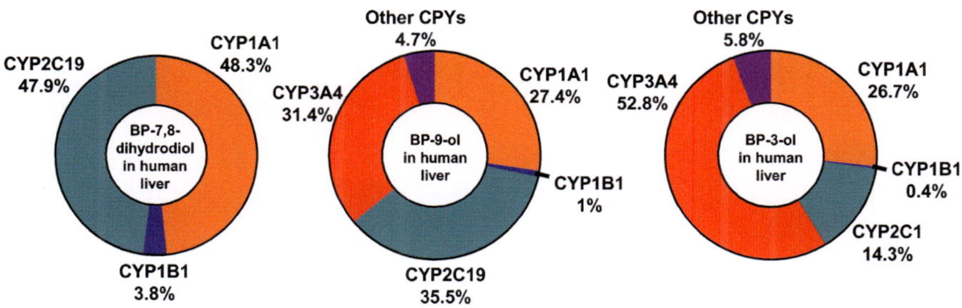

(c)

Figure 5. (a) BP metabolism catalyzed by individual human CYPs in the presence of cytochrome b_5 (Cyb5). Columns show the sum of individual BP metabolites formed by P450s [47]. Data shown are mean ± SD ($n = 3$). (b) Formation of BP metabolites by human P450s in the presence of cytochrome b_5 (Cyb5) [47]. Data shown are mean ± SD ($n = 3$). (c) Contributions of P450 enzymes to the formation BaP-7,8-dihydrodiol, BaP-9-ol and BaP-3-ol in human livers [47]. ND, not detected.

BP oxidation. Whereas some studies found that human CYP1A2 partially oxidized BP [44, 45], other studies reported that human CYP1A2 is not effective in mediating BP hydroxylation [55]. In a recent study comparing the contribution of individual P450 enzymes metabolizing BP, CYP1A1 had the highest BP-metabolizing potency, followed by CYP1B1, 2C19, 3A4, 2B6, 2C9, and 3A5 (Figures 5a and 5b) [47].

Interestingly, another protein of the microsomal P450 monooxygenase system, cytochrome b_5 (compare Figure 3) strongly modulated the activity of some P450 enzymes to metabolize BP, most notably CYP1A1 [46, 47, 51, 52, 56]. The CYP1A1 enzyme system generated the largest amount of BP-3-ol (Figure 5b). Addition of cytochrome b_5 to CYP1A1 increased BP oxidation to this metabolite more than two-fold. The greatest stimulatory effect of cytochrome b_5 was in the formation of BP-3-ol and BP-7,8-dihydrodiol, and to a lesser extent on the other BP metabolites produced [47, 51]. Moreover, increased protein expression of cytochrome b_5, in livers of mouse [50] and rat [52] models suggests that cytochrome b_5 may modulate the P450-mediated bioactivation of BP in organisms, capable of substituting the function of POR [46, 51, 52] as discussed below.

Of the BP metabolites generated by the human P450s, BP-3-ol is formed by most of the P450 enzymes active in BP metabolism as the major BP oxidation product, except CYP1B1 (Figure 5b). BP-3-ol is considered to be a detoxification metabolite [57]. In addition, BP-7,8-dihydrodiol and BP-9-ol were formed (Figure 5b), which are precursors of BPDE and 9-hydroxy-BP-4,5-oxide, respectively, both of which covalently bind to DNA forming adducts (see Figure 4). CYP1A1, 1B1, and 2C19 are capable of forming BP-7,8-dihydrodiol (Figure 5b). The formation of BP-9-ol was also mainly catalyzed by CYP1A1 and 1B1, while CYP2C19 and 3A produced this metabolite to a lesser extent (Figure 5b). CYP1A1 and 1B1 are also the most efficient in catalyzing oxidation of BP to BP-3-ol, although this metabolite was also formed by CYP2B6, 2C9, 2C19, and 3A4/5 (Figure 5b).

Because BP-7,8-dihydrodiol, BP-9-ol, and BP-3-ol are critical intermediates that contribute to DNA damage induced by BP, it is crucial to know which of the P450 enzymes expressed in human liver are most effective in the formation of these metabolites. The liver is the most important organ for the metabolism of xenobiotics including BP in humans. Based on the amounts of BP-7,8-dihydrodiol, BP-9-ol, and BP-3-ol formed by each human P450 enzyme, and the expression levels of these enzymes in human liver [24, 58, 59], the contributions of individual P450s to the formation of these BP metabolites in human liver were established (Figure 5c) [47].

Although CYP1A1 is expressed in human liver at very low levels (<0.7%), it is estimated to form more BP-7,8-dihydrodiol than any other human CYP (~48.3% of all BP-7,8-dihydrodiol synthesis) (Figure 5c). A similar contribution to the formation of BP-7,8-dihydrodiol is attributed to CYP2C19 (~47.9%), whereas CYP1B1 contributed little (~3.8%) to its formation (Figure 5c). Other human hepatic P450s have essentially no impact on the formation of this metabolite.

For BP-9-ol, the highest contribution to its formation is attributed to CYP2C19 (~35.5%), followed by CYP3A4 (~31.4%) and CYP1A1 (~27.4%); in contrast, CYP3A5 (~4.7%) and CYP1B1 (~1%) had minor impacts on its formation (Figure 5c). In human liver, the formation of BP-3-ol, which is the major metabolite formed by most of the tested human P450 enzymes, is attributed to CYP3A4 (~52.8%), followed by CYP1A1 (~26.7%) and CYP2C19 (~14.3%). CYP2C9 (~3.3%), 2B6 (~2.1%), 3A5 (~0.4%), and 1B1 (~0.4%) also participate in the formation of this detoxification BP metabolite in human liver, but their contribution is minor (Figure 5c). Other P450 enzymes expressed in human liver have essentially no ability to form BP-3-ol [47].

Collectively, these recent studies [47, 56] have demonstrated that the degree of activation or detoxification of BP in human liver is affected by both the activities of individual human P450 enzymes to metabolize BP, and the expression levels of these P450 enzymes in human liver. Therefore, modulation of levels and activities of hepatic P450, mediated both by their polymorphisms (or internal regulation), and their induction or inhibition by endogenous and exogenous compounds, determines the genotoxic properties of BP. As presented for the liver here an analogous approach might be utilized for evaluating the contributions of human P450 enzymes to BP metabolism in the lung, a target organ for BP carcinogenicity [40]. However, as P450 expression levels in human lungs have not yet been fully characterized, this remains a challenge for future studies.

The CYP1 family comprises three functional genes in two subfamilies. Located on chromosome 15q24.1 *CYP1A1* and *1A2* are highly conserved. The *CYP1B1* gene can be found on chromosome 2p22.2. The best established transcriptional control of the three *CYP1* genes is *via* the aryl hydrocarbon receptor (AHR) but some reports have also linked CYP1A1 activation to the xenobiotic sensor constitutive androstane receptor (CAR) [60]. As BP is mainly activated by CYP1A1 and 1B1, we now describe the regulation of *CYP1* gene expression *via* AHR in more detail.

As shown in Figure 6, in the absence of a ligand, AHR resides in the cytoplasm bound to two heat-shock proteins 90 (HSP90) [61]. Upon ligand binding, the AHR undergoes a conformation change and the HSP90 molecules are detached. This exposes a nuclear localization sequence, resulting in the translocation of AHR into the nucleus where it

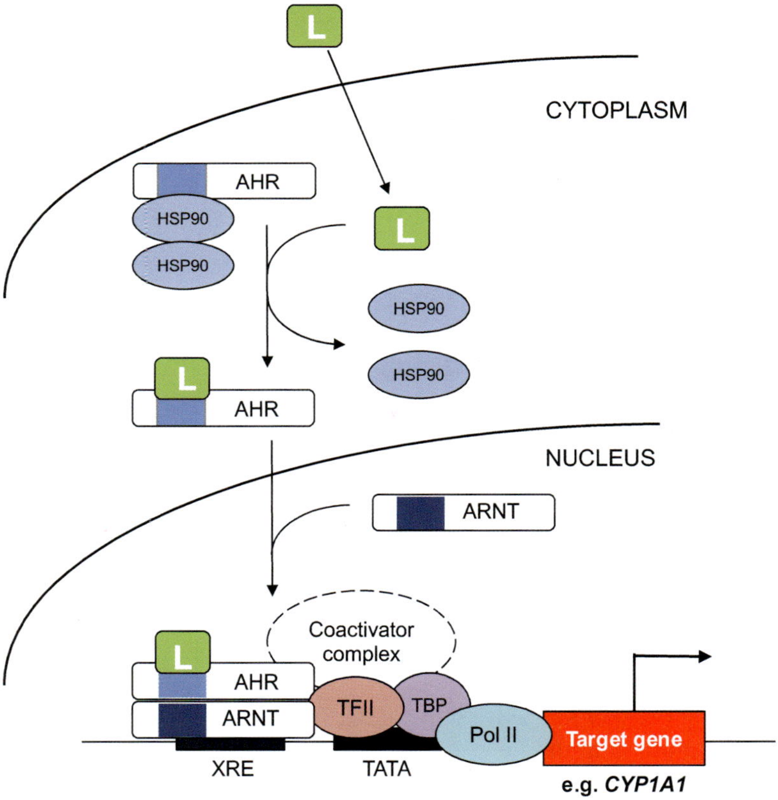

Figure 6. The regulation of *CYP1* gene expression by AHR. See text for details. ARNT, aryl hydrocarbon receptor nuclear translocator; HSP90, heat shock protein 90; Pol II, RNA polymerase II; TATA, gene promoter; TBP, TATA-binding protein; TFII, transcription factor II; XRE, xenobiotic response element.

acquires the AHR nuclear translocator (ARNT) [62, 63]. The formed heterodimer can bind to recognition sequences located in the promoter region of responsive genes. These recognition sequences of the AHR/ARNT complex are referred to as either AHR- or xenobiotic response elements (AREs or XREs) and contain the motif 5′-TNGCGTG-3′ (with N=A, C, G, or T) [64–67]. For example, *CYP1A1* and *1A2* share a 23kb bidirectional promoter that contains at least 13 AHR (XRE) response elements (REs) [68]. The binding of the heterodimer leads to the recruitment of several coactivators (reviewed by [69]) for instance, the histone acetyltransferases p300 and CREB-binding protein (CBP), and components of the general transcription machinery, such as TATA-binding protein (TBP), TFII, and RNA polymerase II to the TATA box. Regulatory components of the signaling pathway such as the AHR repressor (AHRR) are also controlled by AHR generating a negative feedback loop. AHRR can disturb AHR signaling through two mechanisms. It can compete with AHR for ARNT and the AHRR/ARNT heterodimer can also bind to XREs. To avoid prolonged gene activation AHR signaling is quickly terminated. After induction of transcription of its target genes, AHR undergoes nuclear export followed by ubiquitination and degradation by the 26S proteasome [70].

Classical exogenous AHR ligands include dioxin (2,3,7,8-tetrachlorodibenzo-*p*-dioxin [TCDD]) and PAHs. Dioxin has shown the strongest affinity toward the AHR [70] but since PAHs like BP can induce *CYP1* gene expression *via* AHR [71, 72], BP can induce its own metabolic activation thereby promoting BP carcinogenesis.

Recently, a novel pathway has been discovered linking the CYP1A1-mediated bioactivation of BP to p53 function [42, 43]. The tumor suppressor *TP53* gene, which encodes the protein p53, is mutated in approximately 50% of sporadic tumors making it arguably one of the most important cancer genes [41, 73]. p53 can be induced by multiple forms of cellular stress (e.g., DNA damage) and it functions at the crossroads of a network of signaling pathways that are crucial for the regulation of cell growth and apoptosis [74, 75]. Its response to DNA damage is to abrogate DNA synthesis or cell division and this has earned p53 the term "*guardian of the genome*" [76]. Disruption of the normal p53 response by mutation leads to the development of tumors [77]. The best established control of p53 is facilitated *via* the E3 ubiquitin ligase murine double minute 2 (MDM2), which keeps p53 levels low when not required [78]. In normal unstressed cells, MDM2 binds to the transcriptional activation domain (TAD) of p53 and promotes proteasomal degradation of p53 by ubiquitylation. DNA damage (e.g., DNA adducts induced by BP) can cause inhibition of p53-MDM2 binding through phosphorylation of p53 at serine 15 and 20 and subsequently activate p53's transactivation activity. Phosphorylation can be facilitated by mediators such as the kinases ataxia telangiectasia mutated (ATM), ataxia telangiectasia Rad3-related (ATR), and the checkpoint kinases 1 and 2 (CHK1 and CHK2). Further, protein modifications such as methylation and acetylation can add to p53 stabilization [79].

Previously, it has been shown that p53 can induce gene expression of certain *CYPs* by binding to p53REs in the promoter region of these *CYPs* [80]. It was shown that p53 induces the activity of CYP3A4 *via* its binding to p53REs and the subsequent transcriptional enhancement of *CYP3A4*. Recently, using chromatin immunoprecipitation analysis, it was shown that p53 can bind to a p53RE in the regulatory region of the *CYP1A1* gene after BP exposure [42]. Consequently, these results indicated that BP is able to induce CYP1A1 not only by binding as a ligand to AHR promoting *CYP1A1* transcriptional activation but also by induction of p53, which can subsequently trigger *CYP1A1* transcription *via* its binding to p53REs.

6.2. *Aristolochic acid I*

Another example of the dual role of P450 enzymes (activation *vs.* detoxification) in the metabolism of environmental carcinogens is the P450-mediated conversion of the plant carcinogen aristolochic acid I (8-methoxy-6-nitro-phenanthro-(3,4-*d*)-1,3-dioxolo-5-carboxylic acid, AAI), a major component of the natural plant extract AA (Figure 7) [32, 33, 81]. AA has been shown to be the cause of Chinese herb nephropathy [82, 83], now referred to as aristolochic acid nephropathy (AAN) [84, 85]. AAN is a rapidly progressive renal fibrosis with a high risk for patients of developing upper urothelial tract carcinoma (UUC) and, subsequently, bladder urothelial carcinoma [85–87]. AA exposure

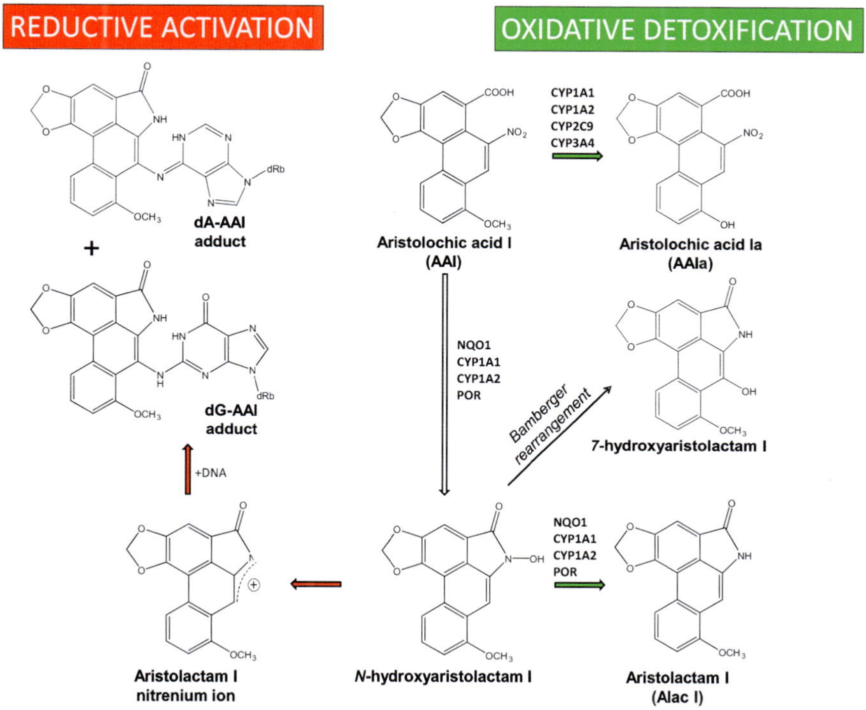

Figure 7. Activation and detoxification pathways of AAI. dA-AAI, 7-(deoxyadenosin-N^6-yl)aristolactam I; dG-AAI, 7-(deoxyguanosin-N^2-yl)aristolactam I; NQO1, NAD(P)H:quinone oxidoreductase.

is also the cause of a similar type of renal disease, Balkan endemic nephropathy (BEN), and its associated occurrence of urothelial malignancy [34, 88–90]. AA and herbal remedies containing AA have been classified as a Group I carcinogen by IARC [91].

Although AAI might directly cause interstitial nephropathy [92], enzymatic activation of AAI to intermediates capable of binding to DNA is a necessary process leading to AA-mediated malignant transformation [32, 34, 84, 85, 88, 93]. Both oxidative and reductive metabolites of AAI are formed in organisms after AAI exposure and these metabolites are excreted in urine and feces (reviewed in [32, 81, 94, 95]). 8-Hydroxyaristolochic acid I (aristolochic acid Ia, AAIa) is the product of oxidative *O*-demethylation of AAI and considered a detoxification metabolite (Figure 7) [96–99]. AAI is reduced to *N*-hydroxyaristolactam I which is either further reduced to aristolactam I (AlacI) or rearranged to 7-hydroxyaristolactam I (Figure 7) [96, 100]. AlacI (Figure 7) is the predominant product of AAI metabolism in humans [96, 101] and should be considered as a detoxification metabolite.

Initial reduction of AAI to *N*-hydroxyaristolactam I is the activation pathway responsible for the carcinogenic effect of AAI. During the reaction, AAI is enzymatically reduced to reactive cyclic acylnitrenium ions that bind to the exocyclic amino groups of adenine and guanine in DNA to form AAI-derived DNA adducts; 7-(deoxyadenosin-N^6-yl)aristolactam I (dA-AAI) and 7-(deoxyguanosin-N^2-yl)aristolactam I (dG-AAI) (Figure 7) [102–107].

Characteristic AA-DNA adducts in renal tissue of AAN and BEN patients are biomarkers of exposure to AA even long after AA exposure has ceased; the dA-AAI adduct is the most abundant adduct formed and the most persistent [86, 90, 104, 108–110]. AA-DNA adduct formation reported in humans is summarized in Table 3. The dA-AAI adduct is an example of an exposure biomarker that has mechanistic relevance as it causes characteristic A to T transversion mutations and such mutations were found in the *TP53* tumor suppressor gene in tumors from AAN and BEN patients [89, 111–113], indicating a molecular mechanism of AA-mediated carcinogenesis [74, 88, 114–117].

P450 enzymes play a dual role in AAI metabolism, as they both activate and detoxify this carcinogen [32, 33, 94, 95, 98, 99, 118–122]. In the case of AAI, two different types of reactions are catalyzed by the P450 enzymes; oxidation (i.e., detoxification) and reduction (i.e., activation) (see Figure 7).

Human CYP1A1 and 1A2 are the major enzymes oxidizing AAI to AAIa under aerobic (i.e., oxidative) conditions [99, 118, 119, 122, 123]. Other P450s such as CYP2C (i.e., CYP2C8/9/19), CYP3A (i.e., CYP3A4/5), 2D6, 2E1, and 1B1 also form AAIa, but their efficiency in catalyzing this reaction is more than one order of magnitude lower compared to CYP1A enzymes [99, 119, 123] (Figure 8a). Cytochrome b_5 does essentially not influence the P450-mediated AAI metabolism [99, 119]. Because the liver is the most important organ responsible for P450-catalyzed xenobiotic biotransformation, the efficiencies of P450s to oxidize AAI (i.e., detoxify

Table 3. DNA adduct levels in human tissue from patients exposed to aristolochic acid.

Country	DNA adduct measured in	DNA adduct detected	DNA adduct level/10^8 nucleotides	Method used for DNA adduct detection	Reference(s)
Belgium	Kidney	dA-AAI	7.0 – 53.0 (n = 6)	TLC[1] ^{32}P-postlabelling	Schmeiser *et al.* 1996 [104]; Bieler *et al.* 1997 [105]
		dG-AAI	0.2 – 1.2 (n = 6)		
		dA-AAII	0.6 – 2.4 (n = 6)		
Belgium	Kidney	dA-AAI	0.1 – 16.5 (n = 38)	TLC ^{32}P-postlabelling	Nortier *et al.* 2000 [86]
		dG-AAI	0.1 – 0.7 (n = 38)		
	Ureter	dA-AAII	0.04 – 0.8 (n = 38)		
		dA-AAI	0.2 – 3.4 (n = 17)		
		dG-AAI	Not reported (n = 17)		
		dA-AAII	Not reported (n = 17)		
Belgium	Kidney	dA-AAI	2.9 – 5.0 (n = 2)	TLC ^{32}P-postlabelling	Arlt *et al.*, 2001 [124]
		dG-AAI	Not detected (n = 2)		
		dA-AAII	0.3 – 0.9 (n = 2)		
Belgium/China	Kidney	dA-AAI	1.8 (n = 1)	TLC ^{32}P-postlabelling	Gillerot *et al.*, 2001 [125]
UK	Kidney	dA-AAI	3.8 (n = 1)	TLC ^{32}P-postlabelling	Lord *et al.*, 2001 [126]; Lord *et al.*, 2004 [111]
	Ureter	dA-AAI	40.0 (n = 1)		
	Bladder	dA-AAI	20.0 (n = 1)		
	Breast tumor	dA-AAI	1.0 (n = 1)		
	Liver tumor[2]	dA-AAI	1.0 (n = 1)		
	Liver (normal)	dA-AAI	16.0 (n = 1)		
Croatia	Kidney	dA-AAI	0.56 – 1.71 (n = 2)	TLC ^{32}P-postlabelling	Arlt *et al.*, 2002 [127]

(Continued)

Table 3. *(Continued)*

Country	DNA adduct measured in	DNA adduct detected	DNA adduct level/10^8 nucleotides	Method used for DNA adduct detection	Reference(s)
Belgium	Kidney	dA-AAI	8.1 (n = 1)	TLC ^{32}P-postlabelling	Nortier *et al.*, 2003 [128]
	Bladder	dA-AAI	Not detectable (n = 1)		
	Liver	dA-AAI	8.7 (n = 1)		
	Pancreas	dA-AAI	0.8 (n = 1)		
	Stomach	dA-AAI	1.9 (n = 1)		
	Lung	dA-AAI	0.2 (n=1)		
	Lymph nodes	dA-AAI	0.5 (n = 1)		
France	Kidney	dA-AAI	0.1 – 5.4 (n = 2)	TLC ^{32}P-postlabelling	Arlt *et al.*, 2004 [129]
	Ureter	dA-AAI	0.3 (n = 1)		
	Liver	dA-AAI	1.8 (n = 1)		
	Lung	dA-AAI	2.2 (n = 1)		
	Stomach	dA-AAI	1.0 (n = 1)		
	Small Intestine	dA-AAI	1.0 (n = 1)		
	Spleen	dA-AAI	2.1 (n = 1)		
	Adrenal	dA-AAI	1.9 (n = 1)		
	Brain	dA-AAI	0.2 (n = 1)		
	Heart	dA-AAI	Not detected (n = 1)		
China	Kidney	dA-AAI	Detected, but not quantified	TLC ^{32}P-postlabelling	Lo *et al.*, 2005 [130]
USA	Kidney (cortex, medulla, and pelvis)	dA-AA	11.0 – 34 (n = 1)	PAGE[3] ^{32}P-postlabelling	Grollman *et al.*, 2007 [89]
		dG-AA	0.2 – 1 (n = 1)	Mass spectrometry	
	Kidney (cortex)	dA-AAI & dA-AAII	Detected, but not quantified (n = 1)		

Croatia	Kidney	dA-AA dG-AA	8 – 59 (n = 4) 2 – 62 (n = 4)	PAGE ^{32}P-postlabelling	Grollman *et al.*, 2007 [89]
Taiwan	Kidney	dA-AA	1.4 – 234 (89/148 [60%])	PAGE ^{32}P-postlabelling	Chen *et al.*, 2012 [113]
Bosnia, Croatia & Serbia	Kidney	dA-AA	0.2 – 19.2 (47/67 [70%]	PAGE ^{32}P-postlabelling	Jelakovic *et al.*, 2012 [131]
Romania	Kidney	dA-AAI	0.3 – 6.5 (n = 7)	TLC ^{32}P-postlabelling	Schmeiser *et al.*, 2012 [109]
Belgium	Kidney	dA-AAI	2 – 22 (n = 11)	TLC ^{32}P-postlabelling	Schmeiser *et al.*, 2014 [110]
	Kidney	dA-AAI	Detected, but not quantified (n = 1)	Mass spectrometry	
Croatia & Serbia	Kidney	dA-AAI	0.2 – 7.0 (n = 15)	Mass spectrometry	Yun *et al.*, 2014 [132]
Belgium	Kidney	dA-AAI	5 (n = 1)	TLC ^{32}P-postlabelling	Roumeguere *et al.* 2015 [133]
Romania	Kidney	dA-AAI	0.7 – 26.8 (n = 14)	Mass spectrometry	Turesky *et al.*, 2016 [134]
Taiwan	Kidney	dA-AAI	0.3 – 258 (39/51 [76%]	Mass spectrometry	Hoang *et al.*, 2016 [135]

[1] Thin-layer chromatography
[2] Metastasis from breast
[3] Polyacrylamide gel electrophoresis

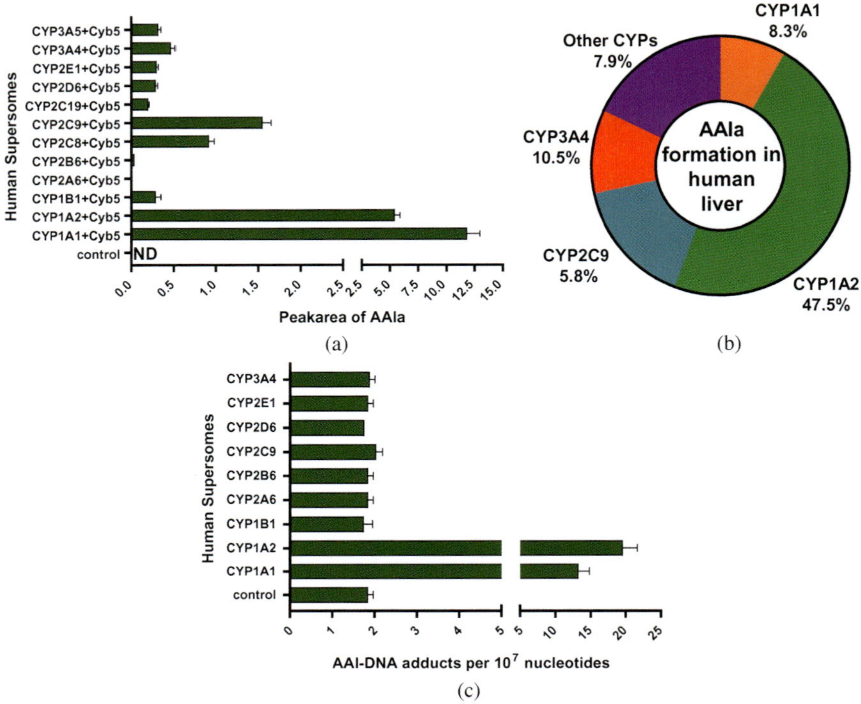

Figure 8. (a) AAIa formation by human recombinant P450s in the presence of cytochrome b_5 (Cyb5). Data shown are mean ± SD ($n = 3$). ND, not detected. Data previously published in [99, 119, 123]. (b) Contributions of P450 enzymes to AAIa formation in human livers. (c). AAI-DNA adduct formation by human P450 enzymes [59, 143].

AAI to AAIa) in human livers were studied in detail [99, 119, 123]. Human CYP1A2 followed by CYP2C9, CYP3A4, and CYP1A1 are the major enzymes contributing to AAI oxidation in human liver (Figure 8b) [123].

O-Demethylation of AAI proceeds *via* the P450-mediated attack of the carbon atom of the methoxy group by oxygen, which leads to the formation of the α-C-hydroxylation product that decomposes to formaldehyde and AAIa. Therefore, the binding orientation of the methoxy group of AAI in the binary complex of AAI with the P450 active sites, which is a prerequisite process for *O*-demethylation of AAI, should dictate the efficiency of individual P450s to catalyze this reaction. Thus, differences among abilities of P450 enzymes to *O*-demethylate AAI might be caused by the affinities of AAI to these enzymes and the binding orientation of the methoxy group of this compound in their active sites.

Using molecular modeling techniques such as *in silico* docking, employing soft-soft, flexible, docking procedure (as outlined in [123]), interactions of AAI with the active site of the P450 Compounds I of human CYP1A1, 1A2, 2C9, and 3A4 were found. P450 Compounds I is the highly reactive P450 intermediate that is responsible for the P450 oxidations, postulated to be an iron(IV)oxo (or ferryl) species with an additional oxidizing equivalent delocalized over the porfyrin and thiolate ligands [136, 137]. These results indicated that CYP1A1, 1A2, 2C9, and 3A4 *O*-demethylate AAI in human liver but with different effectiveness [123]. Every *in silico* docking predicted several binding orientations of the AAI molecule in the P450 active site. Positions showing short distances (3.7–4.9 Å) between the methoxy group of AAI and the activated oxygen atom in Compounds I of P450s was found for all the P450 enzymes examined (Table 4). However, CYP2C9 and 3A4 enzymes were predicted to bind the AAI molecule with a significantly lower affinity than P450s of the CYP1A subfamily as illustrated by the values of binding free energy shown in Table 4. The predicted binding free energy of AAI to human CYP1A1 was slightly lower than that to CYP1A2, but the distance between the carbon in the methoxy group of AAI and the oxygen atom on heme iron in the binary

Table 4. The predicted binding free energies and distances facilitating O-demethylation of AAI bound in selected P450s complexes.

	The most stable productive orientations of AAI in the complex with P450	
	Estimated Free Energy of binding [kcal/mol]	O(Comp I)-OCH_3 (AAI) distance [Å][a]
CYP1A1	−7.0	4.4
CYP1A2	−7.7	4.9
CYP2C9	−5.3	4.3
CYP3A4	−6.0	3.7

[a]Distance between the carbon in the methoxy group of AAI and oxygen atom on heme iron in the complex of an activated P450 enzyme (Compound I) with AAI,

complex of CYP1A1 with AAI was shorter by 0.5 Å (Table 4). The larger distance of the reacting groups would result in a decreased reaction rate during CYP1A2-catalyzed AAI demethylation. Collectively, these results provide an explanation why CYP1A1 and 1A2 are most effective in AAI oxidation, while other P450s (i.e., CYP2C9 and 3A4) are less active in catalyzing this reaction (compare Figures 8a and 8b).

The importance of CYP1A1 and 1A2 enzymes to catalyze the oxidative *O*-demethylation of AAI to AAIa *in vivo* was shown utilizing genetically modified *Cyp1a*-knockout and *CYP1A*-humanized mouse lines [98, 99, 122]. Other studies were conducted in mice where Cyp1a expression was enhanced by inducers [138]. These studies showed that murine Cyp1a1 and 1a2 oxidize AAI to AAIa and protect these animals from AAI-induced acute renal injury [122, 139]. When the animals are not able to detoxify AAI through demethylation to AAIa, higher concentrations of unmetabolized AAI are available for bioactivation to form AAI-DNA adducts, or to cause renal injury. Indeed, high levels of AAI-DNA adducts were found in several organs of *Cyp1a*-knockout mouse lines, predominantly in the kidney [98, 122]. In addition, an increase in AAI-mediated nephrotoxicity in mice lacking Cyp1a was observed in these animals [122]. Furthermore, induction of CYP1A1 and 1A2 in rats led to an increase in AAI detoxification to AAIa, thereby reducing the actual amount of AAI available for reductive activation [120]. This subsequently resulted in lower AAI-DNA adduct levels in the rats *in vivo* demonstrating that of the two opposing CYP1A1/2-catalyzed reactions (i.e., oxidation and reduction of AAI), CYP1A1/2-mediated oxidative detoxification prevails in AAI metabolism *in vivo*, thereby suppressing levels of AAI-DNA adducts (i.e., AAI genotoxicity) [120]. Even more importantly, AAI oxidation to AAIa by human CYP1A1 and 1A2 was also shown *in vivo* using *CYP1A*-humanized mouse lines [99].

Reduction of AAI is catalyzed not only by P450 enzymes, but also by cytosolic enzymes. For example, NAD(P)H:quinone oxidoreductase (NQO1) is the most important cytosolic nitroreductase activating AAI [140–144] (see Figure 7). Of human P450 enzymes, CYP1A1 and 1A2 are the only enzymes that are able to efficiently activate AAI by nitroreduction under anaerobic conditions, while other P450s are almost ineffective in catalyzing this reaction (Figure 8c). In contrast to CYP1A1 and 1A2, the closely related CYP1B1, which is lacking the hydroxyl group containing residues in its active site, is ineffective in catalyzing AAI nitroreductase activity (see Figure 8c) [59, 143, 145–147].

Using molecular modeling [147] and site-directed mutagenesis [121], the possible mechanisms of CYP1A1/1A2/1B1-catalyzed AAI nitroreduction were explained. Molecular modeling predicted that the hydroxyl groups of serine122/threonine124 (Ser122/Thr124) amino acids in the CYP1A1/1A2-AAI binary complexes located near to the nitro group of AAI are mechanistically important as they provide the proton required for the stepwise reduction reaction. These hydroxyl groups in the CYP1A1/1A2-AAI binary complexes are closely placed to the nitro group of the AAI, thereby being able to provide a proton necessary for the stepwise reduction reaction. In contrast, the closely related CYP1B1, which is lacking the hydroxyl group containing residues in its active site, is ineffective in catalyzing AAI nitroreductase activity [143, 147].

Based on these models, mutant forms of CYP1A1 and 1A2 were prepared, where Ser122 and Thr124 were replaced by Ala (CYP1A1-S122A) and Val (CYP1A2-T124V), respectively. Similarly, a CYP1B1 mutant was prepared in which Ala133 was replaced by Ser (CYP1B1-A133S) [121]. Employing these models, S122A and T124V mutations in CYP1A1 and 1A2, respectively, abolished the efficiency of CYP1A1 and 1A2 enzymes to reduce AAI to species generating AAI-DNA adducts *in vitro*. In contrast, the formation of AAI-DNA adducts was catalyzed by CYP1B1 with the A133S mutation. These results fully confirm the importance of the hydroxyl group possessing amino acids in the active center of CYP1A1 and 1A2 for AAI nitroreduction predicted by the theoretical approaches [121, 147].

The role of CYP1A1 and 1A2 in the reductive metabolism of AAI can be explained by the fact that AAI acts as a ligand substrate for human CYP1A1 and 1A2 enzymes, being bound to the heme iron instead of molecular oxygen at low oxygen concentrations, and is reduced instead of being oxidized during the P450-mediated reaction cycle [32, 121, 147]. Under aerobic (i.e., oxidative) conditions, AAI is a classical substrate of CYP1A1 and 1A2, being bound to Compound I of these P450 enzymes. During this process, one atom of molecular oxygen is used to *O*-demethylate the methoxy group of AAI to generate AAIa [32, 123]. These experimental findings [59, 98, 99, 119, 123, 146] are supported by theoretical approaches (i.e., computational studies) [32, 33, 121, 123, 147], indicating that, in addition to CYP1A1 and 1A2 expression levels, the *in vivo* oxygen concentration in tissues might affect the balance between AAI nitroreduction and oxidative *O*-demethylation, which in turn influences tissue-specific toxicity and carcinogenicity.

7. Mechanism of a Monooxygenation Reaction Catalyzed by P450s

The most frequent reaction catalyzed by P450 enzymes is the monooxygenase reaction catalyzing the insertion of one atom of oxygen molecule into the P450 substrate, while the second atom of the oxygen molecule is reduced to water (see equation above). This process is also a typical reaction involved in oxidation of BP and AAI leading to C-hydroxylated and *O*-demethylated metabolites, respectively. However, when a substrate of P450 is bound as a ligand to the heme iron, it can be reduced instead of an oxygen molecule, which is the case during the P450-mediated activation of AAI by nitroreduction [32, 121, 147].

In the active center of P450s, the oxygen is activated by two electrons. Although POR is considered an essential constituent of the electron transport chain toward P450 (Figure 9) [14], its exact role in the P450-mediated reaction cycle is still not clearly established. Namely, the second electron needed for the reduction of P450s in the monooxygenase reaction cycle may also be provided by cytochrome b_5 in the presence of POR or NADH:cytochrome b_5 reductase (compare Figure 3), but cytochrome b_5 plays also additional roles in the monooxygenase system [15, 23, 148–152].

It was demonstrated that a second electron donor, cytochrome b_5, could also modulate (stimulate and/or inhibit) the activity of several P450s both *in vitro* [15, 149, 151, 152] and *in vivo* [152–156]. It is noteworthy that besides the ability of cytochrome b_5 to modulate P450 catalysis by influencing catalytic activities of P450s by donating the second electron to P450 in the catalytic cycle, cytochrome b_5 can also act as an allosteric modifier of the oxygenase (for reviews see [148, 149, 151, 157, 158]). The mechanism(s) underlying such allosteric effects are based on reports that apo-cytochrome b_5 can stimulate P450 catalysis. However, the mechanism(s) remain largely uncertain to date. Nevertheless, it has been shown that cytochrome b_5 binding can cause conformational changes to the substrate access channel and binding pocket in the P450 enzyme [17, 148, 149, 151, 157–160].

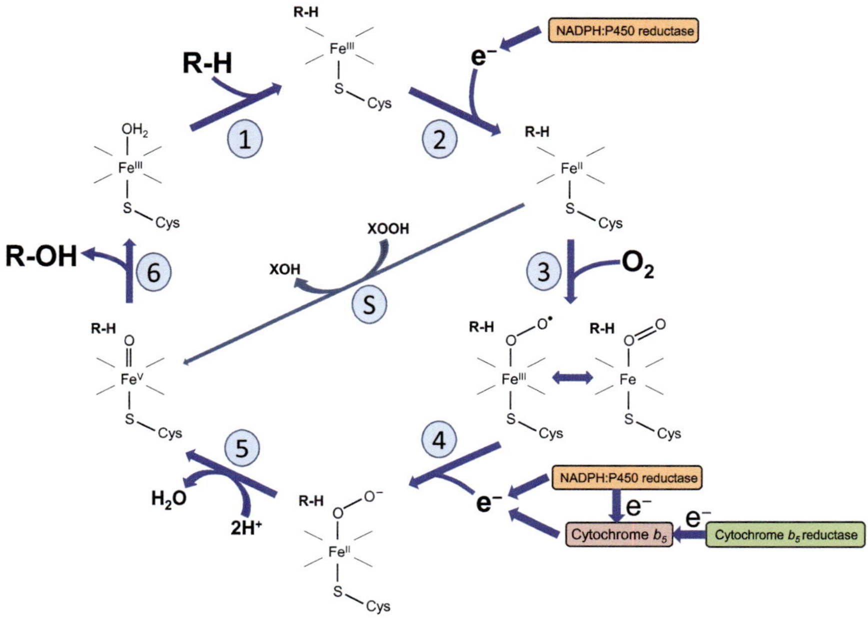

Figure 9. Classical scheme of the reaction cycle of the P450 monoxygenase system, where its first and second reduction is mediated by NADPH:cytochrome P450 reductase (POR). See text for details.

In general, the catalytic cycle of P450s proceeds by reactions shown in Figure 9 as follows:

Step 1. The P450 substrate binds in proximity to the heme iron, on the side opposite to the axial thiolate of the Cysteine (Cys) residue in the active site of the enzyme. Substrate binding produces a conformational change of the active site, often displacing a water molecule from the distal axial coordination position of the heme iron [14, 161] and changing the state of the heme iron from low spin to high spin [162].

Step 2. Substrate binding induces the electron transfer from NADPH *via* POR [163].

Step 3. Molecular oxygen binds to the resulting ferrous heme center at the distal axial coordination position, initially giving a dioxygen adduct (i.e., Fe–O_2).

Step 4. The second electron is transferred either from POR or cytochrome b_5 reducing the Fe–O_2 adduct to give a short-lived peroxo state.

Step 5. The peroxo group formed in Step 4 is rapidly protonated twice, releasing one molecule of water and forming the highly reactive species referred to as P450 Compound I. This highly reactive intermediate was isolated in 2010 [136], P450 Compound 1 is an iron(IV) oxo (or ferryl) species with an additional oxidizing equivalent delocalized over the porphyrin skeleton and thiolate ligands. Evidence for the alternative perferryl iron(V)-oxo [14, 161] is lacking [136].

Step 6. Depending on the substrate and enzyme involved, P450 enzymes can catalyze a wide variety of reactions. After the product has been released from the active site, the enzyme returns to its original state, with a water molecule returning to occupy the distal coordination position of the iron nucleus.

An alternative route for monooxygenation is *via* the "peroxide shunt" (Path "S") in Figure 9. This pathway entails oxidation of the ferric-substrate complex with oxygen-atom donors such as peroxides [14, 15]. A hypothetical peroxide "XOOH" is shown in the diagram (Figure 9).

However, in contrast to this generally accepted mechanism, experiments with human and rat CYP1A1 have shown the involvement of cytochrome b_5 in the NADH-dependent hydroxylation of BP in reconstituted P450-containing systems [51, 52, 164, 165]. Indeed, it has been proven that during the oxidation of BP by CYP1A1, POR as a transporter of electrons from NADPH to CYP1A1 can be substituted by the NADH:cytochrome b_5 reductase/cytochrome b_5/NADH system thereby acting as the sole electron donor for both reduction steps in the CYP1A1-mediated reaction cycle (Figure 10) [51, 52]. These findings demonstrated that the reaction cycle of BP oxidation catalyzed by CYP1A1 can proceed by ways that differ from the abovementioned generally accepted mechanism, where the first reduction of P450 is considered to be catalyzed by POR without cytochrome b_5 [15, 51, 52, 161, 163, 166–168].

Analogous results were found for testosterone 6-β-hydroxylation, a reaction catalyzed by CYP3A4 and used as a marker for CYP3A4 enzyme activity [158, 169]. Further studies have demonstrated the impact of the NADH:cytochrome b_5 reductase/cytochrome b_5/NADH system on the

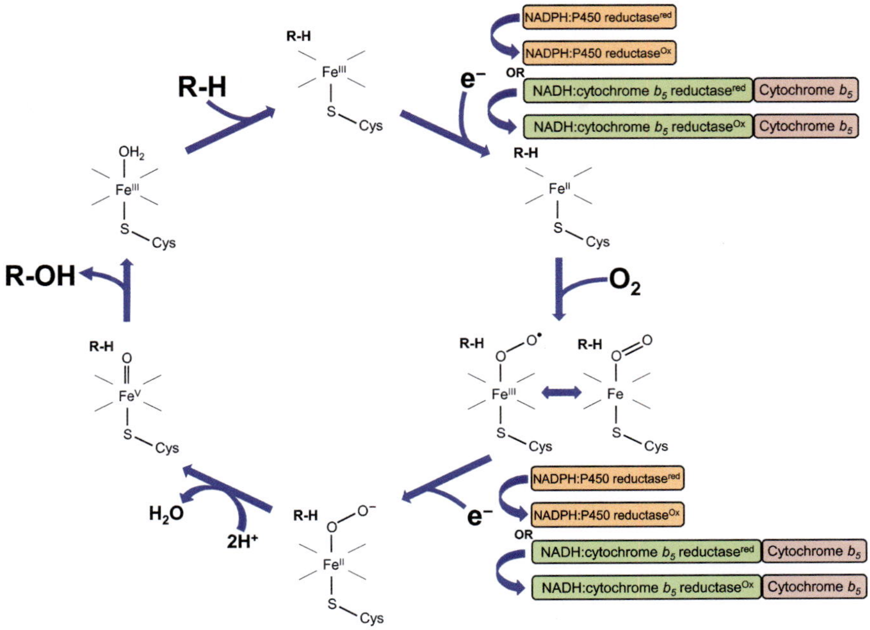

Figure 10. Scheme of the novel reaction cycle of P450, where its first and second reduction is mediated not only by NADPH:cytochrome P450 reductase (POR), but exclusively also by the system of cytochrome b_5 with NADH:cytochrome b_5 reductase. See text for details.

CYP3A4-mediated oxidation of the anticancer drug ellipticine [170]. Collectively, these finding indicate that this mechanism is valid not only for CYP1A1, but also for other P450 enzymes.

Therefore, the actual presence and amounts not only of P450 and POR as the components of the P450 monooxygenase systems, but also NADH:cytochrome b_5 reductase and its substrate cytochrome b_5 can dramatically influence the efficiencies of at least some of the P450s to oxidize their substrates.

8. Concluding Remarks

This chapter provided a summary on the role of P450 enzymes in the metabolism of some environmental carcinogens, and on mechanisms of the reaction cycle of the microsomal P450-monooxygenase system. We showed that the functions of this system in chemical carcinogenesis are complex and that P450 enzymes play a crucial role both in metabolic activation and in detoxification of two human carcinogens highlighted in this chapter. BP and AAI are metabolically activated by P450s to reactive intermediates forming premutagenic DNA adducts, while other P450-mediated reactions lead to the formation of detoxification metabolites that are excreted. Whereas BP is oxidatively metabolized (activated or detoxified) by P450s, AAI can be metabolized both oxidatively (to its detoxification metabolite AAIa) and reductively (to be activated to DNA-binding metabolites). This indicates that metabolism of these carcinogens is dictated by the *in vivo* oxygen concentration in tissues, which can affect both the efficiency of their oxidative reactions and the balance between nitroreduction and oxidative *O*-demethylation of one of them, AAI. This finally influences their tissue-specific toxicity and carcinogenicity. Concerning the mechanisms of reactions catalyzed by the microsomal P450 monooxygenase system reviewed in this chapter, we showed that not only the NADPH/POR reduction system, but also the system consisting of another microsomal reductase, NADH:cytochrome b_5 reductase together with its substrate cytochrome b_5 and the coenzyme NADH, can act as a donor of electrons both for the first and the second reduction of at least some of P450 enzymes in their reaction cycle. These data demonstrate that in the presence of cytochrome b_5 the catalytic cycle of the P450 reactions in the monooxygenase system can proceed by different ways than it was suggested previously for the generally accepted mechanism, namely, the mechanism suggesting that the NADPH/POR system is necessary for the P450 reaction cycle [14, 15]. The finding that cytochrome b_5, together with its reductase, NADH:cytochrome b_5 reductase, can function as exclusive donors of both electrons in the first and second reduction steps of the P450-mediated oxidation of their substrates is a challenge for further research; especially, for studying the mechanism(s) of kinetical and structural features of this novel phenomenon.

Acknowledgments

Work at Charles University is supported by Grant Agency of Czech Republic (Grant GACR 17-12816S). Work at King's College London is supported by Cancer Research United Kingdom (grant number C313/A143329), the Wellcome Trust (grants 101126/Z/13/Z and 101126/B/13/Z), the Natural Environmental Research Council (grant NE/L006782/1), and in part by the National Institute for Health Research Health Protection Research Unit (NIHR HPRU) in Health Impact of Environmental Hazards at King's College London in partnership with Public Health England (PHE). and Imperial College London Laura E. Wohak was supported by a PhD studentship from the Institute of Cancer Research, London, United Kingdom. The views expressed are those of the authors and not necessarily those of the National Health Service, the NIHR, the Department of Health & Social Care, or PHE.

Abbreviations

AA	aristolochic acid
AAI	aristolochic acid I
AAIa	aristolochic acid Ia
AAN	aristolochic acid nephropathy
ABC	ATP-binding cassette
AFB_1	aflatoxin B_1
AHR	aryl hydrocarbon receptor
AKR	aldo-keto reductase
AlacI	aristolactam I
ARE	AHR response element
ARNT	AHR nuclear translocator
ATM	ataxia telangiectasia mutated
ATP	adenosine 5′-triphosphate
ATR	ataxia telangiectasia Rad3-related
BEN	Balkan endemic nephropathy
BP	benzo[*a*]pyrene
CAR	constitutive androstane receptor
CBP	CREB-binding protein
CHK1	checkpoint kinase 1
CHK2	checkpoint kinase 2
CHN	Chinese herbs nephropathy
COMT	catechol-*O*-methyltransferase

COX	cyclooxygenase
CYP	cytochrome P450
dA	deoxyadenosine
dA-AAI	7-(deoxyadenosin-N^6-yl)aristolactam I
dG	deoxyguanosine
dG-AAI	7-(deoxyguanosin-N^2-yl)aristolactam I
EC 1.14.14.1	cytochrome P450 enzymes
FMO	flavin-containing monooxygenases
GST	glutathione *S*-transferase
HAA	heterocyclic aromatic amine
HSP90	heat-shock protein 90
IARC	International Agency for Research on Cancer
*m*EH	microsomal epoxide hydrolase
MDM2	murine double minute 2
MFO	mixed-function oxidase
NADH	nicotinamide adenine dinucleotide reduced
NADPH	nicotinamide adenine dinucleotide phosphate reduced
NAT	*N*,*O*-acetyltransferase
NQO1	NAD(P)H:quinone oxidoreductase 1
PAH	polycyclic aromatic hydrocarbon
POR	NADPH:cytochrome P450 oxidoreductase
RE	response element
SLC	solute carrier
SULT	sulfotransferase
TAD	transcriptional activation domain
TBP	TATA-binding protein
TCDD	2,3,7,8-tetrachlorodibenzo-*p*-dioxin
UGT	uridine diphosphate glucuronosyltransferase
UUC	upper urothelial tract carcinoma
XRE	xenobiotic response element

References

1. Phillips DH, Arlt VM. Genotoxicity: damage to DNA and its consequences. *EXS* 2009; 99: 87–110.

2. Baird WM, Hooven LA, Mahadevan B. Carcinogenic polycyclic aromatic hydrocarbon-DNA adducts and mechanism of action. *Environ Mol Mutagen* 2005; 45: 106–114.
3. Turesky RJ, Le Marchand L. Metabolism and biomarkers of heterocyclic aromatic amines in molecular epidemiology studies: lessons learned from aromatic amines. *Chem Res Toxicol* 2011; 24: 1169–1214.
4. Phillips DH, Venitt S. DNA and protein adducts in human tissues resulting from exposure to tobacco smoke. *Int J Cancer* 2012; 131: 2733–2753.
5. Phillips DH, Arlt VM. The 32P-postlabeling assay for DNA adducts. *Nat Protoc* 2007; 2: 2772–2781.
6. Poirier MC. Linking DNA adduct formation and human cancer risk in chemical carcinogenesis. *Environ Mol Mutagen* 2016; 57: 499–507.
7. Rappaport SM, Li H, Grigoryan H, Funk WE, Williams ER. Adductomics: characterizing exposures to reactive electrophiles. *Toxicol Lett* 2012; 213: 83–90.
8. Walsh AA, Szklarz GD, Scott EE. Human cytochrome P450 1A1 structure and utility in understanding drug and xenobiotic metabolism. *J Biol Chem* 2013; 288: 12932–12943.
9. Coleman MD. *Human Drug Metabolism — An Introduction*, 2nd Edition. Wiley-Blackwell, Chichester, 2010.
10. Rendic S, Guengerich FP. Contributions of human enzymes in carcinogen metabolism. *Chem Res Toxicol* 2012; 25: 1316–1383.
11. Glatt H. Sulfotransferases in the bioactivation of xenobiotics. *Chem Biol Interact* 2000; 129: 141–170.
12. Omiecinski CJ, Vanden Heuvel JP, Perdew GH, Peters JM. Xenobiotic metabolism, disposition, and regulation by receptors: from biochemical phenomenon to predictors of major toxicities. *Toxicol Sci* 2011; 120 Suppl 1: S49–S75.
13. Sim SC, Ingelman-Sundberg M. The Human Cytochrome P450 (CYP) Allele Nomenclature website: a peer-reviewed database of CYP variants and their associated effects. *Hum Genomics* 2010; 4: 278–281.
14. Guengerich FP. Common and uncommon cytochrome P450 reactions related to metabolism and chemical toxicity. *Chem Res Toxicol* 2001; 14: 611–650.
15. Guengerich FP. Cytochrome p450 and chemical toxicology. *Chem Res Toxicol* 2008; 21: 70–83.
16. Wienkers LC, Heath TG. Predicting *in vivo* drug interactions from *in vitro* drug discovery data. *Nat Rev Drug Discov* 2005; 4: 825–833.
17. Porter TD. The roles of cytochrome b5 in cytochrome P450 reactions. *J Biochem Mol Toxicol* 2002; 16: 311–316.

18. Nebert DW, Wikvall K, Miller WL. Human cytochromes P450 in health and disease. *Philos Trans R Soc Lond B Biol Sci* 2013; 368: 20120431.
19. Luch A. Nature and nurture — lessons from chemical carcinogenesis. *Nat Rev* 2005; 5: 113–125.
20. Nebert DW, Dalton TP. The role of cytochrome P450 enzymes in endogenous signalling pathways and environmental carcinogenesis. *Nat Rev* 2006; 6: 947–960.
21. Phillips DH. *Macromolecular Adducts as Biomarkers of Human Exposure to Polycyclic Aromatic Hydrocarbons*. Imperial College Press, London, 2005; 137–169.
22. Phillips DH. DNA adducts as markers of exposure and risk. *Mutat Res* 2005; 577: 284–292.
23. Guengerich FP, Liebler DC. Enzymatic activation of chemicals to toxic metabolites. *Crit Rev Toxicol* 1985; 14: 259–307.
24. Rendic S, Di Carlo FJ. Human cytochrome P450 enzymes: a status report summarizing their reactions, substrates, inducers, and inhibitors. *Drug Metab Rev* 1997; 29: 413–580.
25. Guengerich FP, Shimada T. Oxidation of toxic and carcinogenic chemicals by human cytochrome P-450 enzymes. *Chem Res Toxicol* 1991; 4: 391–407.
26. Besaratinia A, Tommasi S. Genotoxicity of tobacco smoke-derived aromatic amines and bladder cancer: current state of knowledge and future research directions. *Faseb J* 2013; 27: 2090–2100.
27. Hecht SS. Progress and challenges in selected areas of tobacco carcinogenesis. *Chem Res Toxicol* 2008; 21: 160–171.
28. Dohnal V, Wu Q, Kuca K. Metabolism of aflatoxins: key enzymes and interindividual as well as interspecies differences. *Arch Toxicol* 2014; 88: 1635–1644.
29. Hamouchene H, Arlt VM, Giddings I, Phillips DH. Influence of cell cycle on responses of MCF-7 cells to benzo[a]pyrene. *BMC Genomics* 2011; 12: 333.
30. Nebert DW, Shi Z, Galvez-Peralta M, Uno S, Dragin N. Oral benzo[a]pyrene: understanding pharmacokinetics, detoxication, and consequences — Cyp1 knockout mouse lines as a paradigm. *Mol Pharmacol* 2013; 84: 304–313.
31. Luch A, Baird WM. *Metabolic Activation and Detoxification of Polycyclic Aromatic Hydrocarbons*. Imperial College Press, London, 2005; 19–96.
32. Stiborova M, Martinek V, Frei E, Arlt VM, Schmeiser HH. Enzymes metabolizing aristolochic acid and their contribution to the development of aristolochic acid nephropathy and urothelial cancer. *Curr Drug Metab* 2013; 14: 695–705.

33. Stiborova M, Frei E, Arlt VM, Schmeiser HH. Knockout and humanized mice as suitable tools to identify enzymes metabolizing the human carcinogen aristolochic acid. *Xenobiotica* 2014; 44: 135–145.
34. Stiborova M, Arlt VM, Schmeiser HH. Balkan endemic nephropathy: An update on its aetiology. *Arch Toxicol* 2016; 90: 2595–2615.
35. IARC. Some non-heterocyclic polycyclic aromatic hydrocarbons and some related exposures. *IARC Monogr Eval Carcinog Risk Hum* 2010; 92.
36. Phillips DH. Polycyclic aromatic hydrocarbons in the diet. *Mutat Res* 1999; 443: 139–147.
37. Phillips DH. Fifty years of benzo(a)pyrene. *Nature* 1983; 303: 468–472.
38. Perera FP. Environment and cancer: who are susceptible? *Science* (New York, NY) 1997; 278: 1068–1073.
39. Arlt VM, Stiborova M, Henderson CJ, Thiemann M, Frei E, Aimova D, *et al.* Metabolic activation of benzo[a]pyrene *in vitro* by hepatic cytochrome P450 contrasts with detoxification *in vivo*: experiments with hepatic cytochrome P450 reductase null mice. *Carcinogenesis* 2008; 29: 656–665.
40. Arlt VM, Krais AM, Godschalk RW, Riffo-Vasquez Y, Mrizova I, Roufosse CA, *et al.* Pulmonary inflammation impacts on CYP1A1-mediated respiratory tract DNA damage induced by the carcinogenic air pollutant benzo[a] pyrene. *Toxicol Sci* 2015; 146: 213–225.
41. Kucab JE, van Steeg H, Luijten M, Schmeiser HH, White PA, Phillips DH, *et al.* TP53 mutations induced by BPDE in Xpa-WT and Xpa-Null human TP53 knock-in (Hupki) mouse embryo fibroblasts. *Mutat Res* 2015; 773: 48–62.
42. Wohak LE, Krais AM, Kucab JE, Stertmann J, Ovrebo S, Seidel A, *et al.* Carcinogenic polycyclic aromatic hydrocarbons induce CYP1A1 in human cells *via* a p53-dependent mechanism. *Arch Toxicol* 2016; 90: 291–304.
43. Krais AM, Speksnijder EN, Melis JP, Indra R, Moserova M, Godschalk RW, *et al.* The impact of p53 on DNA damage and metabolic activation of the environmental carcinogen benzo[a]pyrene: effects in Trp53(+/+), Trp53(+/–) and Trp53(–/–) mice. *Arch Toxicol* 2016; 90: 839–851.
44. Bauer E, Guo Z, Ueng YF, Bell LC, Zeldin D, Guengerich FP. Oxidation of benzo[a]pyrene by recombinant human cytochrome P450 enzymes. *Chem Res Toxicol* 1995; 8: 136–142.
45. Kim JH, Stansbury KH, Walker NJ, Trush MA, Strickland PT, Sutter TR. Metabolism of benzo[a]pyrene and benzo[a]pyrene-7,8-diol by human cytochrome P450 1B1. *Carcinogenesis* 1998; 19: 1847–1853.
46. Stiborova M, Moserova M, Cerna V, Indra R, Dracinsky M, Sulc M, *et al.* Cytochrome b5 and epoxide hydrolase contribute to benzo[a]pyrene-DNA

adduct formation catalyzed by cytochrome P450 1A1 under low NADPH:P450 oxidoreductase conditions. *Toxicology* 2014; 318: 1–12.

47. Sulc M, Indra R, Moserova M, Schmeiser HH, Frei E, Arlt VM, *et al.* The impact of individual cytochrome P450 enzymes on oxidative metabolism of benzo[a]pyrene in human livers. *Environ Mol Mutagen* 2016; 57: 229–235.
48. Chun YJ, Shimada T, Guengerich FP. Construction of a human cytochrome P450 1A1: rat NADPH-cytochrome P450 reductase fusion protein cDNA and expression in Escherichia coli, purification, and catalytic properties of the enzyme in bacterial cells and after purification. *Arch Biochem Biophys* 1996; 330: 48–58.
49. Jiang H, Gelhaus SL, Mangal D, Harvey RG, Blair IA, Penning TM. Metabolism of benzo[a]pyrene in human bronchoalveolar H358 cells using liquid chromatography-mass spectrometry, *Chem Res Toxicol* 2007; 20: 1331–1341.
50. Arlt VM, Poirier MC, Sykes SE, John K, Moserova M, Stiborova M, *et al.* Exposure to benzo[a]pyrene of Hepatic Cytochrome P450 Reductase Null (HRN) and P450 Reductase Conditional Null (RCN) mice: detection of benzo[a]pyrene diol epoxide-DNA adducts by immunohistochemistry and 32P-postlabelling. *Toxicol Lett* 2012; 213: 160–166.
51. Stiborova M, Indra R, Moserova M, Frei E, Schmeiser HH, Kopka K, *et al.* NADH: cytochrome b5 reductase and cytochrome b5 can act as sole electron donors to human cytochrome P450 1A1-mediated oxidation and DNA adduct formation by benzo[a]pyrene. *Chem Res Toxicol* 2016; 29: 1325–1334.
52. Stiborova M, Indra R, Moserova M, Sulc M, Hodek P, Frei E, *et al.* NADPH- and NADH-dependent metabolism of and DNA adduct formation by benzo[a]pyrene catalyzed with rat hepatic microsomes and cytochrome P450 1A1. *Monatsh Chem* 2016; 147: 847–855.
53. King HW, Thompson MH, Brookes P. The role of 9-hydroxybenzo(a)pyrene in the microsome mediated binding of benzo(a)pyrene to DNA. *Int J Cancer* 1976; 18: 339–344.
54. Fang AH, Smith WA, Vouros P, Gupta RC. Identification and characterization of a novel benzo[a]pyrene-derived DNA adduct. *Biochem Biophys Res Commun* 2001; 281: 383–389.
55. Endo K, Uno S, Seki T, Ariga T, Kusumi Y, Mitsumata M, *et al.* Inhibition of aryl hydrocarbon receptor transactivation and DNA adduct formation by CYP1 isoform-selective metabolic deactivation of benzo[a]pyrene. *Toxicol Appl Pharmacol* 2008; 230: 135–143.
56. Indra R, Moserova M, Kroftova N, Sulc M, Martinkova M, Adam V, *et al.* Modulation of human cytochrome P450 1A1-mediated oxidation of

benzo[a]pyrene by NADPH:cytochrome P450 oxidoreductase and cytochrome b5. *Neuro Endocrinol Lett* 2014; 35 Suppl 2: 105–113.

57. Huang MT, Wood AW, Chang RL, Yagi H, Sayer JM, Jerina DM, *et al.* Inhibitory effect of 3-hydroxybenzo(a)pyrene on the mutagenicity and tumorigenicity of (+/–)-7 beta, 8 alpha-dihydroxy-9 alpha, 10 alpha-epoxy-7,8,9,10-tetrahydrobenzo(a)pyrene. *Cancer Res* 1986; 46: 558–566.
58. Stiborova M, Martinek V, Rydlova H, Hodek P, Frei E. Sudan I is a potential carcinogen for humans: evidence for its metabolic activation and detoxication by human recombinant cytochrome P450 1A1 and liver microsomes. *Cancer Res* 2002; 62: 5678–5684.
59. Stiborova M, Frei E, Hodek P, Wiessler M, Schmeiser HH. Human hepatic and renal microsomes, cytochromes P450 1A1/2, NADPH:cytochrome P450 reductase and prostaglandin H synthase mediate the formation of aristolochic acid-DNA adducts found in patients with urothelial cancer. *Int J Cancer* 2005; 113: 189–197.
60. Tolson AH, Wang H. Regulation of drug-metabolizing enzymes by xenobiotic receptors: PXR and CAR. *Adv Drug Deliv Rev* 2010; 62: 1238–1249.
61. Perdew GH. Association of the Ah receptor with the 90-kDa heat shock protein. *J Biol Chem* 1988; 263: 13802–13805.
62. Hord NG, Perdew GH. Physicochemical and immunocytochemical analysis of the aryl hydrocarbon receptor nuclear translocator: characterization of two monoclonal antibodies to the aryl hydrocarbon receptor nuclear translocator. *Mol Pharmacol* 1994; 46: 618–626.
63. Pollenz RS, Sattler CA, Poland A. The aryl hydrocarbon receptor and aryl hydrocarbon receptor nuclear translocator protein show distinct subcellular localizations in Hepa 1c1c7 cells by immunofluorescence microscopy. *Mol Pharmacol* 1994; 45: 428–438.
64. Dolwick KM, Swanson HI, Bradfield CA. *In vitro* analysis of Ah receptor domains involved in ligand-activated DNA recognition. *Proc Natl Acad Sci USA* 1993; 90: 8566–8570.
65. Hankinson O. The aryl hydrocarbon receptor complex. *Annu Rev Pharmacol Toxicol* 1995; 35: 307–340.
66. Probst MR, Reisz-Porszasz S, Agbunag RV, Ong MS, Hankinson O. Role of the aryl hydrocarbon receptor nuclear translocator protein in aryl hydrocarbon (dioxin) receptor action. *Mol Pharmacol* 1993; 44: 511–518.
67. Shen ES, Whitlock JP, Jr. Protein-DNA interactions at a dioxin-responsive enhancer. Mutational analysis of the DNA-binding site for the liganded Ah receptor. *J Biol Chem* 1992; 267: 6815–6819.

68. Whitlock JP, Jr., Okino ST, Dong L, Ko HP, Clarke-Katzenberg R, Ma Q, *et al.* Cytochromes P450 5: induction of cytochrome P4501A1: a model for analyzing mammalian gene transcription. *FASEB J* 1996; 10: 809–818.
69. Hankinson O. Role of coactivators in transcriptional activation by the aryl hydrocarbon receptor. *Arch Biochem Biophys* 2005; 433: 379–386.
70. Abel J, Haarmann-Stemmann T. An introduction to the molecular basics of aryl hydrocarbon receptor biology. *Biol Chem* 2010; 391: 1235–1248.
71. Hockley SL, Arlt VM, Brewer D, Te Poele R, Workman P, Giddings I, *et al.* AHR- and DNA-damage-mediated gene expression responses induced by benzo(a)pyrene in human cell lines. *Chem Res Toxicol* 2007; 20: 1797–1810.
72. Wang T, Gavin HM, Arlt VM, Lawrence BP, Fenton SE, Medina D, *et al.* Aryl hydrocarbon receptor activation during pregnancy, and in adult nulliparous mice, delays the subsequent development of DMBA-induced mammary tumors. *Int J Cancer* 2011; 128: 1509–1523.
73. Kucab JE, Zwart EP, van Steeg H, Luijten M, Schmeiser HH, Phillips DH, *et al.* TP53 and lacZ mutagenesis induced by 3-nitrobenzanthrone in Xpa-deficient human TP53 knock-in mouse embryo fibroblasts. *DNA Repair* 2016; 39: 21–33.
74. Kucab JE, Phillips DH, Arlt VM. Linking environmental carcinogen exposure to TP53 mutations in human tumours using the human TP53 knock-in (Hupki) mouse model. *FEBS J* 2010; 277: 2567–2583.
75. Vogelstein B, Lane D, Levine AJ. Surfing the p53 network. *Nature* 2000; 408: 307–310.
76. Lane DP. Cancer. p53, guardian of the genome. *Nature* 1992; 358: 15–16.
77. Kenzelmann Broz D, Attardi LD. *In vivo* analysis of p53 tumor suppressor function using genetically engineered mouse models. *Carcinogenesis* 2010; 31: 1311–1318.
78. Brooks CL, Gu W. p53 ubiquitination: Mdm2 and beyond. *Mol Cell* 2006; 21: 307–315.
79. Riley T, Sontag E, Chen P, Levine A. Transcriptional control of human p53-regulated genes. *Nat Rev Mol Cell Biol* 2008; 9: 402–412.
80. Goldstein I, Rivlin N, Shoshana OY, Ezra O, Madar S, Goldfinger N, *et al.* Chemotherapeutic agents induce the expression and activity of their clearing enzyme CYP3A4 by activating p53. *Carcinogenesis* 2013; 34: 190–198.
81. Arlt VM, Stiborova M, Schmeiser HH. Aristolochic acid as a probable human cancer hazard in herbal remedies: a review. *Mutagenesis* 2002; 17: 265–277.

82. Vanherweghem JL, Depierreux M, Tielemans C, Abramowicz D, Dratwa M, Jadoul M, *et al.* Rapidly progressive interstitial renal fibrosis in young women: association with slimming regimen including Chinese herbs. *Lancet* 1993; 341: 387–391.
83. Debelle FD, Vanherweghem JL, Nortier JL. Aristolochic acid nephropathy: a worldwide problem. *Kidney Int* 2008; 74: 158–169.
84. Schmeiser HH, Stiborova M, Arlt VM. Chemical and molecular basis of the carcinogenicity of Aristolochia plants. *Curr Opin Drug Discov Devel* 2009; 12: 141–148.
85. Gokmen MR, Cosyns JP, Arlt VM, Stiborova M, Phillips DH, Schmeiser HH, *et al.* The epidemiology, diagnosis, and management of aristolochic acid nephropathy: a narrative review. *Ann Intern Med* 2013; 158: 469–477.
86. Nortier JL, Martinez MC, Schmeiser HH, Arlt VM, Bieler CA, Petein M, *et al.* Urothelial carcinoma associated with the use of a Chinese herb (*Aristolochia fangchi*). *N Engl J Med* 2000; 342: 1686–1692.
87. Lemy A, Wissing KM, Rorive S, Zlotta A, Roumeguere T, Muniz Martinez MC, *et al.* Late onset of bladder urothelial carcinoma after kidney transplantation for end-stage aristolochic acid nephropathy: a case series with 15-year follow-up. *Am J Kidney Dis* 2008; 51: 471–477.
88. Arlt VM, Stiborova M, vom Brocke J, Simoes ML, Lord GM, Nortier JL, *et al.* Aristolochic acid mutagenesis: molecular clues to the aetiology of Balkan endemic nephropathy-associated urothelial cancer. *Carcinogenesis* 2007; 28: 2253–2261.
89. Grollman AP, Shibutani S, Moriya M, Miller F, Wu L, Moll U, *et al.* Aristolochic acid and the etiology of endemic (Balkan) nephropathy. *Proc Natl Acad Sci USA* 2007; 104: 12129–12134.
90. Arlt VM, Ferluga D, Stiborova M, Pfohl-Leszkowicz A, Vukelic M, Ceovic S, *et al.* Is aristolochic acid a risk factor for Balkan endemic nephropathy-associated urothelial cancer? *Int J Cancer* 2002; 101: 500–502.
91. IARC. Pharmaceuticals: a review of human carcinogens. *IARC Monogr Eval Carcinog Risk Hum* 2012; 100A.
92. Shibutani S, Dong H, Suzuki N, Ueda S, Miller F, Grollman AP. Selective toxicity of aristolochic acids I and II. *Drug Metab Dispos* 2007; 35: 1217–1222.
93. Grollman AP, Aristolochic acid nephropathy: harbinger of a global iatrogenic disease, *Environ Mol Mutagen* 2013; 54: 1–7.
94. Stiborova M, Frei E, Arlt VM, Schmeiser HH. Metabolic activation of carcinogenic aristolochic acid, a risk factor for Balkan endemic nephropathy. *Mutat Res* 2008; 658: 55–67.

95. Stiborova M, Frei E, Schmeiser HH. Biotransformation enzymes in development of renal injury and urothelial cancer caused by aristolochic acid. *Kidney Int* 2008; 73: 1209–1211.
96. Chan W, Cui L, Xu G, Cai Z. Study of the phase I and phase II metabolism of nephrotoxin aristolochic acid by liquid chromatography/tandem mass spectrometry. *Rapid Commun Mass Spectrom* 2006; 20: 1755–1760.
97. Shibutani S, Bonala RR, Rosenquist T, Rieger R, Suzuki N, Johnson F, *et al*. Detoxification of aristolochic acid I by O-demethylation: less nephrotoxicity and genotoxicity of aristolochic acid Ia in rodents. *Int J Cancer* 2010; 127: 1021–1027.
98. Arlt VM, Levova K, Barta F, Shi Z, Evans JD, Frei E, *et al*. Role of P450 1A1 and P450 1A2 in bioactivation versus detoxication of the renal carcinogen aristolochic acid I: studies in Cyp1a1-/-, Cyp1a2-/-, and Cyp1a1/1a2-/- mice. *Chem Res Toxicol* 2011; 24: 1710–1719.
99. Stiborova M, Levova K, Barta F, Shi Z, Frei E, Schmeiser HH, *et al*. Bioactivation versus detoxication of the urothelial carcinogen aristolochic acid I by human cytochrome P450 1A1 and 1A2. *Toxicol Sci* 2012; 125: 345–358.
100. Dong H, Suzuki N, Torres MC, Bonala RR, Johnson F, Grollman AP, *et al*. Quantitative determination of aristolochic acid-derived DNA adducts in rats using 32P-postlabeling/polyacrylamide gel electrophoresis analysis. *Drug Metab Dispos* 2006; 34: 1122–1127.
101. Krumbiegel G, Hallensleben J, Mennicke WH, Rittmann N, Roth HJ. Studies on the metabolism of aristolochic acids I and II. *Xenobiotica* 1987; 17: 981–991.
102. Pfau W, Schmeiser HH, Wiessler M. 32P-postlabelling analysis of the DNA adducts formed by aristolochic acid I and II. *Carcinogenesis* 1990; 11: 1627–1633.
103. Pfau W, Schmeiser HH, Wiessler M. Aristolochic acid binds covalently to the exocyclic amino group of purine nucleotides in DNA. *Carcinogenesis* 1990; 11: 313–319.
104. Schmeiser HH, Bieler CA, Wiessler M, van Ypersele de Strihou C, Cosyns JP. Detection of DNA adducts formed by aristolochic acid in renal tissue from patients with Chinese herbs nephropathy. *Cancer Res* 1996; 56: 2025–2028.
105. Bieler CA, Stiborova M, Wiessler M, Cosyns JP, van Ypersele de Strihou C, Schmeiser HH. 32P-post-labelling analysis of DNA adducts formed by aristolochic acid in tissues from patients with Chinese herbs nephropathy. *Carcinogenesis* 1997; 18: 1063–1067.

106. Stiborova M, Fernando RC, Schmeiser HH, Frei E, Pfau W, Wiessler M. Characterization of DNA adducts formed by aristolochic acids in the target organ (forestomach) of rats by 32P-postlabelling analysis using different chromatographic procedures. *Carcinogenesis* 1994; 15: 1187–1192.
107. Schmeiser HH, Frei E, Wiessler M, Stiborova M. Comparison of DNA adduct formation by aristolochic acids in various *in vitro* activation systems by 32P-post-labelling: evidence for reductive activation by peroxidases. *Carcinogenesis* 1997; 18: 1055–1062.
108. Fernando RC, Schmeiser HH, Scherf HR, Wiessler M. Formation and persistence of specific purine DNA adducts by 32P-postlabelling in target and non-target organs of rats treated with aristolochic acid I. *IARC Sci Publ* 1993: 167–171.
109. Schmeiser HH, Kucab JE, Arlt VM, Phillips DH, Hollstein M, Gluhovschi G, *et al.* Evidence of exposure to aristolochic acid in patients with urothelial cancer from a Balkan endemic nephropathy region of Romania. *Environ Mol Mutagen* 2012; 53: 636–641.
110. Schmeiser HH, Nortier JL, Singh R, da Costa GG, Sennesael J, Cassuto-Viguier E, *et al.* Exceptionally long-term persistence of DNA adducts formed by carcinogenic aristolochic acid I in renal tissue from patients with aristolochic acid nephropathy. *Int J Cancer* 2014; 135: 502–507.
111. Lord GM, Hollstein M, Arlt VM, Roufosse C, Pusey CD, Cook T, *et al.* DNA adducts and p53 mutations in a patient with aristolochic acid-associated nephropathy. *Am J Kidney Dis* 2004; 43: e11–e17.
112. Aydin S, Dekairelle AF, Ambroise J, Durant JF, Heusterspreute M, Guiot Y, *et al.* Unambiguous detection of multiple TP53 gene mutations in AAN-associated urothelial cancer in Belgium using laser capture microdissection. *PLOS ONE* 2014; 9: e106301.
113. Chen CH, Dickman KG, Moriya M, Zavadil J, Sidorenko VS, Edwards KL, *et al.* Aristolochic acid-associated urothelial cancer in Taiwan. *Proc Natl Acad Sci USA* 2012; 109: 8241–8246.
114. Nik-Zainal S, Kucab JE, Morganella S, Glodzik D, Alexandrov LB, Arlt VM, *et al.* The genome as a record of environmental exposure. *Mutagenesis* 2015; 30: 763–770.
115. Poon SL, Pang ST, McPherson JR, Yu W, Huang KK, Guan P, *et al.* Genome-wide mutational signatures of aristolochic Acid and its application as a screening tool. *Sci Transl Med* 2013; 5: 197ra101.
116. Hoang ML, Chen CH, Sidorenko VS, He J, Dickman KG, Yun BH, *et al.* Mutational signature of aristolochic acid exposure as revealed by whole-exome sequencing. *Sci Transl Med* 2013; 5: 197ra102.

117. Olivier M, Hollstein M, Schmeiser HH, Straif K, Wild CP. Upper urinary tract urothelial cancers: where it is A:T. *Nat Rev* 2012; 12: 503–504.
118. Sistkova J, Hudecek J, Hodek P, Frei E, Schmeiser HH, Stiborova M. Human cytochromes P450 1A1 and 1A2 participate in detoxication of carcinogenic aristolochic acid. *Neuro Endocrinol Lett* 2008; 29: 733–737.
119. Levova K, Moserova M, Kotrbova V, Sulc M, Henderson CJ, Wolf CR, *et al.* Role of cytochromes P450 1A1/2 in detoxication and activation of carcinogenic aristolochic acid I: studies with the hepatic NADPH:cytochrome P450 reductase null (HRN) mouse model. *Toxicol Sci* 2011; 121: 43–56.
120. Dracinska H, Barta F, Levova K, Hudecova A, Moserova M, Schmeiser HH, *et al.* Induction of cytochromes P450 1A1 and 1A2 suppresses formation of DNA adducts by carcinogenic aristolochic acid I in rats *in vivo*. *Toxicology* 2016; 344–346: 7–18.
121. Milichovsky J, Barta F, Schmeiser HH, Arlt VM, Frei E, Stiborova M, *et al.* Active site mutations as a suitable tool contributing to explain a mechanism of aristolochic acid I nitroreduction by cytochromes P450 1A1, 1A2 and 1B1. *Int J Mol Sci* 2016; 17: 213.
122. Rosenquist TA, Einolf HJ, Dickman KG, Wang L, Smith A, Grollman AP. Cytochrome P450 1A2 detoxicates aristolochic acid in the mouse. *Drug Metab Dispos* 2010; 38: 761–768.
123. Stiborova M, Barta F, Levova K, Hodek P, Schmeiser HH, Arlt VM, *et al.* A mechanism of O-demethylation of aristolochic acid I by cytochromes P450 and their contributions to this reaction in human and rat livers: experimental and theoretical approaches. *Int J Mol Sci* 2015; 16: 27561–27575.
124. Arlt VM, Pfohl-Leszkowicz A, Cosyns J, Schmeiser HH. Analyses of DNA adducts formed by ochratoxin A and aristolochic acid in patients with Chinese herbs nephropathy. *Mutat Res* 2001: 494(1–2): 143–150.
125. Gillerot G, Jadoul M, Arlt VM, van Ypersele De Strihou C, Schmeiser HH, But PP, Bieler CA, Cosyns JP. Aristolochic acid nephropathy in a Chinese patient: time to abandon the term "Chinese herbs nephropathy"? *Am J Kidney Dis* 2001: 38(5): E26.
126. Lord GM, Cook T, Arlt VM, Schmeiser HH, Williams G, Pusey CD. Urothelial malignant disease and Chinese herbal nephropathy. *Lancet* 2001: 358(9292): 1515–1516.
127. Arlt VM, Ferluga D, Stiborova M, Pfohl-Leszkowicz A, Vukelic M, Ceovic S, Schmeiser HH, Cosyns JP. Is aristolochic acid a risk factor for Balkan endemic nephropathy-associated urothelial cancer? *Int J Cancer* 2002: 101(5): 500–502.

128. Nortier JL, Schmeiser HH, Muniz Martinez MC, Arlt VM, Vervaet C, Garbar CH, Daelemans P, Vanherweghem JL. Invasive urothelial carcinoma after exposure to Chinese herbal medicine containing aristolochic acid may occur without severe renal failure. *Nephrol Dial Transplant* 2003: 18(2): 426–428.
129. Arlt VM, Alunni-Perret V, Quatrehomme G, Ohayon P, Albano L, Gaid H, Michiels JF, Meyrier A, Cassuto E, Wiessler M, Schmeiser HH, Cosyns JP. Aristolochic acid (AA)-DNA adduct as marker of AA exposure and risk factor for AA nephropathy-associated cancer. *Int J Cancer* 2004: 111(6): 977–980.
130. Lo SH, Wong KS, Arlt VM, Phillips DH, Lai CK, Poon WT, Chan CK, Mo KL, Chan KW, Chan A. Detection of Herba Aristolochia Mollissemae in a patient with unexplained nephropathy. *Am J Kidney Dis* 2005: 45(2): 407–410.
131. Jelakovic B, Karanovic S, Vukovic-Lela I, Miller F, Edwards KL, Nikolic J, Tomic K, Slade N, Brdar B, Turesky RJ, Stipancic Z, Dittrich D, Grollman AP, Dickman KG. Aristolactam-DNA adducts are a biomarker of environmental exposure to aristolochic acid. *Kidney Int* 2012: 81(6): 559–567.
132. Yun BH, Yao L, Jelakovic B, Nikolic J, Dickman KG, Grollman AP, Rosenquist TA, Turesky RJ. Formalin-fixed paraffin-embedded tissue as a source for quantitation of carcinogen DNA adducts: aristolochic acid as a prototype carcinogen. *Carcinogenesis* 2014: 35(9): 2055–2061.
133. Roumeguere T, Broeders N, Jayaswal A, Rorive S, Quackels T, Pozdzik A, Arlt VM, Schmeiser HH, Nortier JL. 2015. Bacillus Calmette-Guerin therapy in non-muscle-invasive bladder carcinoma after renal transplantation for end-stage aristolochic acid nephropathy. *Transpl Int* 2015: 28(2):199–205.
134. Turesky RJ, Yun BH, Brennan P, Mates D, Jinga V, Harnden P, Banks RE, Blanche H, Bihoreau MT, Chopard P, Letourneau L, Lathrop GM, Scelo G. Aristolochic acid exposure in Romania and implications for renal cell carcinoma. *Br J Cancer* 2016: 114(1): 76–80.
135. Hoang ML, Chen CH, Chen PC, Roberts NJ, Dickman KG, Yun BH, Turesky RJ, Pu YS, Vogelstein B, Papadopoulos N, Grollman AP, Kinzler KW, Rosenquist TA. Aristolochic acid in the etiology of renal cell carcinoma. *Cancer Epidemiol Biomarkers Prev* 2016: 25(12): 1600–1608.
136. Rittle J, Green MT. Cytochrome P450 compound I: capture, characterization, and C-H bond activation kinetics. *Science* (New York, NY) 2010; 330: 933–937.

137. Jung C, de Vries S, Schunemann V. Spectroscopic characterization of cytochrome P450 Compound I. *Arch Biochem Biophys* 2011; 507: 44–55.
138. Xue X, Xiao Y, Zhu H, Wang H, Liu Y, Xie T, *et al.* Induction of P450 1A by 3-methylcholanthrene protects mice from aristolochic acid-I-induced acute renal injury. *Nephrol Dial Transplant* 2008; 23: 3074–3081.
139. Xiao Y, Ge M, Xue X, Wang C, Wang H, Wu X, *et al.* Hepatic cytochrome P450s metabolize aristolochic acid and reduce its kidney toxicity. *Kidney Int* 2008; 73: 1231–1239.
140. Stiborova M, Frei E, Sopko B, Wiessler M, Schmeiser HH. Carcinogenic aristolochic acids upon activation by DT-diaphorase form adducts found in DNA of patients with Chinese herbs nephropathy. *Carcinogenesis* 2002; 23: 617–625.
141. Stiborova M, Frei E, Sopko B, Sopkova K, Markova V, Lankova M, *et al.* Human cytosolic enzymes involved in the metabolic activation of carcinogenic aristolochic acid: evidence for reductive activation by human NAD(P)H:quinone oxidoreductase. *Carcinogenesis* 2003; 24: 1695–1703.
142. Levova K, Moserova M, Nebert DW, Phillips DH, Frei E, Schmeiser HH, *et al.* NAD(P)H:quinone oxidoreductase expression in Cyp1a-knockout and CYP1A-humanized mouse lines and its effect on bioactivation of the carcinogen aristolochic acid I. *Toxicol Appl Pharmacol* 2012; 265: 360–367.
143. Stiborova M, Frei E, Schmeiser HH, Arlt VM, Martinek V. Mechanisms of enzyme-catalyzed reduction of two carcinogenic nitro-aromatics, 3-nitrobenzanthrone and aristolochic acid I: Experimental and theoretical approaches. *Int J Mol Sci* 2014; 15: 10271–10295.
144. Stiborova M, Mares J, Frei E, Arlt VM, Martinek V, Schmeiser HH. The human carcinogen aristolochic acid i is activated to form DNA adducts by human NAD(P)H:quinone oxidoreductase without the contribution of acetyltransferases or sulfotransferases. *Environ Mol Mutagen* 2011; 52: 448–459.
145. Stiborova M, Frei E, Wiessler M, Schmeiser HH. Human enzymes involved in the metabolic activation of carcinogenic aristolochic acids: evidence for reductive activation by cytochromes P450 1A1 and 1A2. *Chem Res Toxicol* 2001; 14: 1128–1137.
146. Stiborova M, Sopko B, Hodek P, Frei E, Schmeiser HH, Hudecek J. The binding of aristolochic acid I to the active site of human cytochromes P450 1A1 and 1A2 explains their potential to reductively activate this human carcinogen. *Cancer Lett* 2005; 229: 193–204.
147. Jerabek P, Martinek V, Stiborova M. Theoretical investigation of differences in nitroreduction of aristolochic acid I by cytochromes P450 1A1, 1A2 and 1B1. *Neuro Endocrinol Lett* 2012; 33 Suppl 3: 25–32.

148. Schenkman JB, Jansson I. The many roles of cytochrome b5. *Pharmacol Ther* 2003; 97: 139–152.
149. Stiborova M, Indra R, Moserova M, Cerna V, Rupertova M, Martinek V, *et al.* Cytochrome b5 increases cytochrome P450 3A4-mediated activation of anticancer drug ellipticine to 13-hydroxyellipticine whose covalent binding to DNA is elevated by sulfotransferases and N,O-acetyltransferases. *Chem Res Toxicol* 2012; 25: 1075–1085.
150. Stiborova M, Poljakova J, Martinkova E, Ulrichova J, Simanek V, Dvorak Z, *et al.* Ellipticine oxidation and DNA adduct formation in human hepatocytes is catalyzed by human cytochromes P450 and enhanced by cytochrome b5. *Toxicology* 2012; 302: 233–241.
151. Kotrbova V, Mrazova B, Moserova M, Martinek V, Hodek P, Hudecek J, *et al.* Cytochrome b(5) shifts oxidation of the anticancer drug ellipticine by cytochromes P450 1A1 and 1A2 from its detoxication to activation, thereby modulating its pharmacological efficacy. *Biochem Pharmacol* 2011; 82: 669–680.
152. McLaughlin LA, Ronseaux S, Finn RD, Henderson CJ, Roland Wolf C. Deletion of microsomal cytochrome b5 profoundly affects hepatic and extrahepatic drug metabolism. *Mol Pharmacol* 2010; 78: 269–278.
153. Finn RD, McLaughlin LA, Ronseaux S, Rosewell I, Houston JB, Henderson CJ, *et al.* Defining the *in vivo* role for cytochrome b5 in cytochrome P450 function through the conditional hepatic deletion of microsomal cytochrome b5. *J Biol Chem* 2008; 283: 31385–31393.
154. Henderson CJ, McLaughlin LA, Wolf CR. Evidence that cytochrome b5 and cytochrome b5 reductase can act as sole electron donors to the hepatic cytochrome P450 system. *Mol Pharmacol* 2013; 83: 1209–1217.
155. Henderson CJ, McLaughlin LA, Finn RD, Ronseaux S, Kapelyukh Y, Wolf CR. A role for cytochrome b5 in the *in vivo* disposition of anticancer and cytochrome P450 probe drugs in mice. *Drug Metab Dispos* 2014; 42: 70–77.
156. Henderson CJ, McLaughlin LA, Scheer N, Stanley LA, Wolf CR. Cytochrome b5 is a major determinant of human cytochrome P450 CYP2D6 and CYP3A4 activity *in vivo*. *Mol Pharmacol* 2015; 87: 733–739.
157. Yamazaki H, Gillam EM, Dong MS, Johnson WW, Guengerich FP, Shimada T. Reconstitution of recombinant cytochrome P450 2C10(2C9) and comparison with cytochrome P450 3A4 and other forms: effects of cytochrome P450–P450 and cytochrome P450-b5 interactions. *Arch Biochem Biophys* 1997; 342: 329–337.
158. Yamazaki H, Shimada T, Martin MV, Guengerich FP. Stimulation of cytochrome P450 reactions by apo-cytochrome b5: evidence against

transfer of heme from cytochrome P450 3A4 to apo-cytochrome b5 or heme oxygenase. *J Biol Chem* 2001; 276: 30885–30891.

159. Loughran PA, Roman LJ, Miller RT, Masters BS. The kinetic and spectral characterization of the E. coli-expressed mammalian CYP4A7: cytochrome b5 effects vary with substrate. *Arch Biochem Biophys* 2001; 385: 311–321.
160. Zhang H, Myshkin E, Waskell L. Role of cytochrome b5 in catalysis by cytochrome P450 2B4. *Biochem Biophys Res Commun* 2005; 338: 499–506.
161. Meunier B, de Visser SP, Shaik S. Mechanism of oxidation reactions catalyzed by cytochrome p450 enzymes. *Chem Rev* 2004; 104: 3947–3980.
162. Poulos TL, Finzel BC, Howard AJ. High-resolution crystal structure of cytochrome P450cam. *J Mol Biol* 1987; 195: 687–700.
163. Sligar SG, Cinti DL, Gibson GG, Schenkman JB. Spin state control of the hepatic cytochrome P450 redox potential. *Biochem Biophys Res Commun* 90; 1979: 925–932.
164. Lu AY, West SB, Vore M, Ryan D, Levin W. Role of cytochrome b5 in hydroxylation by a reconstituted cytochrome P-450-containing system. *J Biol Chem* 1974; 249: 6701–6709.
165. West SB, Levin W, Ryan D, Vore M, Lu AY. Liver microsomal electron transport systems. II. The involvement of cytochrome b5 in the NADH-dependent hydroxylation of 3,4-benzpyrene by a reconstituted cytochrome P-448-containing system. *Biochem Biophys Res Commun* 1974; 58: 516–522.
166. Lewis DF, Hlavica P. Interactions between redox partners in various cytochrome P450 systems: functional and structural aspects. *Biochim Biophys Acta* 2000; 1460: 353–374.
167. Hlavica P, Schulze J, Lewis DF. Functional interaction of cytochrome P450 with its redox partners: a critical assessment and update of the topology of predicted contact regions. *J Inorg Biochem* 2003; 96: 279–297.
168. Berka K, Hendrychova T, Anzenbacher P, Otyepka M. Membrane position of ibuprofen agrees with suggested access path entrance to cytochrome P450 2C9 active site. *J Phys Chem A* 2011; 115: 11248–11255.
169. Yamazaki H, Nakajima M, Nakamura M, Asahi S, Shimada N, Gillam EM, *et al.* Enhancement of cytochrome P-450 3A4 catalytic activities by cytochrome b(5) in bacterial membranes. *Drug Metab Dispos* 1999; 27: 999–1004.
170. Stiborová M, Indra R, Frei E, Kopečková K, Schmeiser HH, Eckschlager T, *et al.* Cytochrome b_5 plays a dual role in the reaction cycle of cytochrome P450 3A4 during oxidation of the anticancer drug ellipticine. *Monatsh Chem* 2017; 148: 1983–1991.

Chapter 3

DNA Damage by Tobacco Carcinogens

Stephen S. Hecht

Masonic Cancer Center, University of Minnesota
Minneapolis, MN 55455, USA

Abstract

DNA damage is the critical step linking carcinogen exposure from tobacco products to cancer-causing mutations in important growth control genes. This chapter provides an overview of carcinogens in tobacco products and their conversion to metabolites that react readily with DNA bases producing addition products known as DNA adducts found in smokers' DNA. DNA adducts can cause errors during DNA replication leading to permanent changes in DNA sequence. When such mutations occur in critical regions of oncogenes such as *KRAS* and tumor suppressor genes such as *p53*, the result can be loss of normal growth control mechanisms and cancer development. This chapter presents evidence for this important biochemical pathway.

1. Introduction

The relationship between cigarette smoking and cancer is common knowledge but less well known is the massive role that cigarette smoking still plays in overall worldwide cancer mortality. Lung cancer is the major cancer killer caused by cigarette smoking. In 2012, lung cancer killed 1,589,800

Lung	Colorectum
Oral cavity	Liver
Naso, oro, and hypopharynx	Kidney (body and pelvis)
Nasal cavity and accessory	Ureter
Sinuses	Urinary bladder
Larynx	Uterine cervix
Esophagus	Ovary (mucinous)
Stomach	Myeloid leukemia
Pancreas	

Box 1. Cancers caused by cigarette smoking.

people in the world, about three per minute [1]. It is estimated that cigarette smoking caused about 80% of this worldwide death toll in males and about 50% in females. The numbers of deaths from lung cancer are equivalent to those which would occur in about 14 major plane crashes *per day*.

The International Agency for Research on Cancer (IARC), a branch of the World Health Organization, has evaluated the relationship between cigarette smoking and cancers of various organs [2]. As shown in Box 1, cigarette smoking causes multiple types of cancers — to varying extents — in addition to lung cancer. Most of these cancers, as well as lung cancer, are either virtually incurable or involve therapies which can result in significant decreases in quality of life. This is also a fact not widely recognized, particularly when one is regularly exposed to headlines regarding advanced cancer genomics and impressive targeted therapies, which however in most instances do not generally extend life beyond a few years. Worldwide, it is estimated that 1,420,000 cancer deaths per year, about 21% of all cancer deaths, are caused by cigarette smoking while the corresponding figures in the United States are 185,000 cancer deaths per year, and 33% of all cancer mortality [1].

Yet, in the United States there are still about 40,000,000 cigarette smokers [3] while worldwide there are about 1 billion cigarette smokers [4]. These facts attest to the financial power and advertising savvy of tobacco companies.

Cigar and pipe smoking are also causes of cancer, being strongly-related to cancers of the oral cavity, oropharynx, hypopharynx, larynx, and esophagus as well as lung, pancreas, stomach, and urinary bladder [5].

Smokeless tobacco is another popular form of tobacco use. The most common form of smokeless tobacco use at present in the United States is moist snuff, a pinch of which is placed between the cheek and gum.

Worldwide, there is a huge variety of smokeless tobacco products, particularly in India and other parts of southeast Asia, where products range from betel quid containing tobacco to tobacco toothpaste. IARC has evaluated smokeless tobacco as a cause of oral cavity cancer, pancreatic cancer, and esophageal cancer in humans [2]. There are hundreds of millions of smokeless tobacco users worldwide. It is recognized however that smokeless tobacco use, at least in the Western world, is considerably safer than cigarette smoking.

In view of these undeniably frightening facts, why do people continue to use tobacco products? The answer is simple: nicotine, a compound found in all tobacco products, is addictive. People use tobacco products to obtain their required dose of nicotine. This dependence on nicotine results in persistent use of nicotine-containing products and the appearance of symptoms of withdrawal when nicotine exposure is abruptly discontinued. The tobacco industry has designed products, and cigarettes in particular, to maximize their addictive properties as can be achieved through nicotine exposure [6].

2. Cancer-Causing Agents (Carcinogens) are Present in Tobacco Products

Tobacco smoke contains a highly complex mixture of over 7000 characterized chemical compounds, including more than 70 with cancer-causing (carcinogenic) activity, based on studies carried out in laboratory animals or on evidence from human exposure scenarios [7]. Structures of some tobacco smoke carcinogens from representative classes of compounds are illustrated in Box 2, and most of these are considered "carcinogenic to humans" by IARC.

Benzo[*a*]pyrene (BaP) and benz[*a*]anthracene (BaA) are typical representatives of the class of compounds known as polycyclic aromatic hydrocarbons (PAHs). Multiple PAHs are present in tobacco smoke. These are combustion products formed when organic material (the tobacco leaf) is burned. Because combustion to CO_2 and H_2O is incomplete during this process (due to insufficient oxygen being present), some intermediate carbon radicals combine to form these PAHs. BaP is a particularly strong carcinogen, readily inducing tumors at the site of application as well as at distant sites when tested in laboratory mice and rats [8]. BaP is the classic representative of the PAH group of compounds; it has been associated with multiple cancer induction scenarios in humans, resulting from exposure to tars, soot, and polluted air [8]. BaA is a weaker carcinogen than BaP but

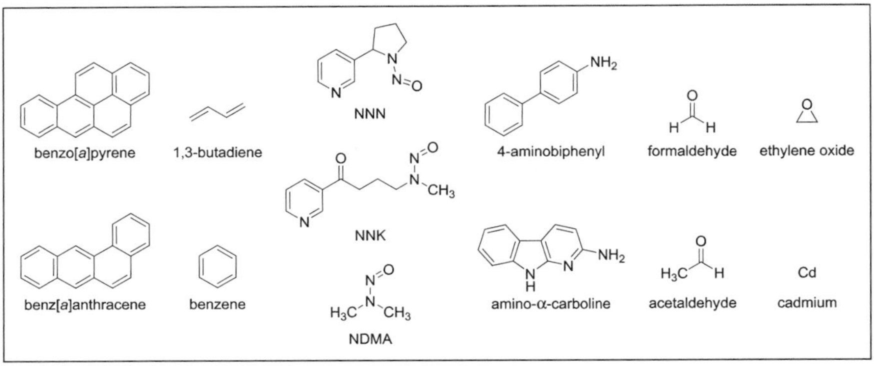

Box 2. Representative carcinogens in cigarette smoke.

is generally present in higher amounts. There are at least 15 other characterized PAH carcinogens in tobacco smoke as well as a considerable number that are more poorly defined in terms of carcinogenicity.

1,3-Butadiene and benzene are two of the important volatile hydrocarbon carcinogens in tobacco smoke, both occurring in considerable quantities [7,9]. Both are multiorgan carcinogens in laboratory mice, and benzene in particular is well established as a cause of leukemia in humans. These two carcinogens are formed during the burning of tobacco, similar to the PAH.

N'-Nitrosonornicotine (NNN) and 4-(methylnitrosamino)-1-(3-pyridyl)-1-butanone (NNK, for nicotine-derived nitrosamino ketone) are formed during the curing and related processing of tobacco, which converts the green leaves of the tobacco plant to the brown and palatable material used in making cigarettes. These compounds, also known as "tobacco-specific nitrosamines," are present in virtually all cigarette tobacco and are transferred to the smoke during smoking [10,11]. They are both strong carcinogens and tend to affect specific organs, independent of the route of administration. NNN causes oral cavity and esophageal tumors in rats while NNK targets the lung in a variety of animal species. Another nitrosamine commonly found in cigarette smoke, albeit at lower levels than NNN and NNK, is *N*-nitrosodimethylamine (NDMA).

4-Aminobiphenyl and amino-α-carboline are aromatic amine carcinogens, which like the PAH and volatile hydrocarbons, are formed during the combustion process when cigarettes are smoked. 4-Aminobiphenyl is a bladder carcinogen in rodents and humans while amino-α-carboline targets the liver in mice and has also shown some activity in the colon [9,12].

Formaldehyde and acetaldehyde are present in considerable quantities in cigarette smoke [7]. Formaldehyde causes nasopharyngeal cancer while acetaldehyde is believed to be a significant cause of cancers associated with alcohol consumption [9]. Ethylene oxide and cadmium are additional constituents of smoke that cause cancer in humans and are likely involved in the carcinogenic properties of tobacco smoke [7,9].

Carcinogens in smokeless tobacco include the tobacco-specific nitrosamines and metals, with considerably lower amounts or none of the volatile carcinogens and combustion products that are found in cigarette smoke.

3. DNA Adducts are Central in Carcinogenesis by Tobacco Products

The currently accepted central mechanism by which cigarette smoking causes lung cancer (and most likely other cancers as well) is summarized in Figure 1 [6]. People start smoking cigarettes, usually when they are teenagers, due to peer pressure and other societal factors. They may not realize the addictive power of nicotine. They may be aware of the dangers of smoking but think they can stop at some time in the future. Unfortunately, it is very difficult to overcome nicotine addiction. Even the best smoking cessation programs which combine counseling and nicotine replacement therapy (using nicotine patches or related methods of delivery of nicotine alone) have 6-month cessation rates of only 20%–25% [13]. *Nicotine is not a carcinogen*, but the cigarette is a disastrous nicotine delivery device because the smoke of each cigarette delivers to the smoker over 70 carcinogens, as noted above, along with the nicotine. The chronic exposure to this mixture of carcinogens, multiple times per day for many years, is a recipe for cancer induction.

Most cigarette smoke carcinogens are relatively unreactive chemicals that the human body recognizes as foreign. The body has enzyme systems

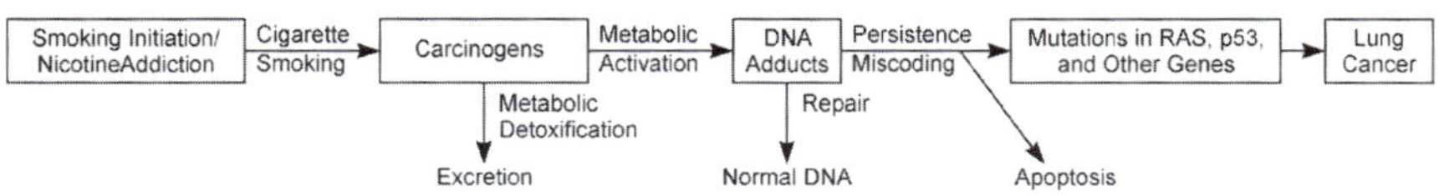

Figure 1. Accepted central mechanism by which cigarette smoking causes lung cancer.

that act on such foreign compounds or on therapeutic drugs — as another example — to convert them to metabolites that are more readily excreted. These enzymes belong to several families, but the most important are the cytochrome P450 enzymes that generally catalyze the addition of an oxygen atom to the foreign compound resulting in the formation of a metabolite, which is more polar and more water soluble than the parent compound, and therefore more readily excreted [14]. This part of the process is illustrated as "Metabolic Detoxification" in Figure 1. There are additional enzymes, such as UDP-glucuronosyl transferases and glutathione-*S*-transferases, that often participate in this process. In some cases, the biotransformation that occurs during metabolism by cytochrome P450s produces chemically reactive (usually electrophilic) metabolites that can react with nucleophilic sites on the DNA bases. The resulting covalent DNA addition products are commonly referred to as "DNA adducts." Some of these reactions are discussed in the next section.

The body has DNA repair enzymes that can remove these DNA adducts and return DNA to normal [15]. But if that process is overwhelmed or not completely efficient, the DNA adducts can persist. These persistent DNA adducts are critical in the carcinogenic process because they can cause miscoding during DNA replication, thus permanently altering the DNA code. Sometimes, cells with adducted DNA are removed by apoptosis, or programmed cell death. The importance of DNA adducts is discussed in more detail in a subsequent section of this chapter, but it is well established that these miscoding events can occur in critical growth control genes such as *RAS* and *p53*. If alteration of the DNA code — referred to as a mutation — occurs in a critical region of one of these genes, the result can be loss of normal growth control mechanisms and, ultimately, cancer development. We know from many experimental studies that blocking the metabolic activation process and consequently decreasing or eliminating DNA adduct formation will result in decreased tumor formation [16].

4. DNA Adducts are Formed by Reaction of Carcinogenic Metabolites with DNA

As noted in the previous section, some of the metabolites formed from tobacco smoke carcinogens, in the intended process of detoxification, are

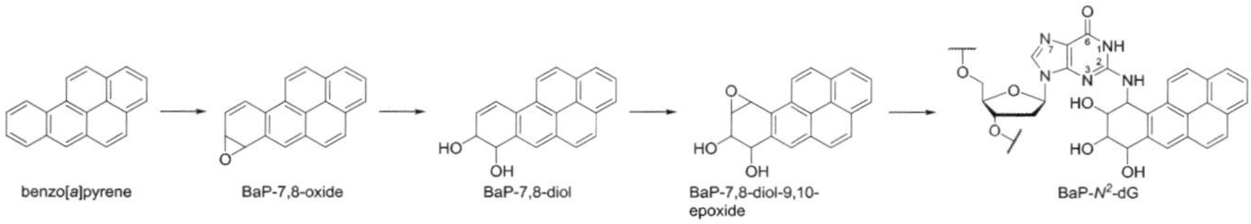

Figure 2. Metabolism of benzo[*a*]pyrene leading to formation of a DNA adduct, BaP-N^2-dG.

actually quite reactive and can react chemically with DNA, producing covalent addition products called DNA adducts. In this section, the process will be illustrated for several carcinogens.

The pathway to DNA adduct formation from BaP is illustrated in Figure 2 [8]. BaP is converted to BaP-7,8-epoxide by cytochrome P450 enzymes (mainly P450s 1A1, 1A2, and 1B1). This intermediate then undergoes reaction with H_2O catalyzed by epoxide hydrolase to produce BaP-7,8-diol. This metabolite is then further converted to BaP-7,8-diol-9,10-epoxide in another reaction catalyzed by cytochrome P450 enzymes. This diol epoxide metabolite is highly reactive with DNA, resulting in the formation of multiple adducts among which the most well-studied is BaP-N^2-dG in which the exocyclic amino group at the 2-position of deoxyguanosine (dG) in DNA has reacted with the epoxide ring to form the adduct illustrated.

Typical pathways of DNA damage by 1,3-butadiene are illustrated in Figure 3 [17,18]. Cytochrome P450 2E1 catalyzes the addition of oxygen to this volatile hydrocarbon, producing 1,2-epoxy-3-butene. As in the case of BaP-7,8-epoxide, this can be acted upon by epoxide hydrolase, forming 3-butene-1,2-diol. Further oxidation of this metabolite produces 1,2-dihydroxy-3,4-epoxybutane. Both 1,2-epoxy-3-butene and 1,2-dihydroxy-3,4-epoxybutane react with DNA with the production of multiple adducts,

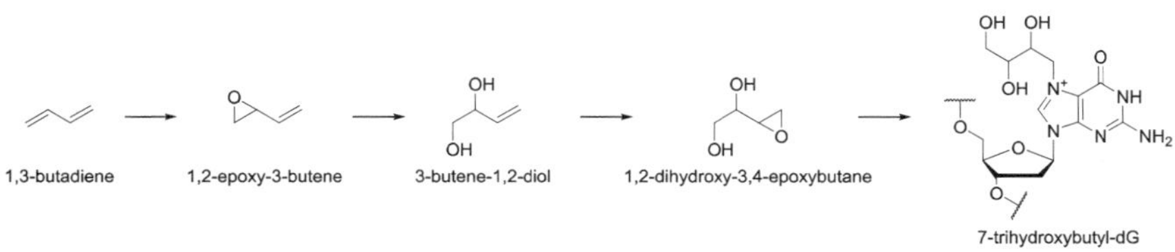

Figure 3. Metabolism of 1,3-butadiene leading to formation of a DNA adduct, 7-trihydroxybutyl-dG.

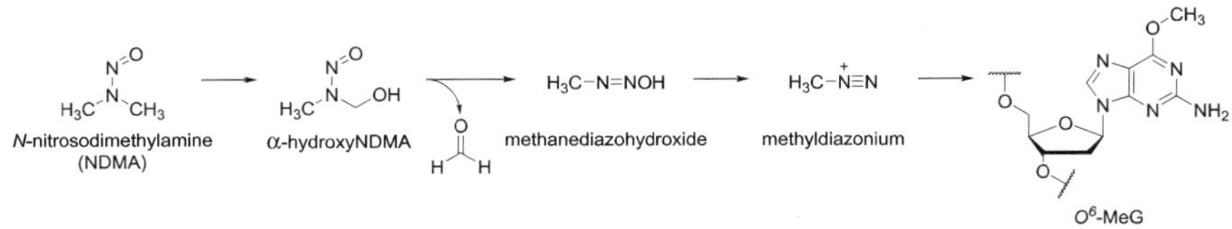

Figure 4. Metabolism of *N*-nitrosodimethylamine leading to formation of a DNA adduct, O^6-MeG.

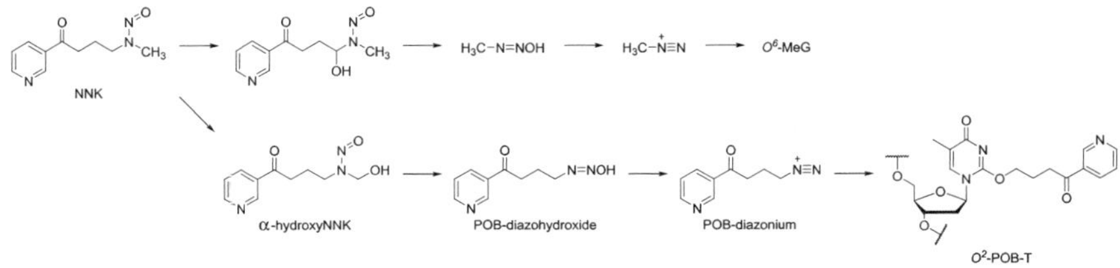

Figure 5. Metabolism of NNK leading to formation of the DNA adducts O^6-MeG and O^2-POB-T.

among which the one illustrated in Figure 3, 7-trihydroxybutyl-dG, results from the reaction of 1,2-dihydroxy-3,4-epoxybutane with the 7-position of dG.

The mode of DNA interaction of NDMA is illustrated in Figure 4 [19]. Cytochrome P450 2E1 catalyzes the hydroxylation of this carcinogen resulting in α-hydroxyNDMA. This metabolite is unstable and spontaneously decomposes producing formaldehyde and methanediazohydroxide as illustrated. Methanediazohydroxide spontaneously converts to the methyl diazonium ion intermediate which reacts with DNA to give O^6-MeG (shown in Figure 4) as well as a variety of other adducts in which various positions of the DNA bases have been methylated. The ratio of 7-MeG to O^6-MeG is initially about 10:1, but O^6-MeG is biologically more important for the reasons discussed in the next section. DNA repair systems which can specifically remove O^6-MeG, returning the DNA to its normal structure, have evolved to manage this kind of damage.

A similar metabolism pathway is operative for NNK, as illustrated in Figure 5 [20]. Hydroxylation of the methylene group next to the *N*-nitroso-nitrogen, catalyzed by cytochrome P450 2A13 and other enzymes, produces an α-hydroxyNNK intermediate which spontaneously decomposes to the same methyl diazonium ion as that formed from NDMA, with consequent production of the same methyl DNA adducts, among which is

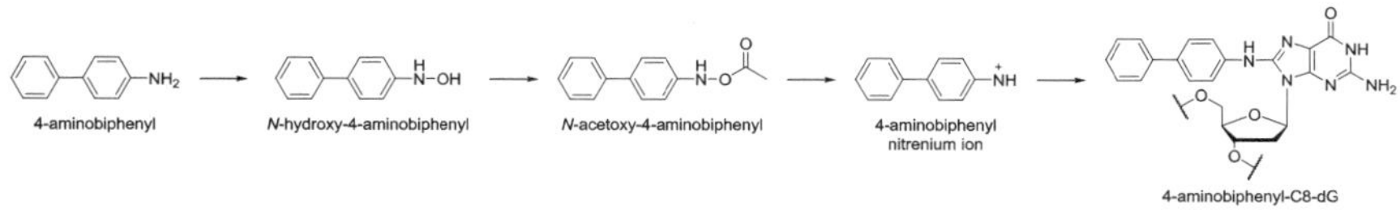

Figure 6. Metabolism of 4-aminobiphenyl leading to formation of a DNA adduct, 4-aminobiphenyl-C^8-dG.

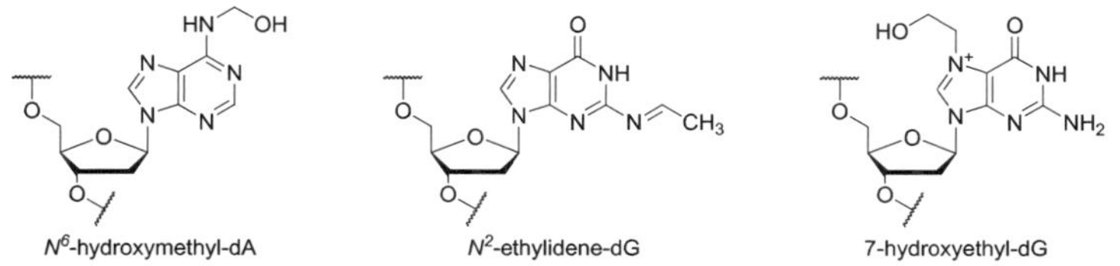

Figure 7. Structures of DNA adducts formed from formaldehyde (N^6-hydroxymethyl-dA), acetaldehyde (N^2-ethylideine-dG) and ethylene oxide (7-hydroxyethyl-dG).

O^6-MeG. Hydroxylation of the methyl group of NNK similarly produces pyridyloxobutyl (POB) diazohydroxide and POB-diazonium ion which react with various positions on DNA forming POB-DNA adducts among which one of the most prevalent is O^2-POB-thymidine (O^2-POB-T).

An important pathway of DNA adduct formation from 4-aminobiphenyl is summarized in Figure 6 [21]. This aromatic amine undergoes *N*-hydroxylation, catalyzed by cytochrome P450 1A2, producing *N*-hydroxy-4-aminobiphenyl, which is further metabolized by *N*-acetyltransferase enzymes, yielding *N*-acetoxy-4-aminobiphenyl. This compound spontaneously decomposes to a nitrenium ion intermediate which reacts with the C-8 position of dG, producing 4-aminobiphenyl-C8-dG, illustrated.

Not all tobacco smoke carcinogens require metabolic activation to produce electrophilic intermediates that react with DNA. Some of these carcinogens are inherently reactive; typical examples are formaldehyde, acetaldehyde, and ethylene oxide. Formaldehyde reacts with DNA to produce a variety of adducts, among which N^6-hydroxymethyl-dA has been identified in smokers' white blood cells [22]. Acetaldehyde reacts directly with the N^2-position of dG, producing N^2-ethylidene-dG, which is also formed when people consume alcohol [23]. Ethylene oxide reacts directly with dG to form N^7-hydroxyethyl-dG [24]. These DNA adducts are illustrated in Figure 7.

5. Specific Adducts are Found in Smokers' DNA

In the previous section, we have seen that carcinogens undergo certain metabolic changes that produce reactive products capable of covalently binding to DNA. Most of the studies described in that section were based on experiments carried out in the laboratory. Can the same DNA adducts also be found in DNA isolated from tissues of cigarette smokers? The answer is yes, but the data are still somewhat fragmentary, particularly when one considers detection of structurally specific DNA adducts. There is solid evidence for the presence in smokers' tissues or urine of DNA adducts derived from most of the compounds shown in Figure 1. However, some of the studies are small, while in others a significant difference between smokers and nonsmokers has not been observed, indicating that the source of the DNA adducts may not be tobacco smoke [25]. There is massive evidence for DNA adduct formation in multiple tissues of smokers, at levels higher than in nonsmokers, from many studies carried out using nonspecific DNA adduct detection methods. These include immunoassay and ^{32}P-postlabeling, techniques with great sensitivity and/or ease of application that allow the analysis of DNA adducts in very small samples of DNA of the type that are typically available in clinical studies. However, these methods, as applied in such studies, do not provide information on the chemical structures of the DNA adducts which are detected [26]. Mass spectrometry is the method of choice for analysis and quantitation of structurally specific DNA adducts in tissues or body fluids from smokers. This technology has now reached a level of sensitivity which promises its application in larger and more convincing studies for the analysis of structurally specific DNA adducts in human tissues.

6. DNA Adducts Cause Mutations

The centrality of DNA adducts in carcinogenesis results from their ability to cause miscoding during DNA replication. This process has been studied in detail using a wide variety of sophisticated approaches in which a given DNA adduct has been inserted into DNA at only one specific position, and then replication has been allowed to proceed [27,28]. By using this approach, multiple groups have obtained consistent results which only vary somewhat based on the system in which the experiments were performed. As an example, one can consider the mutagenic consequences of O^6-MeG formation. Normally, DNA polymerase enzymes would catalyze the insertion of deoxycytidine (C) opposite deoxyguanosine (G). However, experimentation has clearly demonstrated that O^6-MeG is recognized by DNA polymerases as deoxyadenosine (dA). Consequently, a thymidine (dT) is inserted opposite O^6-MeG, resulting in an O^6-MeG→T base pair in the DNA. In the next round of replication, a dA is inserted opposite dT, resulting in conversion of what was originally a strand containing dG (and then O^6-MeG at that position) to one containing dA. This is known as a G→A transition mutation (purine to purine; purine to pyrimidine or vice versa is called a transversion mutation). Thus, a DNA strand that contained a dG that was modified by a carcinogen such as NDMA, to form O^6-MeG (Figure 4), now contained an A at that position. This mutation is permanent. In similar experiments, as examples, BaP-N^2-dG (Figure 2) has caused mainly G→T transversion mutations, O^2-POB-T caused T→A transversions [27–29], and O^6-POB-G produced both G→A transition mutations and G→T trasnversions [30]. Many tobacco smoke-related carcinogens and DNA adducts have been associated with G→T transversion mutations [24].

7. The Predicted Mutations are Found at Specific Sites in Critical Genes

As indicated in Figure 1, mutations are commonly found in certain critical genes in tissues of smokers, the most extensive studies having been carried out in lung tumors. Two of the most commonly mutated genes are the *KRAS* oncogene and the *p53* tumor suppressor gene. The *KRAS* oncogene codes for a protein involved in signal transduction and cell growth. When the gene is mutated, this growth pathway is locked in the "on" position. The *p53* tumor suppressor gene is one of the most commonly mutated genes in cancer. When mutated, it loses its ability to suppress tumor growth by a variety of mechanisms.

As an example, when mice were treated with a chemically activated analogue of the carcinogen NDMA (Figure 4), they developed lung tumors (the analogue is called acetoxymethylmethylnitrosamine [AMMN], and does not require cytochrome P450-catalyzed metabolism). When these lung tumors were analyzed for mutations in the *KRAS* gene, tumors from all treated mice had G→A transition mutations at one specific location in the gene, namely the second position of codon 12. The G→A transition mutation is consistent with the miscoding properties of O^6-MeG, discussed above, as this is one of the DNA adducts produced by this particular compound. There are many other dG's in the *KRAS* gene, including one in the first position of codon 12, but the mutations were only seen at the second position of codon 12. This may be due to preferential reactivity of this position or selection during the process of tumorigenesis. Nevertheless, the observed selectivity is quite remarkable. When the mice were treated with the tobacco carcinogen NNK, their lung tumors similarly had almost exclusively G→A mutations in the second position of codon 12, probably due to the DNA methylating ability of NNK, as indicated in Figure 5 (top part). When the mice were treated with a chemically activated analogue of NNK which goes through the POB-diazonium pathway of Figure 5 (bottom part), a variety of mutations were observed in codon 12, including GGT to GAT, GGT to TGT, and GGT to GTT [31]. Mouse models prone to lung tumor development have also been shown to develop mutations at codon 12 [32]. Overall, about 20%–30% of tumors from the lungs of smokers contain mutations in codons 12 and 13 of the *KRAS* gene, and the most frequent mutation is a G→T transversion at the first position of codon 12 [33].

The *p53* gene provides another example of specificity in interaction with compounds found in cigarette smoke. BaP-7,8-diol-9,10-epoxide (Figure 2) preferentially forms DNA adducts at codons 156, 157, and 158 of exon 5; codon 248 of exon 7; and codon 273 of exon 8 in the *p53* gene in normal human bronchial epithelial cells [24,34]. Another cigarette smoke constituent, acrolein, gives a very similar pattern, preferentially forming adducts at codons 152, 154, 156, 157, and 158 in exon 5; codons 248 and 249 in exon 7; and codons 273 and 282 in exon 8 [34]. A remarkable aspect of these observations is that there are specific hotspots in the distribution of mutations, primarily G→T transversions, along the *p53* gene in lung cancer, and these mutational hotspots are essentially identical to the patterns of reactivity of BaP-7,8-diol-9,10-epoxide (and related diol epoxides from other PAH) and acrolein noted above [24]. This provides convincing evidence that PAH diol epoxides and acrolein are causative agents in tobacco smoke for the observed mutations.

8. Thousands of Mutations are Present in Smokers' Lung Tumors

Modern sequencing techniques allow the interrogation of point mutations in lung tumors [25]. In one study, this technology was used to investigate DNA sequence in a small cell lung cancer cell line [35]. A total of more than 22,000 somatic substitutions were identified including 134 in coding exons. In another study, a nonsmall cell lung cancer from a smoker was analyzed [36]. More than 50,000 single nucleotide variants were observed. In a third study of lung adenocarcinomas, more than 1000 mutations were identified including 915 point mutations, 12 dinucleotide mutations, 29 insertions, and 57 deletions. More than 25 significantly mutated genes were identified, including most commonly *p53* and *KRAS*[37]. These results are consistent with the multiple carcinogens present in tobacco smoke, and their ability to form DNA adducts, leading to miscoding and mutations in critical genes, as summarized in Figure 1.

9. Summary

We have seen that there are more than 70 carcinogenic compounds in tobacco smoke, including PAH, volatile hydrocarbons such as 1,3-butadiene and benzene, tobacco-specific nitrosamines such as NNN and NNK, aromatic amines such as 4-aminobiphenyl, other volatiles such as formaldehyde and acetaldehyde, and a variety of metals. Some of these carcinogens, including formaldehyde and acetaldehyde, react directly with DNA to form adducts, while others, including PAH and tobacco-specific nitrosamines, require a metabolic activation process generally catalyzed by cytochrome P450 enzymes. Structurally distinct DNA adducts are formed when the carcinogens or their metabolites react with DNA at various nucleophilic sites of the DNA bases dG, dA, dT, and dC. Typical modes of formation and structures of these DNA adducts have been summarized in Figures 2–7. We have seen that many of these specific DNA adducts have been found in DNA isolated from tissues of cigarette smokers and that, particularly in the case of more generalized analytical techniques such as ^{32}P-postlabeling and immunoassay, levels of DNA adducts are consistently higher in tissues of smokers than nonsmokers. We have seen that specific DNA adducts can cause mutations during DNA replication. The mutations result from the placement of an incorrect base opposite to the DNA adduct during replication, and this incorrect information is carried forward permanently. We have also seen that the result of these processes can be quite specific with respect to the observed mutations in critical genes such as *KRAS* and *p53*. The multiple mutations seen in smokers' lungs are fully consistent with the formation of multiple DNA adducts from multiple carcinogens in tobacco smoke.

Abbreviations

AMMN	acetoxymethylmethylnitrosamine
dG-C8-2ABP	N-(deoxyguanosin-8-yl)-4-aminobiphenyl
BaA	benz[*a*]anthracene
BaP	benzo[*a*]pyrene
BaP 7,8-diol-9,10-epoxide	7β,8α-dihydroxy-9α,10α-epoxy-7,8,9,10-tetrahydrobenzo[*a*]pyrene
BaP-N^2-dG	*r*7,*t*8,*t*9-trihydroxy-*c*-10-(N^2-deoxyguanosyl)-7,8,9,10-tetrahydrobenzo[*a*]pyrene
dG	deoxyguanosine
dA	deoxyadenosine
dC	deoxycytidine
dT	thymidine
NDMA	*N*-nitrosodimethylamine
NNK	(for nicotine-derived nitrosamino ketone), 4-(methylnitrosamino)-1-(3-pyridyl)-1-butanone
NNN	*N′*-nitrosonornicotine
O^6-Me-dG	O^6-methyldeoxyguanosine
O^2-POB-T	O^2-pyridyloxobutylthymidine
PAH	polycyclic aromatic hydrocarbons
POB	pyridyloxobutyl

References

1. Stewart BW, Wild CP. *World Cancer Report 2014*. IARC, Lyon, 2014.
2. International Agency for Research on Cancer. *Personal Habits and Indoor Combustions*. IARC, Lyon, 2012.
3. Jamal A, Homa DM, O'Connor E, Babb SD, Caraballo RS, Singh T, *et al.* Current cigarette smoking among adults-United States, 2005–2014. *MMWR Morb Mortal Wkly Rep* 2015; 64: 1233–1240.
4. Shafey O, Eriksen MP, Ross H, Mackay J. *The Tobacco Atlas*, 3rd Edition. American Cancer Society and World Lung Foundation, Atlanta, 2009.
5. International Agency for Research on Cancer. *Tobacco Smoke and Involuntary Smoking. IARC Monographs on the Evaluation of Carcinogenic Risks to Humans*, Vol. 83. IARC, Lyon, 2004.
6. United States Department of Health and Human Services. *The Health Consequences of Smoking: 50 Years of Progress. A Report of the Surgeon General*. U. S. Department of Health and Human Services, Centers for Disease Control and Prevention, National Center for Chronic Disease Prevention and Health Promotion, Office on Smoking and Health, Atlanta, 2014.
7. Hecht SS. Research opportunities related to establishing standards for tobacco products under the Family Smoking Prevention and Tobacco Control Act. *Nicotine Tob Res* 2012; 14: 18–28.
8. International Agency for Research on Cancer. *Some Non-Heterocyclic Polycyclic Aromatic Hydrocarbons and Some Related Exposures. IARC Monographs on the Evaluation of Carcinogenic Risks to Humans*, Vol. 92. IARC, Lyon, 2010, pp. 35–818.
9. International Agency for Research on Cancer. *Chemical Agents and Related Occupations. IARC Monographs on the Evaluation of Carcinogenic Risks to Humans*, Vol. 100F. IARC, Lyon, 2012, pp. 249–294.
10. Hecht SS. It is time to regulate carcinogenic tobacco-specific nitrosamines in cigarette tobacco. *Cancer Prev Res* 2014; 7: 639–647.
11. International Agency for Research on Cancer. *Smokeless Tobacco and Tobacco-Specific Nitrosamines. IARC Monographs on the Evaluation of Carcinogenic Risks to Humans*, Vol. 89. IARC, Lyon, 2007, pp. 41–583.
12. Pathak KV, Bellamri M, Wang Y, Langouet S, Turesky RJ. 2-Amino-9H-pyrido[2,3-b]indole (AaC) adducts and thiol oxidation of serum albumin as potential biomarkers of tobacco smoke. *J Biol Chem* 2015; 290: 16304–16318.

13. Fiore MC, Jaen CR, Baker TB, Bailey WC, Benowitz NL, Curry SJ, *et al.* *Treating Tobacco Use and Dependence: 2008 Update. Clinical Practice Guideline*. U. S. Department of Health and Human Services, Public Health Service, Rockville, MD, 2008.
14. Guengerich FP. Cytochrome p450 and chemical toxicology. *Chem Res Toxicol* 2008; 21: 70–83.
15. Hecht SS, Belinsky SA, Bode AM, Christiani D, Dennis PA, Dong Z, *et al.* Cancer. In *How Tobacco Causes Disease — The Biology and Behavioral Basis for Smoking-Attributable Disease: A Report of the Surgeon General*, Sidransky D, ed., 2006.
16. Hecht SS, Kassie F, Hatsukami DK. Chemoprevention of lung carcinogenesis in addicted smokers and ex-smokers. *Nat Rev Cancer* 2009;9:476–488.
17. Sangaraju D, Villalta PW, Wickramaratne S, Swenberg J, Tretyakova N. NanoLC/ESI$^+$ HRMS3 quantitation of DNA adducts induced by 1,3-butadiene. *J Am Soc Mass Spectrom* 2014; 25: 1124–1135.
18. Koc H, Tretyakova NY, Walker VE, Henderson RF, Swenberg JA. Molecular dosimetry of N-7 guanine adduct formation in mice and rats exposed to 1,3-butadiene. *Chem Res Toxicol* 1999; 12: 566–574.
19. Singer B, Grunberger D. *Molecular Biology of Mutagens and Carcinogens*. Plenum Press, New York, 1983.
20. Hecht SS. Progress and challenges in selected areas of tobacco carcinogenesis. *Chem Res Toxicol* 2008; 21: 160–171.
21. Penning TM. *Chemical Carcinogenesis*. Humana Press, New York, 2011.
22. Wang M, Cheng G, Balbo S, Carmella SG, Villalta PW, Hecht SS. Clear differences in levels of a formaldehyde-DNA adduct in leukocytes of smokers and non-smokers. *Cancer Res* 2009; 69: 7170–7174.
23. Balbo S, Meng L, Bliss RL, Jensen JA, Hatsukami DK, Hecht SS. Kinetics of DNA adduct formation in the oral cavity after drinking alcohol. *Cancer Epidemiol Biomarkers Prev* 2012; 21: 601–618.
24. Pfeifer GP, Denissenko MF, Olivier M, Tretyakova N, Hecht SS, Hainaut P. Tobacco smoke carcinogens, DNA damage and p53 mutations in smoking-associated cancers. *Oncogene* 2002; 21: 7435–7451.
25. Hecht SS. Lung carcinogenesis by tobacco smoke. *Int J Cancer* 2012; 131: 2724–2732.
26. Phillips DH, Venitt S. DNA and protein adducts in human tissues resulting from exposure to tobacco smoke. *Int J Cancer* 2012; 131: 2733–2753.
27. Basu AK, Essigmann JM. Site-specifically modified oligodeoxynucleotides as probes for the structural and biological effects of DNA-damaging agents. *Chem Res Toxicol* 1988; 1: 1–18.

28. Delaney JC, Essigmann JM. Biological properties of single chemical-DNA adducts: a twenty year perspective. *Chem Res Toxicol* 2008; 21: 232–252.
29. Weerasooriya S, Jasti VP, Bose A, Spratt TE, Basu AK. Roles of translesion synthesis DNA polymerases in the potent mutagenicity of tobacco-specific nitrosamine-derived O^2-alkylthymidines in human cells. *DNA Repair (Amst)* 2015; 35: 63–70.
30. Pauly GT, Peterson LA, Moschel RC. Mutagenesis by O^6-[4-oxo-4-(3-pyridyl)butyl]guanine in *Escherichia coli* and human cells. *Chem Res Toxicol* 2002; 15: 165–169.
31. Ronai Z, Gradia S, Peterson LA, Hecht SS. G to A transitions and G to T transversions in codon 12 of the Ki-*ras* oncogene isolated from mouse lung tumors induced by 4-(methylnitrosamino)-1-(3-pyridyl)-1-butanone (NNK) and related DNA methylating and pyridyloxobutylating agents. *Carcinogenesis* 1993; 14: 2419–2422.
32. Jackson EL, Willis N, Mercer K, Bronson RT, Crowley D, Montoya R, *et al.* Analysis of lung tumor initiation and progression using conditional expression of oncogenic K-ras. *Genes Dev* 2001; 15: 3243–3248.
33. Califano R, Landi L, Cappuzzo F. Prognostic and predictive value of K-ras mutations in non-small cell lung cancer. *Drugs* 2012; 72 Suppl 1: 28–36.
34. Feng Z, Hu W, Hu Y, Tang M-S. Acrolein is a major cigarette-related lung cancer agent. Preferential binding at *p53* mutational hotspots and inhibition of DNA repair. *Proc Natl Acad Sci USA* 2006; 103: 15404–15409.
35. Pleasance ED, Stephens PJ, O'Meara S, McBride DJ, Meynert A, Jones D, *et al.* A small-cell lung cancer genome with complex signatures of tobacco exposure. *Nature* 2010; 463: 184–190.
36. Lee W, Jiang Z, Liu J, Haverty PM, Guan Y, Stinson J, *et al.* The mutation spectrum revealed by paired genome sequences from a lung cancer patient. *Nature* 2010; 465: 473–477.
37. Ding L, Getz G, Wheeler DA, Mardis ER, McLellan MD, Cibulskis K, *et al.* Somatic mutations affect key pathways in lung adenocarcinoma. *Nature* 2008; 455: 1069–1075.

Chapter 4

Measurement of Polycyclic Aromatic Hydrocarbon-DNA Adducts and Studies of Human Cancer Risk Using Antibody-Based Methods

Regina M. Santella

Department of Environmental Health Sciences,
Mailman School of Public Health of Columbia University,
New York, NY 10032, USA

Abstract

Polyclonal and monoclonal antibodies have been developed that recognize the guanine adducts formed by the reactive intermediates of benzo(a)pyrene, a ubiquitous environmental carcinogen. The antibodies cross-react with the guanine adducts of several other polycyclic aromatic hydrocarbons (PAH) and thus can detect a wide range of PAH diol epoxide guanine adducts. These antibodies have been used in highly sensitive enzyme-linked immunosorbent and immunohistochemical assays to monitor human exposure to PAHs. Occupational, lifestyle, and environmental exposures to PAHs have been detectable by measurement of white blood cell DNA adduct levels. Other studies have used the assays to investigate the role of PAHs in cancer induction. In the last 30 years, data produced using antibody-based methods have been essential for much of the progress in our understanding of the hazards of PAHs.

1. Introduction

Polycyclic aromatic hydrocarbons are a class of ubiquitous environmental pollutants produced during the combustion of organic materials including tobacco and fossil and biomass fuel [1]; they are also present on char-broiled and smoked foods [2]. Benzo(a)pyrene (BP) has been extensively studied as a model carcinogenic PAH, and its metabolism *in vivo* to 7β,8α-dihydroxy-9α,10α-epoxy-7,8,9,10-tetrahydrobenzo[a]-pyrene (BPDE) that can bind to DNA, primarily on guanine, causing mutations ultimately leading to cancer is well established (Figure 1) [1, 3]. Many early studies of the formation and repair of BP-DNA adducts were carried out in cell culture or in animals treated with [^{14}C] or [^{3}H]-labeled BP [4], but because it involves radioactivity, this methodology clearly could not be used to investigate the role of PAH exposure in human populations.

In the 1960s, a method was developed for preparation of purine and pyrimidine-containing antigens using ribonucleosides [5]. Periodate oxidation of the adjacent hydroxides on the sugar ring was followed by reaction of the product with the amino groups in protein, and the linkage was stabilized

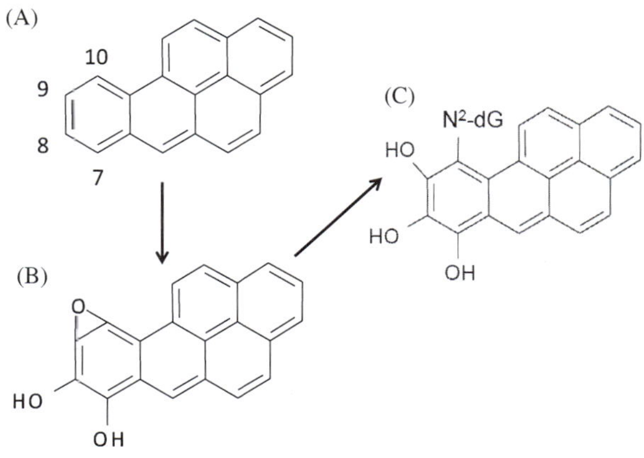

Figure 1. Structures of (A) BP, the parent compound; (B) BPDE, the ultimate metabolite which binds to DNA; and (C) the major BPDE-deoxyguanosine adduct, BPdG. BP, benzo(a) pyrene; BPDE, 7β,8α-dihydroxy-9α,10α-epoxy-7,8,9,10-tetrahydrobenzo[a]-pyrene; BPdG, r7,t8 ,t9-trihydroxy-c-10-(N^2-deoxyguanosyl)-7,8,9,10-tetrahydrobenzo[*a*]pyrene.

by reduction with $NaBH_4$. Using this methodology, highly specific antibodies were generated against guanine, adenine, thymidine, cytosine, and uracil, which reacted best with denatured DNA [5]. The development of these antibodies provided the impetus for the subsequent preparation of antibodies to carcinogen-modified DNA. The first antibody against carcinogen-modified DNA, developed in 1977, recognized the DNA adducts of *N*-2-acetylaminofluorene, a model carcinogen that had been used in early studies of chemical carcinogenesis [6]. Success with these antibodies led to the development of antibodies and immunoassays to other carcinogen-DNA adducts (reviewed in [7–9]) including BP. Furthermore, the antisera elicited against DNA containing the major BP-DNA adduct, the r7,t8,t9-trihydroxy-c-10-(N^2-deoxyguanosyl)-7,8,9,10-tetrahydrobenzo[*a*]pyrene (BPdG), provided a methodology to quantitate PAH-DNA adducts in humans.

2. Antibodies and Methodologies for Measuring PAH-DNA Adducts

Two types of antigens have been used for the development of antibodies to PAH-DNA adducts. One method involves the modification of DNA with BPDE such that at least 1% of the bases are BPdG [10–12]. The DNA is subsequently electrostatically complexed with methylated bovine serum albumin (BSA) before immunization. In the second approach, the BPdG adduct was coupled to BSA, a carrier protein, using the periodate method mentioned earlier. Both polyclonal [10, 11] and monoclonal [12] antibodies were developed. The polyclonal antibodies are easier to obtain but an advantage of monoclonal antibodies is that they can be made available to other investigators. Monoclonal 5D11, generated against BPDE-DNA, and 8E11 generated against BPG-BSA, are licensed by several commercial companies (Hycult Biotech, R&D Systems, Trevigen).

Using these antibodies, different types of assays have been developed for DNA adduct quantitation in biologic samples. While initially radioimmunoassays [10] were used, recently the more sensitive competitive enzyme-linked immunosorbent assay (ELISA) has become the method of choice. Variations in ELISA endpoints have included color [13], fluorescence [14], or chemiluminesence [15] detection of enzyme activity. A representative protocol for a competitive ELISA with chemiluminescence (CIA) detection is illustrated in Figure 2A. With the most sensitive assays, approximately

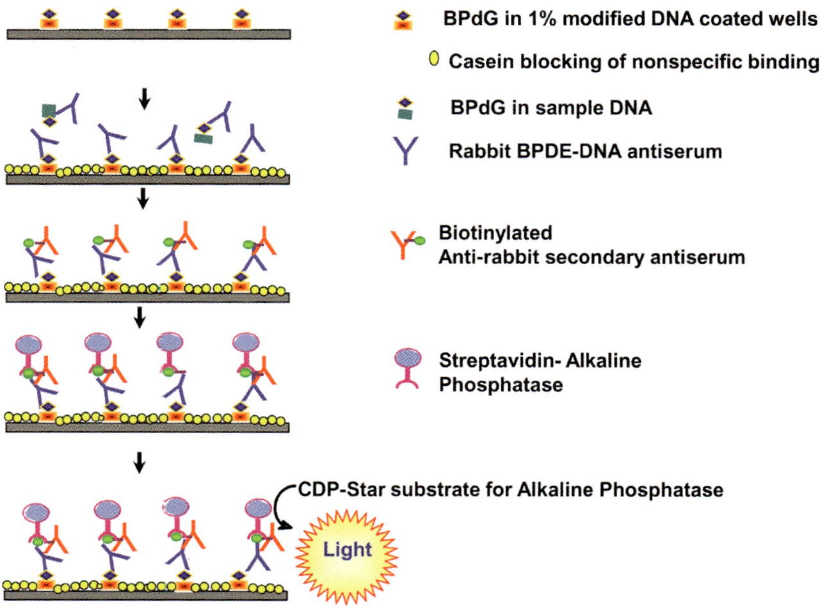

(A)

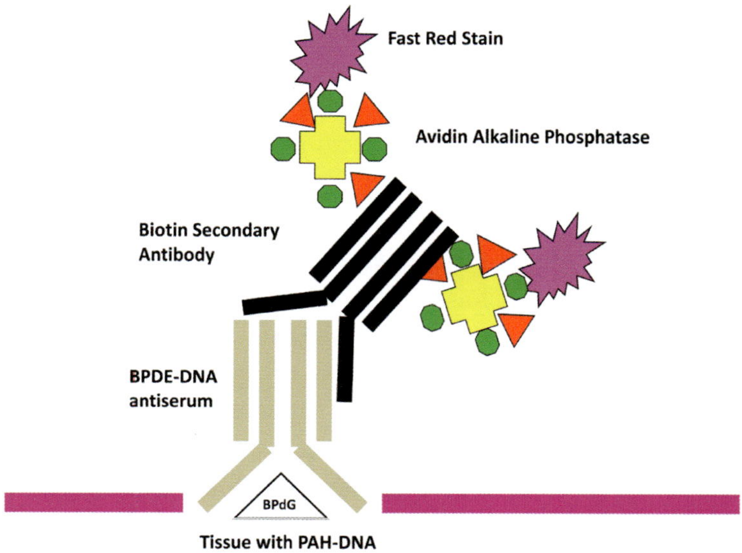

(B)

Figure 2. Methodologies for (A) ELISA with a chemiluminescence endpoint for detection of PAH-DNA adducts in DNA isolated from biological tissues; (B) immunohistochemical detection of PAH-DNA adducts in cells or sectioned tissues. ELISA, enzyme linked immunosorbent assay; PAH, polycyclic aromatic hydrocarbons.

1 fmol (10^{-15}) of adduct can be detected [15]. In addition to antibody affinity to the adduct, the detection limit is also dependent upon the amount of DNA that can be put into a single well of a 96-well plate in a competitive ELISA. For maximum sensitivity, the DNA is sonicated and denatured before analysis. Levels ranging from 5 to 50 μg/well have been used. Depending on the standard curve, this translates into adduct levels of 0.3–3 adducts/10^8 nucleotides. More recently, a direct sandwich chemiluminescence assay, which fragments the DNA with restriction enzyme digestion, captures the modified fragments onto the ELISA plate with an anti-BPDE-DNA antiserum, and then quantitates the bound DNA with an anti-single-stranded DNA antibody, has been developed [16]. A limit of sensitivity of 3 adducts/10^9 nucleotides was reported using 5-μg DNA/well.

An important aspect of antibody-based studies comprises the differences in antibody affinity observed with differing levels of DNA modification. For example, most adducted DNA immunogens are modified to the extent of 1–3 bases/100 nucleotides or 1%–3%. However, most biological samples are modified to the extent of 1 adduct/10^6 nucleotides or lower. Three-dimensionally these species will be very different, and antibody recognition for the 1% modified immunogen DNA will be much stronger than for the biological DNA [17]. To surmount this difference and obtain an accurate value by ELISA, we typically use a plate in which the high-modified (1%) DNA is coated on the wells but the standard curve contains DNA modified to 1 adduct/10^6 nucleotides or lower. When evaluating human DNA where the PAH-DNA adduct composition is unknown, the standard used is low-modified BPDE-DNA but the results, being semi-quantitative, are expressed as PAH-DNA adducts.

Various methods have been used to validate the assays. In our initial study, keratinocytes were treated with [^{3}H]BP and modification levels in the extracted DNA determined by ELISA and radioactivity measurements [17]. Excellent correlations were observed. Later calf thymus DNA modified with [^{3}H] BPDE was assayed by radiolabeling, [^{32}P]-postlabeling, a dissociation-enhanced lanthanide fluoroimmunoassay, and the CIA [15]. All assays gave similar results. Liver DNAs from mice exposed to 0.5 and 1.0 mg [^{3}H] BP were assayed by the same four assays. A dose–response was observed but the values by radiolabeling were higher than by the other assays. This may be due to some DNA adducts, which are produced in liver cells, that are not recognized by the antiserum. In addition, early studies tested for cross-reactivity of the BPDE-DNA antiserum with DNA samples modified with other carcinogens. No recognition of DNA modified with adducts of acetylaminofluorene [12], nitropyrene [12], 8-methoxypsoralen [11], or aflatoxin B_1[11] was observed. Several years later, chemists synthesized the reactive diol-epoxides of most of the known carcinogenic PAHs, and DNA was modified with these compounds [18]. All of these diol-epoxides predominately formed N^2 guanine adducts, as does BPDE. Cross-reactivity was observed for all the diol-epoxide-modified DNAs with both polyclonal and monoclonal antibodies. For example, for polyclonal antibody #29 generated against BPDE-DNA, 50% inhibition was 30 fmol for BPDE-DNA, 18 fmol for chrysene-1,2-diol-3,4-epoxide-DNA, 42 fmol for benz(a)anthracene-8,9-diol-10,11-epoxide-DNA, and 114 fmol for benz(a)anthracene-3,4-diol-1,2-epoxide-DNA [19]. Similar results were observed with another polyclonal antibody [20] as well as with our monoclonal antibody 5D11 [21]. Thus, the antibodies elicited against DNA modified with BP have a broader specificity than just for BPDE-DNA, because they recognize a family of carcinogenic PAH diol-epoxide adducts formed on guanine in DNA. It is important to recognize that PAHs which do not bind to DNA are typically not carcinogenic.

Accurate quantitation for DNA adduct formation is possible using these antibodies in cell culture or animal studies where a single PAH is used for dosing and the appropriate standard curve is used. However, humans are exposed to complex mixtures of PAHs the composition of which may differ based on the source. Therefore, it is impossible to design a standard that would have the appropriate ratio of adducts for each exposure scenario. Typically, we use BPDE-DNA modified to biological levels for the standard curve, but recognizing the lack of absolute accuracy, the data are considered to be equivalents of BPDE-DNA adducts that would cause a similar percent inhibition in the assay, and the final numbers are termed PAH-DNA adducts. All of the PAH-DNA adducts recognized by this antiserum are mutagenic and potentially carcinogenic, so the assay provides an extremely useful method for determining relative levels of DNA damage resulting from carcinogenic PAH exposure.

The lack of absolute quantitation for a single specific PAH-DNA adduct can be considered a major limitation of the ELISA. However, the broad antibody specificity does have several advantages. Compared to some other methods of DNA adduct analysis, it is relatively easy and inexpensive to carry out. Over 1800 samples were assayed in one study [22]. It does not involve the use of radioactivity as does [^{32}P] postlabeling, and equipment costs are much lower than for mass spectrometric methods. Essential equipment includes a 96-well plate reader and a 96-well plate washer. Furthermore, the DNA can be recovered from the ELISA plate and directly used for genotyping, sequencing, or recovery if needed for other downstream applications.

In addition to analysis of adducts in isolated DNA, the antibodies have been used for evaluation of adducts in nuclei of fixed cells or tissues by immunohistochemistry [23]. A major advantage is the retention of tissue morphology allowing visualization of adducts in individual cells. In addition, when only small amounts of cells or tissue are available, it may not be possible to isolate sufficient DNA for sensitive analysis by ELISA. Similar to the ELISA, when immunohistochemistry is used to quantitate PAH-DNA adducts in humans, absolute levels cannot be determined. The method involves treatment of paraffin-embedded or frozen tissue sections with RNase and proteinase K to remove RNA adducts that might cross-react with the antibody and to remove nuclear proteins making the DNA more accessible to the antibody. Finally, the DNA is denatured *in situ* with either acid or base to further enhance antibody binding. Color [24], immunofluorescence [25], and immunoperoxidase [26] methods have been developed. Figure 2B illustrates a typical protocol for staining with alkaline phosphatase-labeled secondary antiserum and a fast red endpoint. While early studies evaluated the staining qualitatively, later studies used microscopic systems that allowed semi-quantitation of staining. An automated system has also been developed that allows rapid quantitation of staining and conversion to PAH-DNA adduct levels with a BPDE-DNA standard curve (reviewed in [23]). Immunohistochemistry appears to be about 10-fold less sensitive for detection of PAH-DNA adducts than the most sensitive competitive ELISA, but localization of adducts within particular cells of a complex tissue can also provide valuable information regarding the process of cancer induction.

3. Application of Antibodies to Detection of PAH-DNA Adducts in Human Populations

3.1. *Enzyme-linked immunosorbent assays*

The first study using polyclonal antiserum #29 to evaluate whether its use in a competitive ELISA could quantitate levels of PAH-DNA adducts in humans was a pilot study [27], which showed detectable levels of PAH-DNA adducts in five tumor and/or nontumor lung tissues from lung cancer patients (Table 1). Another early study evaluated whether exposure to cigarette smoke could be detected in placental tissue using the same ELISA but with both polyclonal antibody #29 and monoclonal antibody 5D11 [28]. While PAH-DNA adduct levels were higher in smokers compared to nonsmokers with both the monoclonal and polyclonal assays, the differences were not statistically significant. These disappointing results raised concerns about the validity of the assay in humans since values were near the limit of assay sensitivity. Further evaluation of the assay employed blood samples from subjects with very high exposures to PAHs, and they included psoriasis patients treated clinically with the application of coal tar [29], as well as foundry [30, 31] and coke oven [32] workers with high occupational PAH exposures. Samples from these subjects gave values clearly well above the limit of assay sensitivity and provided more confidence in the results generated. Table 1 gives additional examples of some of the quantitative data obtained from the analysis of DNA samples. It is difficult to compare data across studies given the use of different primary adduct-specific antibodies and endpoint detection systems. However, within a study the comparisons between exposure group or disease status are valid.

In molecular epidemiological studies of cancer, it is ideal to measure DNA adducts as the biomarker of choice in the target tissue. However, because this is clearly not feasible in many situations, blood is frequently used as a surrogate, since it is readily and repeatedly available. Various studies (Table 1) have used both total white blood cells (WBC), and specific cell fractions. WBC are most frequently used, but studies that employ [^{32}P] postlabeling for DNA adduct detection have observed that the longer lived mononuclear cells have the highest levels of PAH-DNA adducts [33]. Since mononuclear cells are only about 15% of total WBC, larger

Table 1. PAH-DNA adducts in human samples measured by ELISA or immunohistochemistry

Sample	Exposure	Number of subjects	Antibody	Adducts/10^8 nucleotides	Reference
ELISA					
Lung DNA	Unknown	15	#29	ND-5.6	27
Placenta	Smokers	16	#29	18.0 ± 4.6	28
	Nonsmokers	14		12.0 ± 2.3	
Placenta	Smokers	17	5D11	54.0 ± 15	28
	Nonsmokers	14		31.0 ± 11	
Mononuclear cells	Smokers	63	#29	4.5 ± 4.3	
	Nonsmokers	27		1.4 ± 4.0	14
WBC	Foundry worker Low	13	#29	5.1 ± 4.1	30
	Medium	10		6.1 ± 4.3	
	High	15		9.6 ± 8.1	
WBC	Coke oven workers	31	#29	15.3	34
	Local controls	15		13.0	
	Countryside controls	13		2.3	
WBC	Countryside summer	22	#29	3.0 ± 3.3	46
	Countryside winter	21		11.0 ± 22.6	
	Exposed summer	21		4.2 ± 13.5	
	Exposed winter	20		30.4 ± 1.7	
WBC	Controls	48	#29	2.0 ± 3.9	47
	Rickshaw drivers	46		8.5 ± 21.4	
WBC	Mexico City		#29		
	Dry season	113		1.1 ± 0.3	48
	Rainy season	92		1.0 ± 0.3	
WBC	Firefighters		#29		49
	No charbroiled meat	19		41.0 ± 16.6	
	Charbroiled meat 3 times per month	7		84.0 ± 72.0	

(*Continued*)

Table 1. (*Continued*)

Sample	Exposure	Number of subjects	Antibody	Adducts/10^8 nucleotides	Reference
WBC	US Army soldiers June-Germany	22	#29	3.1 ± 4.1	50
	August — Kuwait	22		1.6 ± 1.8	
	October — Germany	22		3.9 ± 3.5	
WBC	Feeding volunteers		#29		51; 52
	Before charbroiled meat consumption	4		1.0–3.0	
	During charbroiled meat consumption	4		3.0–15.0	
Immunohistochemistry				**Relative staining level**	
Oral cells	Smokers	20	#1	684 ± 107	42
	Nonsmokers	20		370 ± 83	
Urothelial cells	Smokers	11		689 ± 72	42
	Nonsmokers	11		495 ± 57	
Mononuclear cells	Coke oven workers	16	#1	1.4 ± 1.1	53
	Ambient environment	16		1.2 ± 0.8	
	Countryside	16		0.8 ± 0.6	
Esophagus	United States	6	#29	0.3 ± 0.1	24
	Linxian, China	5		1.8 ± 0.7	
Liver	Control	36	#1	301 ± 123	41
	Hepatocellular cancer	96		368 ± 164	

amounts of blood are needed to isolate sufficient DNA to specifically evaluate this cell type. Many studies have routinely collected 30, or even 50 mL of blood to ensure adequate amounts of mononuclear cell DNA for analysis.

One of the earliest occupational studies was carried out in Finnish foundry workers known to be at elevated exposure to PAHs and risk of lung cancer. Thus, they were an ideal population for the purpose of further validating the PAH-DNA assay in humans. Workers were categorized by low, medium, and high exposure to BP (<0.05, 0.05–0.2, and >0.2 μg BP/m^3) based on air monitoring levels [31]. Compared to non-worker controls (2.2 PAH-DNA adducts/10^8 nucleotides), PAH-DNA adducts increased with increasing exposure to BP with values of 8, 21, and 50 PAH-DNA adducts/10^8 nucleotides for low, medium, and high exposures, respectively. Subsequently, a study of Polish coke oven workers, which also included people who lived in the surrounding community with high levels of air pollution, and those living in a rural region of Poland with low levels of PAH exposure, was performed [34]. The environmentally exposed individuals had PAH-DNA adduct levels (13/10^8 nucleotides) almost as high as workers in the plant (15/10^8 nucleotides) while controls from the countryside (2.3/10^8 nucleotides) were comparable to the Finnish controls [34]. Many other exposed subjects, for which detectable levels of PAH-DNA adducts were documented in either total WBCs or mononuclear cells (Table 1), include roofing and aluminum plant workers, firefighters, coal tar exposed psoriasis patients, and smokers. Ambient environmental exposures and dietary exposures, including charbroiled meat consumption, have also induced measurable levels of detectable PAH-DNA in some studies (reviewed in [9, 23, 35]). Cord blood has provided material for examination of environmental exposures and their effects on birth outcome [36]. Furthermore, PAH-DNA adducts have been used as intermediate endpoints in a smoking cessation study [37] and an antioxidant intervention study in smokers [38].

In addition to detection of environmental exposures, measurement of DNA adducts in blood has been used to examine associations between PAH-DNA adducts and cancer risk. In one study, PAH-DNA adducts were measured in total WBC DNA adducts in 119 lung cancer cases and 98 noncancer controls [39] and among the current smokers, adducts were higher in cases compared to controls. In the largest study to date, PAH-DNA adducts were measured in mononuclear cells of 873 breast cancer cases and 941 controls [22], and detectable adducts were significantly associated with increased breast cancer risk, but there was no dose–response.

3.2. *Immunohistochemical localization of PAH-DNA adducts*

The first study using immunohistochemical analysis of PAH-DNA adducts in human tissues detected adducts in bronchial epithelial cells, cervical epithelium, oocytes, and several other tissues; however, this study lacked appropriate controls [40]. Our own initial studies used highly exposed patients undergoing a modified Goeckerman therapy of topical coal tar application followed by ultra-violet (UV) radiation. Patients provided 4-mm punch biopsies for evaluation of skin levels of PAH-DNA adducts [25]. Given the small amounts of tissue available, it was not possible to isolate sufficient amounts of DNA for analysis by ELISA. Thus, an immunohistochemical method was used with fluorescence-labeled secondary antibodies. Positive nuclear staining was observed in coal tar treated patients but not in untreated controls, and given the high levels of direct skin exposure these results were not unexpected.

While initial studies evaluated the immunohistochemical data qualitatively, later studies used methods that allowed semiquantitation of PAH-DNA staining. Table 1 gives examples of some of this semiquantitative data. In a study of liver tissues from hepatocellular cancer patients and controls, quantitation of immunoperoxidase staining demonstrated significant differences in relative staining intensity between case and control tissues [41]. Oral mucosa [26] and bladder [42] cells of smokers were also found to contain higher levels of adducts than those from nonsmokers. An example of peroxidase staining of oral mucosa samples of smokers and nonsmokers is illustrated in Figures 3A and 3B. Elevated staining was observed in the nuclei of the smoker compared to less intense staining in the nuclei of the nonsmoker.

Studies by the Poirier laboratory used an automated cell imaging system to routinely quantitate staining in specific cells or tissue regions (reviewed in [23, 35]). A standard curve was prepared from cells treated with increasing concentrations of BPDE, extracted for DNA and stained by immunohistochemistry with the same antiserum. Levels of the BPdG adduct were evaluated in the treated cells by CIA and simultaneously by semiquantitative immunohistochemistry. The resulting standard curve, comparing CIA data with immunohistochemistry staining values, could be used to convert staining intensity to level of comparable PAH-DNA adducts. Using this method, nuclear PAH-DNA staining was determined

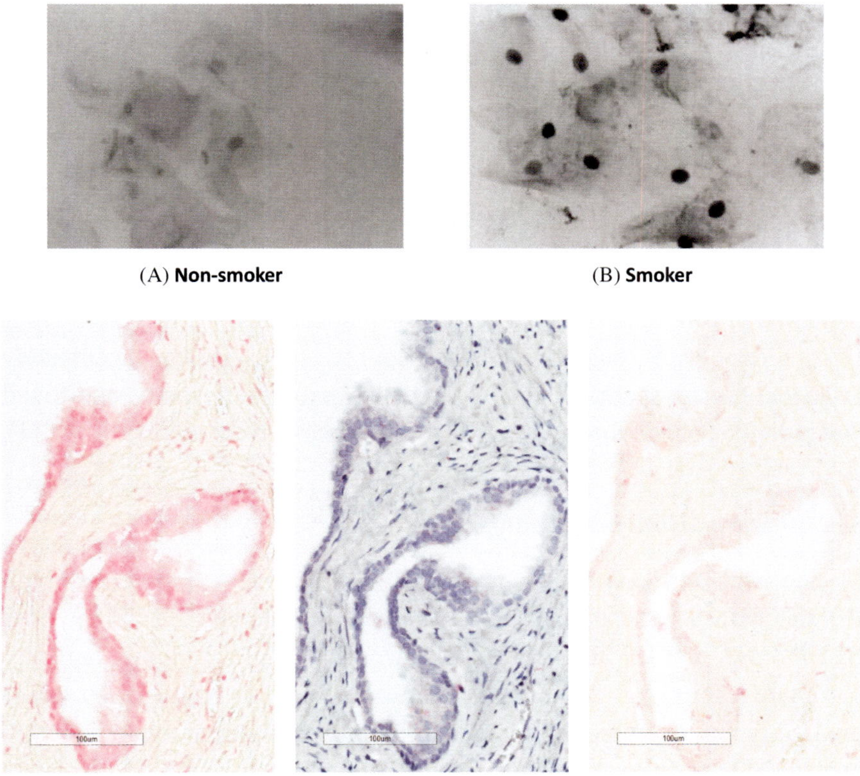

(A) **Non-smoker** (B) **Smoker**

(C) **Specific Serum** (D) **Hematoxylin** (E) **Absorbed Serum**

Figure 3. Examples of PAH-DNA immunohistochemical staining of human cells by immunoperoxidase staining of oral mucosa samples from a (A) nonsmoker and (B) smoker using polyclonal antibody #29 [26] where the darkest cells contain the most PAH-DNA; C,D,E Immunohistochemical staining of human prostate from a middle-aged man in the UK (courtesy of F. Martin and K. John) showing specific PAH-DNA staining in nuclei of endothelial lining cells (C), hematoxylin to localize nuclei (D), and the absorbed serum negative control (E).

in typically 100–2000 cells in esophagus, prostate, placenta, vulva, and cervix [23]. Levels as high as 2 adducts/10^5 nucleotides were observed, but the values were semiquantitative and subject to bias, because the investigator selected cells for inclusion from the regions with the highest adduct levels. Figures 3C–E illustrates PAH-DNA adduct staining in ductal endothelial cells of human prostate. Studies by ourselves and others (reviewed in [35]) have detected adducts in numerous human cells and tissues including breast, prostate, esophagus, cervix, placental fetal tissues, sperm, and embryos.

The relative ease of application of the immunohistochemical method to stored paraffin blocks has resulted in widespread use. For example, a molecular epidemiology study of 1000 prostatectomy samples using a nested case-control study design in a historical cohort of 6692 men with histopathologically benign prostate specimens showed a significantly increased prostate cancer risk in African-American men with elevated PAH-DNA adduct levels, but not for Caucasian men [43]. Another recent study evaluated human sperm bulky DNA adducts and semen quality, and found that individuals with the highest adducts had the poorest sperm quality [44].

3.3. *Immunoaffinity chromatography*

Monoclonal antibody 8E11 generated from animals immunized with BPDE-G-BSA was found to recognize with equal sensitivity both BPDE-DNA and BPdG [12]. This allowed for its use in immunoaffinity chromatography for isolation of adducts prior to analysis by other methods. For example, immunoaffinity purification of enzymatically digested DNA followed by [^{32}P] postlabeling was found to eliminate some contaminants previously observed; however, this method gave variable recovery of different PAH adducts [45].

4. Summary

The development of polyclonal and monoclonal antibodies as well as ELISA and immunohistochemical methods for determination of PAH-DNA adducts has opened up new areas of research and greatly enhanced the field of molecular epidemiology. Studies on the relationship between ambient levels of PAHs and PAH-DNA adducts in blood have led to new understanding of the hazards of pollution. Case-control studies have provided information on the role of PAHs in cancer induction, and molecular dosimetry studies have confirmed that human exposure to carcinogenic PAHs is ubiquitous and induces widespread DNA damage in multiple human tissues.

Abbreviations

BP	benzo(a)pyrene
BPDE	7β,8α-dihydroxy-9α,10α-epoxy-7,8,9,10-tetrahydro benzo[a]-pyrene
BPdG	r7,t8,t9-trihydroxy-c-10-(N^2-deoxyguanosyl)-7,8,9,10-tetrahydrobenzo[*a*]pyrene
BSA	bovine serum albumin
CIA	chemiluninescence immuno assay
ELISA	enzyme-linked immunosorbent assay
PAH	polycyclic aromatic hydrocarbons
WBC	white blood cells

References

1. Moorthy B, Chu C, Carlin DJ. Polycyclic aromatic hydrocarbons: from metabolism to lung cancer. *Toxicol Sci* 2015; 145: 5–15, PubMed Central PMCID: PMC4408964.
2. Phillips DH. Polycyclic aromatic hydrocarbons in the diet. *Mutat Res* 1999; 443: 9.
3. Phillips DH. Fifty years of benzo(a)pyrene. *Nature* 1983; 303: 468–472.
4. Ivanovic V, Geacintov NE, Yamasaki H, Weinstein IB. DNA and RNA adducts formed in hamster embryo cell cultures exposed to benzo[a]pyrene. *Biochemistry* 1978; 17: 1597–1603.
5. Erlanger BF, Beiser SM. Antibodies specific for ribonucleosides and ribonucleotides and their reaction with DNA. *Proc Natl Acad Sci USA* 1964; 52: 68–74, PubMed Central PMCID: PMC300575.
6. Poirier MC, Yuspa SH, Weinstein IB, Blobstein S. Detection of carcinogen-DNA adducts by radiommunoassay. *Nature* 1977; 270: 186–188.
7. Santella RM. Application of new techniques for the detection of carcinogen adducts to human population monitoring. *Mutat Res* 1988; 205: 271–282.
8. Poirier MC, Weston A. Human DNA adduct measurements: state of the art. *Environ Health Perspect* 1996; 104 Suppl 5: 883–893, PubMed Central PMCID: PMCPMC1469680.
9. Poirier MC, Santella RM, Weston A. Carcinogen macromolecular adducts and their measurement. *Carcinogenesis* 2000; 21: 353–359.
10. Poirier MC, Santella R, Weinstein IB, Grunberger D, Yuspa SH. Quantitation of benzo(a)pyrene-deoxyguanosine adducts by radioimmunoassay. *Cancer Res* 1980; 40: 412–416.

11. Mumford JL, Williams K, Wilcosky TC, Everson RB, Young TL, Santella RM. A sensitive color ELISA for detecting polycyclic aromatic hydrocarbon-DNA adducts in human tissues. *Mutat Res* 1996; 359: 171–177.
12. Santella RM, Lin CD, Cleveland WL, Weinstein IB. Monoclonal antibodies to DNA modified by a benzo[a]pyrene diol epoxide. *Carcinogenesis* 1984; 5: 373–377.
13. Santella RM, Hsieh LL, Lin CD, Viet S, Weinstein IB. Quantitation of exposure to benzo[a]pyrene with monoclonal antibodies. *Environ Health Perspect* 1985; 62: 95–99, PubMed Central PMCID: PMC1568684.
14. Santella RM, Grinberg-Funes RA, Young TL, Dickey C, Singh VN, Wang LW, *et al.* Cigarette smoking related polycyclic aromatic hydrocarbon-DNA adducts in peripheral mononuclear cells. *Carcinogenesis* 1992; 13: 2041–2045.
15. Divi RL, Beland FA, Fu PP, Von Tungeln LS, Schoket B, Camara JE, *et al.* Highly sensitive chemiluminescence immunoassay for benzo[a]pyrene-DNA adducts: validation by comparison with other methods, and use in human biomonitoring. *Carcinogenesis* 2002; 23: 2043–2049.
16. Georgiadis P, Kovacs K, Kaila S, Makedonopoulou P, Anna L, Poirier MC, *et al.* Development and validation of a direct sandwich chemiluminescence immunoassay for measuring DNA adducts of benzo[a]pyrene and other polycyclic aromatic hydrocarbons. *Mutagenesis* 2012; 27: 589–597, PubMed Central PMCID: PMC3422811.
17. Santella RM, Weston A, Perera FP, Trivers GT, Harris CC, Young TL, *et al.* Interlaboratory comparison of antisera and immunoassays for benzo[a]pyrene-diol-epoxide-I-modified DNA. *Carcinogenesis* 1988; 9: 1265–1269.
18. Weston A, Manchester DK, Poirier MC, Choi JS, Trivers GE, Mann DL, *et al.* Derivative fluorescence spectral analysis of polycyclic aromatic hydrocarbon-DNA adducts in human placenta. *Chem Res Toxicol* 1989; 2: 104–108.
19. Santella RM, Li Y, Zhang YJ, Young TL, Stefanidis M, Lu XQ, *et al.* Immunological methods for the detection of polycyclic aromatic hydrocabon-DNA and protein adducts. In *Genetic Toxicology of Complex Mixtures,* Waters MD, ed., Plenum Press, New York, 1990, p. 291.
20. Weston A, Trivers G, Vahakangas K, Newman M, Rowe M, Mann D, *et al.* Detection of carcinogen-DNA adducts in human cells and antibodies to these adducts in human sera. *Prog Exp Tumor Res* 1987; 31: 76–85.
21. Santella RM, Gasparo F, Hsieh LL. Quantitation of carcinogen-DNA adducts with monoclonal antibodies. *Prog Exp Tumor Res* 1987; 31: 63–75.

22. Gammon MD, Sagiv SK, Eng SM, Shantakumar S, Gaudet MM, Teitelbaum SL, *et al.* Polycyclic aromatic hydrocarbon-DNA adducts and breast cancer: a pooled analysis. *Arch Environ Health* 2004; 59: 640–649.
23. Pratt MM, John K, MacLean AB, Afework S, Phillips DH, Poirier MC. Polycyclic aromatic hydrocarbon (PAH) exposure and DNA adduct semi-quantitation in archived human tissues. *Int J Environ Res Public Health* 2011; 8: 2675–2691, PubMed Central PMCID: PMC3155323.
24. van Gijssel HE, Divi RL, Olivero OA, Roth MJ, Wang GQ, Dawsey SM, *et al.* Semiquantitation of polycyclic aromatic hydrocarbon-DNA adducts in human esophagus by immunohistochemistry and the automated cellular imaging system. *Cancer Epidemiol Biomarkers Prev* 2002; 11: 1622–1629.
25. Zhang YJ, Li Y, DeLeo VA, Santella RM. Detection of DNA adducts in skin biopsies of coal tar-treated psoriasis patients: immunofluorescence and 32P postlabeling. *Skin Pharmacol* 1990; 3: 171–179.
26. Zhang YJ, Hsu TM, Santella RM. Immunoperoxidase detection of polycyclic aromatic hydrocarbon-DNA adducts in oral mucosa cells of smokers and nonsmokers. *Cancer Epidemiol Biomarkers Prev* 1995; 4: 133–138.
27. Perera FP, Poirier MC, Yuspa SH, Nakayama J, Jaretzki A, Curnen MM, *et al.* A pilot project in molecular cancer epidemiology: determination of benzo[a]pyrene–DNA adducts in animal and human tissues by immunoassays. *Carcinogenesis* 1982; 3: 1405–1410.
28. Everson RB, Randerath E, Santella RM, Cefalo RC, Avitts TA, Randerath K. Detection of smoking-related covalent DNA adducts in human placenta. *Science* 1986; 231: 54–57.
29. Santella RM, Perera FP, Young TL, Zhang YJ, Chiamprasert S, Tang D, *et al.* Polycyclic aromatic hydrocarbon-DNA and protein adducts in coal tar treated patients and controls and their relationship to glutathione S-transferase genotype. *Mutat Res* 1995; 334: 117–124.
30. Santella RM, Hemminki K, Tang DL, Paik M, Ottman R, Young TL, *et al.* Polycyclic aromatic hydrocarbon-DNA adducts in white blood cells and urinary 1-hydroxypyrene in foundry workers. *Cancer Epidemiol Biomarkers Prev* 1993; 2: 59–62.
31. Perera FP, Hemminki K, Young TL, Brenner D, Kelly G, Santella RM. Detection of polycyclic aromatic hydrocarbon-DNA adducts in white blood cells of foundry workers. *Cancer Res* 1988; 48: 2288–2291.
32. Hemminki K, Perera FP, Phillips DH, Randerath K, Reddy MV, Santella RM. Aromatic deoxyribonucleic acid adducts in white blood cells of foundry and coke oven workers. *Scand J Work Environ Health* 1988; 14 Suppl 1: 55–56.

33. Savela K, Hemminki K. DNA adducts in lymphocytes and granulocytes of smokers and nonsmokers detected by the 32P-postlabelling assay. *Carcinogenesis* 1991; 12: 503–508.
34. Hemminki K, Grzybowska E, Chorazy M, Twardowska-Saucha K, Sroczynski JW, Putman KL, *et al.* DNA adducts in human environmentally exposed to aromatic compounds in an industrial area of Poland. *Carcinogenesis* 1990; 11: 1229–1231.
35. Santella RM. Immunological methods for detection of carcinogen-DNA damage in humans. *Cancer Epidemiol Biomarkers Prev* 1999; 8: 733–739.
36. Whyatt RM, Jedrychowski W, Hemminki K, Santella RM, Tsai WY, Yang K, *et al.* Biomarkers of polycyclic aromatic hydrocarbon-DNA damage and cigarette smoke exposures in paired maternal and newborn blood samples as a measure of differential susceptibility. *Cancer Epidemiol Biomarkers Prev* 2001; 10: 581–588.
37. Mooney LA, Santella RM, Covey L, Jeffrey AM, Bigbee W, Randall MC, *et al.* Decline of DNA damage and other biomarkers in peripheral blood following smoking cessation. *Cancer Epidemiol Biomarkers Prev* 1995; 4: 627–634.
38. Jacobson JS, Begg MD, Wang LW, Wang Q, Agarwal M, Norkus E, *et al.* Effects of a 6-month vitamin intervention on DNA damage in heavy smokers. *Cancer Epidemiol Biomarkers Prev* 2000; 9: 1303–1311.
39. Tang D, Santella RM, Blackwood AM, Young TL, Mayer J, Jaretzki A, *et al.* A molecular epidemiological case-control study of lung cancer. *Cancer Epidemiol Biomarkers Prev* 1995; 4: 341–346.
40. Shamsuddin AK, Gan R. Immunocytochemical localization of benzo(a) pyrene-DNA adducts in human tissue. *Hum Pathol* 1988; 19: 309–315.
41. Chen SY, Wang LY, Lunn RM, Tsai WY, Lee PH, Lee CS, *et al.* Polycyclic aromatic hydrocarbon-DNA adducts in liver tissues of hepatocellular carcinoma patients and controls. *Int J Cancer* 2002; 99: 14–21.
42. Hsu TM, Zhang YJ, Santella RM. Immunoperoxidase quantitation of 4-aminobiphenyl- and polycyclic aromatic hydrocarbon-DNA adducts in exfoliated oral and urothelial cells of smokers and nonsmokers. *Cancer Epidemiol Biomarkers Prev* 1997; 6: 193–199.
43. Tang D, Kryvenko ON, Wang Y, Jankowski M, Trudeau S, Rundle, A, *et al.* Elevated polycyclic aromatic hydrocarbon-DNA adducts in benign prostate and risk of prostate cancer in African Americans. *Carcinogenesis* 2013; 34: 113–120, PubMed Central PMCID: PMC3534199.
44. Ji G, Yan L, Wu S, Liu J, Wang L, Zhang S, *et al.* Bulky DNA adducts in human sperm associated with semen parameters and sperm DNA fragmentation in infertile men: a cross-sectional study. *Environ Health* 2013; 12: 82, PubMed Central PMCID: PMC4015644.

45. Randerath K, Sriram P, Moorthy B, Aston JP, Baan RA, van den Berg PTM, *et al.* Comparison of immunoaffinity chromatography enrichment and nuclease P1 procedures for ^{32}P-postlabelling analysis of PAH-DNA adducts. *Chemico-Biological Interactions* 1988; 110: 85–105.
46. Perera FP, Hemminki K, Gryzbowska E, Motykiewicz G, Michalska J, Santella RM, *et al.* Molecular and genetic damage in humans from environmental pollution in Poland. *Nature* 1992; 360: 256–258.
47. Rahman MH, Arslan MI, Chen Y, Ali S, Parvin T, Wang LW *et al.* Polycyclic aromatic hydrocarbon-DNA adducts among rickshaw drivers in Dhaka City, Bangladesh. *Int Arch Occup Environ Health* 2003; 76: 533–538.
48. Garcia-Suastegui WA, Huerta-Chagoya A, Carrasco-Colin KL, Pratt MM, John K, Petrosyan P, *et al.* Seasonal variations in the levels of PAH-DNA adducts in young adults living in Mexico City. *Mutagenesis* 2011; 26: 385–391, PubMed Central PMCID: PMCPMC3081332.
49. Liou SH, Jacobson-Kram D, Poirier MC, Nguyen D, Strickland PT, Tockman MS. Biological monitoring of fire fighters: sister chromatid exchange and polycyclic aromatic hydrocarbon-DNA adducts in peripheral blood cells. *Cancer Res* 1989; 49: 4929–4935.
50. Poirier MC, Weston A, Schoket B, Shamkhani H, Pan CF, McDiarmid MA, *et al.* Biomonitoring of United States Army soldiers serving in Kuwait in 1991. *Cancer Epidemiol Biomarkers Prev* 1998; 7: 545–551.
51. Kang DH, Rothman N, Poirier MC, Greenberg A, Hsu CH, Schwartz BS, *et al.* Interindividual differences in the concentration of 1-hydroxypyrene-glucuronide in urine and polycyclic aromatic hydrocarbon-DNA adducts in peripheral white blood cells after charbroiled beef consumption. *Carcinogenesis* 1995; 16: 1079–1085.
52. Rothman N, Poirier MC, Baser ME, Hansen JA, Gentile C, Bowman ED, *et al.* Formation of polycyclic aromatic hydrocarbon-DNA adducts in peripheral white blood cells during consumption of charcoal-broiled beef. *Carcinogenesis* 1990; 11: 1241–1243.
53. Motykiewicz G, Malusecka E, Grzybowska E, Chorazy M, Zhang YJ, Perera FP, *et al.* Immunohistochemical quantitation of polycyclic aromatic hydrocarbon-DNA adducts in human lymphocytes. *Cancer Res* 1995; 55: 1417–1422.

Chapter 5

Chemical Exposures, DNA Damage, and Human Lung Cancer Risk

Ainsley Weston* and Miriam C. Poirier†

**National Institute for Occupational Safety and Health, Centers for Disease Control and Prevention (CDC), Morgantown, West Virginia 26505-2888, USA*

†Laboratory of Cancer Biology and Genetics, Center for Cancer Research, National Cancer Institute, National Institutes of Health (NIH), Bethesda, Maryland 20892-4255, USA

Abstract

Currently lung cancer accounts for 1.7 million deaths worldwide and more than 150,000 deaths in the United States yearly. The molecular events responsible for the development of lung carcinogenesis are becoming elucidated with continuing research. Carcinogens in tobacco, tobacco smoke, vehicle exhaust, and industrial effluents are inhaled mostly and act either directly or following biotransformation to damage DNA. DNA damage leads to dysregulation of cell growth and renewal, which in turn is followed by mutagenesis, hyperplasia, and malignancy. This chapter describes the multistage process of human chemical carcinogenesis through the example of lung cancer. Major players in the carcinogenic process leading to human lung cancer include chemicals of the classes: (1) polycyclic aromatic hydrocarbons (PAHs), (2) tobacco-specific nitrosamines (TSNs), and (3) diesel engine exhaust fumes (DEEF). Evidence of exposure to these classes of chemicals can be found through DNA adduct formation in normal lung tissue, mutational spectra in lung tumor DNA, and changes in gene expression specific to lung tumor pathology. Knowledge of the dysregulated biological pathways involved in carcinogenesis, including proliferative mechanisms, tyrosine kinase upregulation, motility, differentiation, angiogenesis, and metastasis, have led to development of individualized or precision therapies. In the future, early detection of lung cancer will likely involve low-dose computed tomography, which will allow successful implementation of personized therapeutic approaches.

1. Introduction

Worldwide, lung cancer is the most common cause of cancer-associated mortality, accounting for an estimated 1.7 million deaths in 2015 [1]. The etiology of most human lung cancers contains a tobacco smoking component, because usually only a small percentage of lung cancer patients are non- or never-smokers. Between 1992 and 2013, lung cancer rates in the United States began to decline slightly [2]. However, the National Cancer Institute estimated that almost a quarter of a million new lung cancer cases would be diagnosed (more than 13% of all new cancers) and over 158,000 people would die from this disease in 2016 in the United States alone [2, 3]. The poor prognosis of this disease is indicated by the 20% 5-year survival rate.

Lung cancer induction typically involves chronic exposures to complex chemical mixtures, and is a long-term, multistage process. Tobacco-related lung cancer is considered largely due to the presence of carcinogenic chemicals and irritants in tobacco smoke; however, lung carcinogens are also encountered in the environment or through diet, lifestyle, and occupation (Table 1). Tobacco smoke is responsible for cancers of the lung, esophagus, larynx, bladder, head and neck, and others [4]. Many chemical carcinogens, including some nitrosamines and alkylating agents (methylnitrosourea), can react directly with intracellular components, but others, including polycyclic aromatic hydrocarbons (PAHs), aromatic amines, and some tobacco-specific nitrosamines (TSNs) must be

Table 1. Examples of human lung and head and neck carcinogens that act via DNA damaging mechanisms.

Exposure domain	Source of agents carcinogenic in humans	Active chemical carcinogens
Ambient Environment	Diesel engine exhaust fume	PAHs and Nitro-PAHs Aromatic amines
	Secondhand tobacco smoke	N-nitrosamines PAHs and Nitro-PAHs Tobacco-specific Nitrosamines Aromatic amines Metals
Lifestyle	Tobacco smoke	N-nitrosamines and Tobacco-specific nitrosamines PAHs and Nitro-PAHs Aromatic amines Metals
	Smokeless tobacco (snuff/snus)	Tobacco specific nitrosamines/PAHs
	e-Cigarettes (ENDS)	N-nitrosamines, flavorings, PAHs
	Betel quid	Same as tobacco smoke plus slaked lime
Occupation	Steelworks/Foundries/Roadwork	PAH
	Asbestos[L]	Fibers and cleavage fragments
	Uranium[L]	Ionizing radiation (oxygen free radicals)
	Mining/Fracking/Counter-tops	Silica
	Metals	Beryllium, chromium, cadmium
Diet	Food mutagens	Heterocyclic amines
	Char-broiled meat/fish	PAHs
	Cured meat	N-nitrosamines, nitrites

metabolically bioactivated to carcinogenic and mutagenic electrophiles [5] (also see chapter by Hecht S.S.). Toxin metabolism has evolved to achieve the solubilization and excretion of endogenous compounds; however, electrophilic chemical species resulting from metabolism are naturally attracted to nucleophiles like DNA and protein. Covalent bonding of carcinogenic chemicals with DNA, otherwise known as *DNA adduct* (*note:* words and phrases in italics are defined in the Glossary) formation, results in genetic dysfunction that can include induction of permanent mutations, DNA single strand and double strand breaks, and *aneuploidy* (chromosome losses or gains) [5]. All of these mechanisms contribute to the carcinogenic process, as DNA damage produces mutations in genes critical for growth control.

Internalized xenobiotic carcinogens, subject to competing processes of metabolic activation and detoxification, are processed with considerable interindividual variation among humans. Mechanisms for metabolic processing include activation or covalent binding to informational macromolecules, detoxification or metabolism to execrable forms, mechanisms for removal or repair of DNA damage, and cellular growth control (see chapter by Stiborova, M.). These are the primary factors responsible for interindividual variability in human chemical cancer risk, and they form the basis for gene–environment interactions, defined as alterations in risk of environmentally induced cancers by heritable traits. For example, only 7%–10% of tobacco smokers develop lung cancer, presumably because some individuals possess endogenous resistance factors (https://www.verywell.com/what-percentage-of-smokers-get-lung-cancer-2248868) [5–7]. These individuals may, however, be subject to other fatal conditions, including emphysema, chronic obstructive pulmonary disease, stroke, and heart disease, all of which can be caused in part by tobacco use.

2. Relevance of Initiation and Promotion in Human Lung Cancer Risk

Within the conceptual framework of multistage carcinogenesis, the first step is termed *tumor initiation* and consists of metabolism of the xenobiotic compound to metabolites that bind to DNA, forming DNA adducts, and/or other DNA damage. Some examples of carcinogen-DNA adduct levels measured in human lung are shown in Table 2. Initiated cells are irreversibly altered and have an elevated risk of *malignant conversion*, which largely occurs because DNA adducts induce mutations in genes critical for growth control (e.g., *driver gene mutations* and *tumor suppressor gene* mutations). The next step in this chemical carcinogenic continuum is *tumor promotion,* caused by compounds which facilitate the clonal expansion of the initiated cells largely by proliferation, inflammation, and *epigenetic* mechanisms,

Table 2. DNA adduct levels in human blood and lung from smokers.

Carcinogen DNA adduct (Method)	DNA adducts measured in	Subjects	DNA adduct level /10^8 nucleotides	Reference
PAH-DNA (ELISA)[c]	Lung tissue and Lung tumor	Tobacco smoke-exposed	4.7–6.0	(34)
		Unexposed	ND[b]	
Bulky adducts (^{32}P-Postlabeling)[c]	Lung	Smokers	1.5–34.3	(35)
		Non-smokers	1.9–3.3	
		Former smokers (>5 yrs)	1.7–4.9	
Bulky adducts (^{32}P-Postlabeling,[d] DRZ)	WBC[a]	Smokers	0.8–407.0	(37)
		Former smokers	0.6–135.0	
		Non-smokers	0.3–52.3	
		Unknown smoking	0.6–12.3	
Putative BPDE (spot in DRZ)	WBC	Smokers	ND – 382.0	
		Former smokers	ND – 128.0	
		Non-smokers	ND – 71.0	
		Unknown smoking	ND – 0.6	
PAH-DNA (^{32}P-Postlabeling)	WBC	lung cancer patients	1.5–150.0	
(ELISA)	Lung	lung cancer patients	1.9–21.1	
Bulky Adducts (^{32}P-Postlabeling) and	PBL[a]	Smokers	6.0	(38)
		Non-smokers	4.6	
PAH-DNA (ELISA)	Lung tumor	Smokers	4.7	
			3.7	
		Non-smokers	2.8	
			2.6	
	Normal lung	Smokers	11.0	
			6.2	
		Non-smokers	4.7	
			5.1	
	Normal Bronchus	Smokers	11.5	
		Non-smokers	4.6	
Bulky Adducts (^{32}P-Postlabeling)	Normal Bronchus	Smokers	7.5–9.4	(39)
		Non-smokers	4.0–6.0	

[a] PBL, peripheral blood lymphocytes; WBC, white blood cells.

[b] ND, not detectable.

[c] ELISA, enzyme-linked immunosorbent assay.

[d] ^{32}P-Postlabeling DRZ, diagonal reactive zone.

defined as functional expression of the genome without changes in the DNA base sequence. Tumor promotion may confer a selective growth advantage on a focus of preneoplastic cells, so that the cells represent an enlarging target for the further action of chemical carcinogens, and are more vulnerable to tumorigenesis. Not only do these cells proliferate more rapidly than normal cells, but also they continue to accumulate genetic and epigenetic preneoplastic changes [5, 8].

3. Gene–Environment Interaction in Human Lung Carcinogenesis

Fundamental to human chemical carcinogenesis is the concept of gene–environment interaction, where normal human interindividual polymorphic variation in metabolism may lead to a greater or lesser risk of cancer (Figure 1). Thus, polymorphic variations in DNA sequence may result in subtle interindividual variation that contributes to chemical carcinogen susceptibility (see *genetic polymorphisms*). Whereas some individuals have a high propensity for carcinogen metabolic activation to DNA binding agents and poor ability to detoxify/eliminate carcinogenic species, yet others have lower propensity for metabolic activation and greater ability to detoxify carcinogenic species [9]. Pivotal genes involved in the chemical carcinogenic cascade can be assigned to classes called *caretaker genes* (e.g., carcinogen metabolism and DNA repair genes) and *gatekeeper genes* (cell cycle checkpoint and *apoptosis* or programmed cell death genes) [8, 10, 11]. As driver mutations accumulate (these include activated oncogenes [see point mutations, and fusion genes], inactivated tumor suppressor genes [see point mutations, and deletions], and damaged DNA repair genes [see point mutations, deletions, and insertions]) genomic instability increases. The resulting *mutator phenotype* [12, 13] leads to an acceleration in the genetic and epigenetic changes taking place (including *passenger mutations*). This scenario is followed by *malignant conversion*, *tumor progression*, and *metastasis*. Molecular mechanisms of human chemical carcinogenesis are being increasingly understood, and such

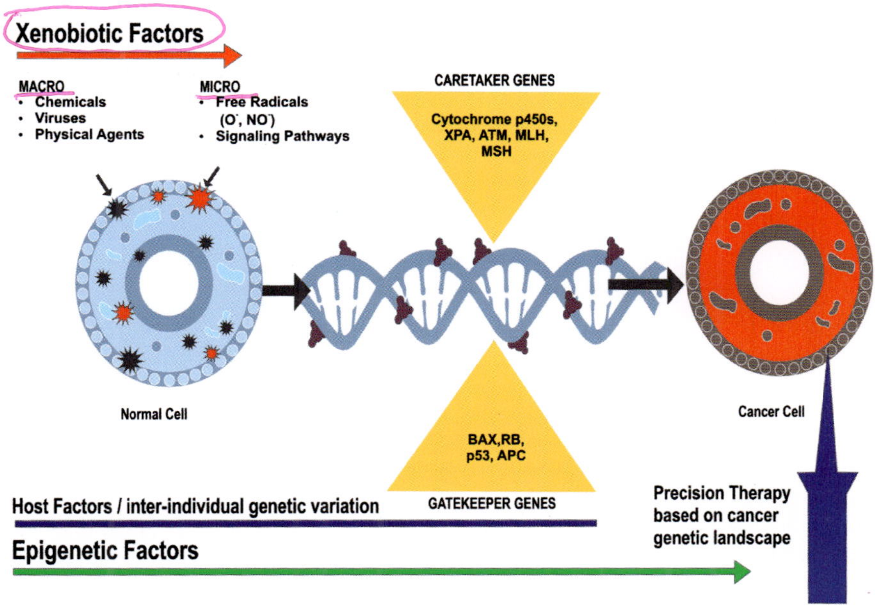

Figure 1. Depiction of the complexity of multistage chemical carcinogenesis that takes into account gene-environment interactions. The environment of the normal cell can be affected by xenobiotic chemicals, viruses, and physical agents that cause the formation of DNA adducts, other DNA damage, and DNA mutations. The cellular environment is also affected by intrinsic factors, including signaling pathways (hormones, growth factors, and apoptosis). Inter-individual host factors (caretaker genes) come into play through polymorphisms in carcinogen metabolism genes, DNA repair genes, and genes involved in cell growth and proliferation. Similarly, gatekeeper genes can be polymorphic. Furthermore, epigenetic factors interact in all stages of the carcinogenic process, resulting in dysregulation of gene expression by 5-me-dC hypomethylation and hypermethylation, as well as histone modifications. Elucidation of all these events has allowed the development of precision therapy in lung cancer treatment.

awareness assists in generating better methods to investigate human cancer risk, susceptibility, and treatment. The results of these studies are intended to frame strategies for prevention, intervention, and cure. As an example, understanding the normal physiology of *gatekeeper genes*, like the tumor suppressor *TP53*, has provided an opportunity to develop new, targeted, or precision-therapeutic approaches [8, 10, 11].

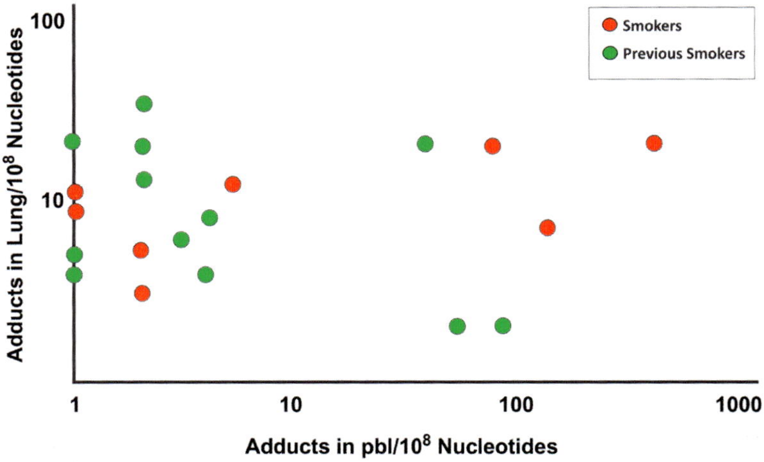

Figure 2. Correlation between bulky DNA adducts detected in peripheral blood lymphocytes (PBL) and in normal peripheral lung tissue from the same cancer patient. Both assays were conducted by ^{32}P-postlabeling, (from van Schooten *et al.* [37]). Here DNA adduct levels were plotted as integers/10^8 nucleotides, where smokers are shown by red circles ($r^2 = 0.27$, $p = 0.19$) and previous smokers are shown in green circles ($r^2 = 0.15$, $p = 0.21$).

4. Classes of Chemical Carcinogen Known to Damage Human Lung DNA

4.1. *Polycyclic aromatic hydrocarbons*

PAHs are the products of incomplete combustion of fossil fuels, which can include cooked food, especially charbroiled meat and fish, and vegetal matter, including tobacco and industrial oils. PAHs are ubiquitous, meaning that they are found in air, soil, cooked foods, water, and some work environments. Human exposure to PAHs, compounds composed of multiple fused benzene rings, is unavoidable. Although they are essentially inert, PAHs are metabolized for excretion from human tissues. The initial metabolic steps are epoxidations occurring at double bonds, catalyzed by cytochrome p450s CYP1A1, CYP1B1, and CYP1A2, to produce an arene oxide. In a second reaction, the arene oxide is hydrolyzed by epoxide hydrolase to form a dihydrodiol. In the case of benzo[a]pyrene (BP), metabolic activation occurs through the further oxidation of the remaining olefinic double bond forming a highly unstable vicinal dihydrodiol-epoxide

7β,8α-dihydroxy-9α,10α-epoxy-7,8,9,10-tetrahydrobenzo[*a*]-pyrene (BPDE), which spontaneously converts to a carbocation at the 10 position and binds covalently to the exocyclic amino group of deoxyguanosine forming r7,t8,t9-trihydroxy-c-10-(N^2-deoxyguanosyl)-7,8,9,10-tetrahydrobenzo [*a*]pyrene (BPdG), the major stable DNA adduct of BP in human tissues [5], (see also chapters by Santella R.M. and Hecht S.S. for further information on this pathway).

4.2. *Aromatic amines*

Aromatic amines are found in cigarette smoke and dyestuffs, and exposure to aromatic amines is associated with lung and bladder cancer in smokers, and bladder cancer in dye workers. These chemicals are metabolized by competing pathways, and the slow-acetylator phenotype is associated with an increased risk of bladder cancer. Therefore, it follows that smokers with slow-acetylator metabolism are likely to be at higher risk of bladder cancer, than smokers with fast-acetylator metabolism. Conversely, smokers who are rapid acetylators are at increased risk of colon cancer [14]. In the case of aromatic amines found in tobacco smoke, a well-studied example is 4-aminobiphenyl (4-ABP). The metabolic pathway resulting in binding to DNA involves an initial N-oxidation of the amine group by cytochrome p4501A2, followed by an acid catalyzed covalent reaction to produce N-hydroxy-4-ABP. Further activation by acetyl coenzyme A-dependent O-acetylase or an O-sulfotransferase gives rise to the N-sulfoonxy- or N-acetoxy-arylamine, which forms N-(deoxyguanosin-8-yl)-2-aminobiphenyl (dG-C8-ABP) in DNA [5].

4.3. *Diesel engine exhaust fumes*

In 2012, diesel engine exhaust fumes (DEEF) were upgraded from Group 2A (probably carcinogenic in humans) to Group 1 (carcinogenic in humans) by the International Agency for Research on Cancer (IARC) [15–17]. The components of DEEF include PAHs, aromatic amines, and other nitroaromatic compounds (1-nitropyrene, 3-nitropyrene, 6-nitrobenzo{a]pyrene, and dinitro-aromatic compounds) [18]. Nitration of PAHs occurs readily, and because nitro-PAH and nitro-biphenyls are components of vehicle emissions they are widespread in the environment.

Some of these chemicals (e.g., 1-nitropyrene, found at levels of 2 mg/kg to >2 g/kg in diesel engine exhaust) can cause genotoxic damage without biotransformation. However, a complex metabolic pathway for most nitrated-hydrocarbons is known [18–20]. 1-Nitropyrene is carcinogenic in a mouse lung model, although insufficient evidence for carcinogenesis to man was reported in 1989 (IARC Group 3) [18]. 1-Nitropyrene metabolism proceeds through nitrogen reduction to 1-aminopyrene by a number of possible enzymes (cytochrome p450, aldehyde oxidase, xanthine oxidase, and DT-diaphorase). Then various oxides (1-nitropyrene-3,4-oxide and 1-nitropyrene-10,11-oxide), hydroxylated (1-hydroxypyrene, 8-hydroxy-1-nitropyrene, 3,4-hydroxy-1-nitropyrene), and N-acetylated metabolites are possible [19].

Since DEEF was first recognized as an IARC Group 1 carcinogen [15–17], numerous other reports have appeared that describe DNA damage, chromosome damage, and mutagenic properties that are the result of exposure to DEEF [21–29]. DEEF is composed of three main fractions gaseous species, a liquid fraction, and particulate materials [21]. Epidemiological studies have shown a positive correlation between levels of particulate air pollution and adverse health effects both systemic (cardiovascular and circulatory) as well as local to the lungs (cancer and other respiratory diseases) [21]. Both reactive oxygen species and reactive nitrogen species can be formed at particle surfaces, but many of the gaseous components of diesel exhaust can contribute to oxidative and proinflammatory damage. It is also well known that metabolic conversion of PAHs and nitrogenated-PAHs leads to redox cycling and gene mutations [22, 23]. Adducts of 8-hydroxy-deoxyguanosine (8-OH-dG) have also been documented in nontobacco smoking traffic policemen (n = 148), and 135 controls, adduct levels were detected by liquid chromatography and electrochemical analysis [24]. Data from this study suggested that stress from vehicular traffic increased oxidative damage, related to cytochrome P450 and glutathione [s] transferase (GST)-polymorphisms, and resulted in higher 8-OH-dG adduct levels [24]. Similarly, among 117 DEEF exposed workers and 112 non-DEEF exposed workers, higher monohydroxylated-PAHs in urine were associated with promotor-region hypomethylation in DNA damage response (DDR) genes (*p16*, *RASSF1A*, and *MGMT*) [25]. In short, these results indicate that DEEF exposure and increased DNA damage are associated with hypomethylation of DDR genes. In a similar DEEF study, increased urinary levels of lipid peroxidation products (malondialdehyde, 4-OH-nonenal, and etheno-DNA adducts) were associated with DEEF exposure [26]. Two other epidemiologic studies of 86 and 101 DEEF exposed workers and 99 and 106 nonexposed controls, respectively, reported elevated urinary levels of monohydrolated PAHs, etheno-DNA adducts, and urinary 1-OH-pyrene [27, 28].

The assessment of genotoxicity in an *in vivo* micronucleus assay has the advantage of including the action of modification (activation and detoxication) enzymes and other physiological parameters. Corrêa *et al.* [29] assessed the genotoxicity of DEEF in a *Vicia faba* (broad bean) root cell *in vivo* micronucleus assay. These investigators observed that both metal ions, as well as organic chemical extracts of diesel particulate materials were genotoxic. These elements of DEEF were shown to cause *clastogenic* events, micronucleus formation, and toxic events that inhibited cell division (by observing reduced mitotic indices) [29].

4.4. *Nitrosamines*

Nitrosamines, a very important class of chemical carcinogens, are found in processed foods, alcoholic beverages, cosmetics, cutting oils, hydraulic fluid, rubber, and tobacco products, including electronic cigarettes or electronic nicotine delivery systems (ENDS), snuff, and *snus*. The tobacco-specific N-nitrosamines, for example, 4-(methylnitrosoamino)-1-(3-pyridyl)-1-butanone (NNK) and N-nitrosodimethylamine (NDMA), are very important because exposure can occur in the absence of combustion, when individuals are using chewing tobacco or ENDS, which provide exposure to tobacco extract and *e-juices*. People who use smokeless tobacco are at high risk of a variety of cancers (head and neck, oral pharynx, larynx, lip, gum, esophagus, and possibly stomach) [4]. TSNs are asymmetric and therefore give rise to a variety of both small and large adducts. For example, cytochrome p450 catalyzed hydroxylation of NNK results in two positively charged species, a relatively large pyridyloxobutyl ion and a relatively small methyl diazonium ion. Covalent reaction of these chemical species with DNA can cause methylation to produce a large adduct, the O^2-pyridyloxobutyl-deoxyguanosine, and a small adduct, the O^6-methyl-deoxyguanosine (O^6-me-dG) [20] (see chapter by Hecht S.S.).

4.5. *Alkylating agents*

Another class of carcinogens, the alkylating agents, can add methyl, ethyl or propyl groups at many sites in DNA, either directly from the action of N-alkylureas (N-methyl-N-nitrosourea), or through metabolism, as is the case of NNK explained above. Protonated alkyl-functional groups generally form covalent lesions at nucleophilic centers within the DNA bases, including adenine (N1, N3, N7), cytosine (N3), guanine (N2, O6, N7), and thymine (O2, N3, O4). Some lesions are known to be successfully repaired such as O^6-me-dG, whereas others, such as N7-methyl-deoxyguanosine (N7-me-dG), are removed spontaneously and are refractory to DNA repair. This explains why O^6-me-dG is a promutagenic lesion, while N7-me-dG is not [30].

5. Additional DNA-Altering Mechanisms: Epigenetics and Reactive Oxygen Species

Other changes in DNA that impact mutagenesis, and potentially carcinogenesis, include epigenetic modifications and reactive oxygen species. We will discuss each in turn. Epigenetic modifications control chromatin conformation and gene expression by either modifying chromosome proteins or methylating DNA at the 5-position of cytosine (see chapter by Pogribny I.P.). For example, 5-methylcytosine (5-meC) is an epigenetic marker that controls transcription, but can also contribute to mutagenesis. The 5-meC marker typically silences transcription. However, if 5-meC becomes deaminated at a CpG dinucleotide, a TpG mismatch forms where 5-me-C is replaced by thymine, and can be incorrectly repaired. Deamination of cytosine can therefore result in CpG pairing with a TpA. Furthermore, cytosine deamination can form a C→T transition if glycosylation and G→T mismatch repair is in error [5].

It is well known that *xenobiotic metabolism* leads to the formation of oxygen free radicals and reactive oxygen species, which can produce oxidative DNA damage. This is especially common when the metabolism involves cytochrome P450s or Fenton chemistry in the presence of metals. Lipid metabolism can generate oxidative stress, and free radicals of other types can be generated from inflammation and immune responses. Oxidative DNA damage, the most common example of which is 8-oxo-deoxyguanosine

(8-oxo-dG), has been shown to form in DNA of biological tissues through Fenton chemistry, after exposure to asbestos, because of the endogenous mineral matrix [31]. Silica is also a source of free radical activity, particularly in hydrologic fracturing (fracking), sandblasting, construction, and other occupations where tiny sand particles are used. Nanoparticles, by their size (1–100 μm) in at least one dimension, also generate reactive oxygen species [32]. Many sources of free radicals can pose a threat to DNA molecules because they are able to covalently react with DNA, not only to form 8-oxo-dG, but also other types of DNA damage [33].

6. Human DNA Adduct Formation in Lung, Bronchus, and Surrogate White Blood Cells

A critical step in the lung cancer process, the formation of carcinogen-DNA adducts, is considered necessary but not sufficient for initiating the process of carcinogenesis. Table 2 summarizes some of the classic studies in which DNA damage has been quantified in human lung, bronchus, and white blood cells (WBCs). The first study, published in 1982 [34], involved analysis of lung tissue obtained at surgery or autopsy, and showed undetectable values of PAH-DNA adducts in lungs from individuals with no smoking exposure, and values between 4.7 and 6.0 adducts/10^8 nucleotides in lungs from individuals with smoking exposure. Subsequently, Phillips *et al.* [35] showed that the range of bulky DNA adduct formation in human lung extended to much higher levels in smokers, compared to former smokers or nonsmokers. Table 2 also shows studies by van Schooten *et al.* [36, 37] in which both lung tissue and surrogate peripheral blood lymphocytes (PBLs) had bulky DNA adduct levels that were higher in smokers > former smokers > nonsmokers. These investigators showed that surrogate blood cell DNA adducts may not always reflect DNA adducts in the target tissue, and that there was no correlation between DNA adduct levels in PBLs and in lung tissue from the same lung cancer patients. Table 2 presents data by Gyórffey *et al.* [38], who used both enzyme-linked immunosorbent assay (ELISA) and ^{32}P-postlabeling to show that the DNA adduct levels in both the target tissue and surrogate (PBL) tissues of smokers were higher than those in nonsmokers. Consistent with other studies, DNA adduct levels in former smokers and nonsmokers were shown not to be indicative of lung cancer risk [38]. Bronchus tissue from lung cancer patients was studied by Gyorffy *et al.* [38] and Ozawa *et al.* [39], and both studies showed that bulky adduct levels in smokers were higher than those in lifelong nonsmokers. In recent years, a great many laboratories have documented carcinogen-DNA adducts in human lung tissue and WBC from the same individual, and a comprehensive review of studies performed between 1982 and 2004 can be found in a review by Philips [40].

6.1. *Risk of lung cancer associated with high levels of carcinogen-DNA adducts*

The association between human DNA adduct formation and human cancer risk has been difficult to validate because large human populations must be studied using appropriate epidemiologic study designs (case/control, or nested prospective case/control). Table 3 shows some studies in which carcinogen-DNA adducts were measured in human tissues of smokers and nonsmokers, and the same individuals were evaluated with respect to lung cancer risk. In 1995, Tang *et al.*, measuring PAH-DNA adducts in peripheral blood leukocytes by ELISA [41], found significantly higher risk of lung cancer (odds ratio [OR] 7.7), and higher PAH-DNA adduct levels, in lung cancer cases compared to controls. Subsequently, the same authors studied smokers and nonsmokers by gender, with regard to DNA adduct levels, and found that lung cancer risk for female smokers was double that for male smokers [42]. Furthermore, they showed that among current smokers, PAH-DNA adduct levels were predictive of lung cancer risk (OR = 3.0) [43].

A meta-analyses of lung cancer studies performed using ^{32}P-postlabeling for bulky DNA adduct determination by Veglia *et al.* [44, 45] support results obtained in studies evaluating PAH-DNA adducts by ELISA. These investigators identified four studies showing that bulky DNA adducts in current smokers are predictive of lung cancer risk using the weighted mean difference, and showing a 15.5-fold increase in smokers compared to nonsmokers [44]. Interestingly, DNA adduct levels in former smokers nor nonsmokers did not show an increase in cancer risk. A second paper published, after 5 years of follow-up [45], showed an 8.4-fold increase in lung cancer risk for current smokers, compared to never-smokers (Table 3).

Table 3. Associations between carcinogen-DNA adducts and lung cancer risk.

Carcinogen DNA adduct (Method)	DNA adducts measured in	Subjects	DNA adduct levels/10^x nucleotides	RR or OR (CI)[a] for Lung cancer	Reference
PAH-DNA	PBL[b]	Cases	3.9 – $10.8/10^{11}$	7.7 (1.7–34.0)	(41)
(BPDE-DNA ELISA)		Controls	3.3 –$5.2/10^{11}$		
PAH-DNA	PBL	All Smokers		2.3 (1.1–7.0)	(42)
(BPDE-DNA ELISA)		All non-Smokers		0.9 (0.2–4.3)	
		♀ Smokers		3.0 (1.1–8.4)	
		♀ non-Smokers		3.1 (0.3–40)	
		♂ Smokers		1.4 (0.5–4.1)	
		♂ non-Smokers	Low adducts	0.3 (0.0–4.0)	
		GSTM1 (+)	Low adducts	1.0 (reference)	
		GSTM1 null	High adducts	2.3 (1.1–5.0)	
		GSTM1 (+)	High adducts	6.9 (1.5–31.4)	
		GSTM1 null		16.2 (2.8–93.9)	
Bulky adducts	PBL		Low vs. high adducts	3.0 (1.1–8.4)	(43)
(^{32}P-Postlabeling)		Current smokers	5.6 vs $11.0/10^8$	2.2 (1.0–4.6)	
		Former smokers	9.8 vs $7.0/10^8$	0.9 (0.6–1.5)	
		Never smokers	5.8 vs $5.0/10^8$	0.8 (0.2–2.8)	
Bulky adducts	WBC[b]	Current Smokers		8.4 (6.1–11.4)	(45)
(^{32}P-Postlabeling)		Former smokers		2.0 (1.4 – 2.7)	
		Never-smokers		1.0 (reference)	

[a]OR, Odds Ratio; RR, Relative Risk; CI, Confidence Interval. [b]PBL, peripheral blood lymphocytes; WBC, white blood cells.

6.2. *Mutations found in different types of lung tumors*

Early studies on the genetics of lung cancer began with tumors having defined *karyotypes*. Beginning in the 1960s, this approach was of limited use because the karyotypes of lung neoplasms were highly abnormal and varied greatly from tumor to tumor. Although some common regions were identified, consistent changes in specific genes remained elusive [46]. In the 1970s and 1980s, cancer research investigators began a quest to look for genes that had been somehow changed, either mutated or lost, during lung carcinogenesis. Initial studies focused on lung cancer cell lines but because of the possibility that mutations might arise during manipulation of cells during establishment of cell lines, work progressed to the isolation of transforming genes from solid human tumors [47–49]. This work is important because mutations in lung tumors provide attractive druggable targets for new therapies.

The first connection to a specific gene found to be mutated in human lung cancer was the *KRAS* gene [47]. Although the initial evidence linked the cellular transforming gene isolated from the human lung cancer to the activated form of the Kirsten-mouse sarcoma virus, it was noted that the results were also consistent with the hypothesis that neoplasms induced by certain carcinogens may be associated with the activation of an oncogene [48–51]. Indeed, a series of experiments were performed a number of years later that demonstrated the activation of the *HRAS* protoncogene *in vitro* through exposure to the ultimate carcinogenic metabolite of (BP), namely BPDE [50]. During the 1980s, attention turned to a technique known as loss of heterozygosity (LOH). This technology involved examination of chromosomal deletions, identified by karyotyping, in childhood tumors like retinoblastoma (chromosome 13q) and Wilms' tumor (chromosome 11p) to search for specific areas of deletion that might have been the location of a tumor suppressor gene [52–54]. The LOH technology was soon applied to adult human cancers like colon cancer, lung cancer, and others [55–57]. These studies led to the identification of *TP53* in a variety of human cancers including lung cancer. It further became clear that TP53 could have the role of loss of tumor suppression function or the role of an activated oncogene or both roles at the same time [5, 8, 58–60].

6.3. *Lung adenocarcinoma*

Adenocarcinoma of the lung accounts for about 35%–40% of lung cancers, and approximately one third are found in never tobacco smokers (defined as having smoked less than 100 cigarettes) [1, 2]. It arises out of peripheral glandular/ductal tissue and secretes mucus. Common gene mutations that occur in lung adenocarcinomas are in *TP53, ALK, ROS1, RET, EGRF, HER2, KRAS, BRAF, PIK3CA, KEAP1*, and *MET* (Table 4). Several of these genes are kinases and can be treated with tyrosine kinase inhibitors [49, 61–63]. Mutations in the epidermal growth factor receptor (EGFR) are found in 15%–20% of lung adenocarcinomas. EGFR is a transmembrane protein with a ligand binding domain that lies outside of the cell and a tyrosine kinase domain within the cell [64, 65]. Mutations of EGFR, either point mutations or in-frame deletions, occur principally in exons 18–21 which affects the ATP-dependent tyrosine kinase region. Because the catalytic region of the protein was disrupted, it provides a target for tyrosine kinase inhibitors (e.g., gefitinib and erlotinib). Through this mechanism, tyrosine kinase inhibitors block mutant EGFR activation of the RAS-RAF-MEK-ERK pathway, which signals survival, migration, metastasis, and proliferation. In addition, the PI3K-PIP3-Akt pathway that has a similar effect to the RAS pathway is also blocked [66, 67].

Anaplastic lymphoma receptor tyrosine kinase (AKL) forms a fusion gene by rearrangement with EML4 or ROS1 or RET in 3%–7% of non–small cell lung cancers, principally adenocarcinomas [66–68]. Similarly, ROS1 and RET can participate in other rearrangements with CD74 and KIF5B or CCDC6, respectively, however, other ROS1 and RET fusions are possible (Table 4) [67].

The gene *HER2* (human epidermal growth factor 2) can be activated by point mutations and insertions, which are drivers, in the kinase domain. These mutations occur in approximately 2%–5% of lung adenocarcinomas exclusive of mutations in EGFR, KRAS, NRAS, BRAF, PI3KCA, MEK1, and AKT or the ALK fusion gene [69]. These lung adenocarcinomas activated by HER2 sometimes occur in a relatively high proportion of nonsmokers (up to 10% in European populations and 90% in Asian populations) [70, 71]. In terms of targeted treatments, some success has been obtained using tyrosine kinase inhibitor drugs that counteract EGFR,

Table 4. Lung ADENOCARCINOMA genes activated by mutation, and/or fusion with a partner gene, and genes ablated by deletion and mutation.

Gene	Class or mode of action	Occurrence (% of tumors)
ALK	Anaplastic lymphoma receptor tyrosine kinase (fusion gene mostly with EML4)	5
EGFR	Transmembrane protein kinase (fusion and point mutation)	15–20
EML4	Echinoderm microtubule associated protein 4 (fusion gene with ALK, in non- or light smokers)	4–8
RET[a]	Transmembrane tyrosine kinase (fusion gene mostly with KIF5B or CCDC6)	1–2
ROS1[b]	Transmembrane tyrosine kinase (fusion gene mostly with CD74)	1–2
CCDC6	Coiled-coil domain-containing protein (Wild-type May function as tumor suppressor)	1
KIF5B	Kinesin with a motor domain moves along microtubules (supports mitosis, meiosis and protein transport)	1
CD74	HLA class II invariant chain (Ii) (supports formation and transport of MHC II protein)	1–2
TP53	Tumor suppressor gene, can assume role of driver oncogene if mutated (point mutations and deletions)	45–50
KRAS	Driver oncogene not usually present with fusion genes (point mutations, usually at codons 12, 13 or 61)	33
HER2	AKA: ERBB/neu: tyrosine kinase receptor (driver point mutations and insertions in the kinase domain)	2–5
BRAF	Protooncogene B-Raf and v-Raf, serine/threonine protein kinase (driver oncogene when mutated)	1–3

(Continued)

Table 4. (*Continued*)

Gene	Class or mode of action	Occurrence (% of tumors)
PIK3CA	Phosphatidylinositol-4,5-bisphosphate 3-kinase (P3K) mutations in exons 9 and 20 (driver point mutations or increase in copy number)	4–6
KEAP1	senses stress/negative regulator of Nrf2 which regulates antioxidant response (preventing oxidative and inflammatory stress). In the quiescent state Nrf2 expression stabilizes normal cell metabolism, one important target gene of Nrf2 is glutathione, major antioxidant. (Epigenetic down regulation by promotor region methylation ~50% lung tumors).	15
MET	hepatocyte growth factor receptor (HGFR) tyrosine kinase, amplified in up to 20% of lung squamous cell carcinomas and adenocarcinomas. Driver of proliferation through phosphorylation HGFR, also imparts resistance to EGFR-specific tyrosine kinase inhibitors.	1–4
NFE2L2	or Nrf2 is a leucine zipper and a transcription factor that protects against oxidative damage caused by injury and inflammation.	

[a]In a minority of tumors other coiled-coil proteins that form fusion genes with RET are: NCOA, a steroid receptor coactivator; TRIM33, a transcriptional corepressor; CUX1, a DNA binding and regulation of gene expression; KIAA1468, which is a ubiquitously expressed cytoplasmic protein.

[b]Other proteins fused with ROS1 are: EZR, a protein tyrosine kinase substrate; SLC34A2, an Na-dependent phosphate transport; TPM3, tropomtosin-3, an actin binding coiled coil protein dimer; SDC4, syndecan, a transmembrane proteoglycan; LRIG3, which has rich membranous content in alveolar epithelium; FIG, a Golgi associated PDZ and CCDC protein; CLTC, clathrin, which facilitates vesicle formation for nutrient and signaling transport by endocytosis and exocytosis; and, CCDC6 (see above) [47].

HER2, and HER4 [72]. However, some HER2 mutants are drug resistant. An example is mutant HER2 that has a 12 base pair insertion in the tyrosine kinase domain at codon 776 (exon 20). This insertion gives rise to the amino acid sequence YVMA so that a nine codon region of the HER2 tyrosine kinase domain becomes synonymous with that of EGFR. Thus, this HER2 mutant is able to operate in the absence of ERBB ligands since activation can occur through the action of epidermal growth factor; moreover, the YVMA mutant activates EGFR and confers resistance to the tyrosine kinase inhibitors erlotinib and gefitinib [72]. When the monoclonal antibody trastuzumab (Herceptin®) was tested for its effectiveness in blocking HER2-YVMA mutant cell proliferation *in vitro,* it was effective, but not against cell lines overexpressing wild-type HER2 [72]. A different antibody, 2C4, was effective against cells expressing wild-type HER2 *in vitro* and *in vivo.* It was concluded from these results that trastuzumab, an IgG1, does not have specificity for EGF or the EGFR/HER2-YVMA heterodimer, and that 2C4 was effective at arresting growth of wild-type HER2-overexpressing cells through steric hindrance of the heregulin-mediated ERB3-ERB2 signaling [72].

6.4. *Lung squamous cell carcinoma*

Squamous cell lung carcinoma accounts for about 20%–25% of lung cancers, and approximately one twentieth to one tenth (1%–2%) of these are found in never tobacco smokers [1, 2]. Common gene mutations that occur in squamous cell lung carcinomas are in *TP53, TP63, SOX2, FGFR1, NFE2L2, NOTCH1*, and *KEAP1.* A comprehensive genomic characterization of squamous cell lung cancers was published in 2012 [73]. This report recurrent mutations in the 10 genes listed in Table 5.

Although both adenocarcinomas of the lung and squamous cell carcinomas of the lung display mutations in p53, a greater proportion of squamous cell carcinomas 35% versus 20% of adenocarcinomas ($p = .006$) have them. In addition, the p53 mutational spectra are somewhat different, where adenocarcinoma mutations occur in codons 248 and 249 of p53 and squamous cell carcinoma p53 mutations are clustered in exons 5–8 as well as a hotspot in codon 267 [74].

The p53 analogue ΔNp63 (TP63) was positively expressed in 100% of squamous cell carcinomas, in addition p40 immunoreactivity which

Table 5. Lung SQUAMOUS CELL CARCINOMA genes activated by mutation, and/or amplification, and genes ablated by deletion and mutation.

Gene	Class or mode of action	Occurrence (% of tumors)
TP53	See Table 4	35
TP63	Tumor protein 63 (3q28) is a member of the TP53, TP63 and TP73 gene family.	100
P40	DeltaNp63 overexpressed, and functions as both a positive and negative transcription regulator and blocks in vitro differentiation	100
SOX2	Transcription factor (3q26), essential for self-renewal. In lung controls branching morphogenesis. Over-expression leads to hyperplasia and ultimately SCC	70
FGFR1	Fibroblast growth factor receptor, amplified in human lung SCC. Promotes carcinogenesis by knockdown of tumor suppressor genes	13
FGFR2	Fibroblast growth factor receptor 2	3
DDR2	Tyrosine kinase, kinase domain mutations appear to be oncogenic driver events	4
NFE2L2	Nrf2 is a leucine zipper and a transcription factor that protects against oxidative damage caused by injury and inflammation.	45
NOTCH1	Regulates/coordinates proliferation, self-renewal and differentiation, dysregulation of *NOTCH1* can cause imbalance of these functions that leads to cell transformation.	30
KEAP1	Senses stress and is a negative regulator of Nrf2 which regulates antioxidant response (preventing oxidative and inflammatory stress). In the quiescent state Nrf2 expression stabilizes normal cell metabolism, one important target gene of Nrf2 is glutathione, which is a major antioxidant.	40

also recognizes ΔNp63 also revealed 100% of squamous cell carcinomas but only 3% of adenocarcinomas [75]. Amplification of the chromosomal region 3q26.33 results in overexpression of *SOX2*, which is a transcription factor that is essential for self-renewal. Overexpression of SOX2 cooperates with *FOXE1* and *FGFR2* causing hyperplasia and ultimately squamous cell carcinomas in lung and esophageal tissues [76].

Mutations in the kinase domains of *FGFR1*, *FGFR2*, and *DDR2* appear to be driver oncogenic events in a proportion of squamous cell carcinomas of the lung (overall approximately 20%). Such mutations are targetable by tyrosine kinase inhibitors like dasatinib and erlotinib [77].

MicroRNAs (miRNA) are also potential classifiers of squamous cell carcinoma prognosis. Starting with 61 squamous cell lung cancer samples and 10 adjacent normal tissue samples, miRNA expression was evaluated by Raponi *et al.* [78]. In this study, they identified 15 unique miRNAs differentially expressed in squamous cell lung cancer which had over 2000 target genes. Genetic pathways that were involved included extracellular signal-regulated kinase/mitogen activated protein kinase (*MAPK*) signaling and axonal guidance. These factors contributed to prediction of prognosis.

6.5. *Small cell and large cell lung carcinoma*

Small cell lung carcinoma also accounts for about 15%–25% of lung cancers, with only a tiny fraction being among never-tobacco smokers [1, 2]. Common gene mutations and amplifications that occur in small cell lung carcinomas are in *RB1, p53, RLF-MYCL1, MYCL1, MYCN, MYC, SOX2, FGFR1,* and *SRSF1* [79–82]. *SRSF1* is a serine/arginine-rich splicing factor, and has a role in preventing exon skipping to ensure accuracy of splicing and regulation of alternative splicing. This gene is also an oncoprotein because it controls the splicing of tumor suppressor genes, kinases, and kinase receptors, and since it is overexpressed in approximately 40% of small cell lung cancers, it is a target for therapy [80].

Large cell lung carcinoma accounts for about 5%–10% of lung cancers, and approximately one fifth being among never-tobacco smokers [1, 2]. Common gene mutations that occur in large cell lung carcinomas are in the neurotrophic tyrosine kinases *(NTRK)* gene family, that is, *NTRK1, NTRK2*, and *NTRK3* [83]. These genes are concerned with neuronal and nerve development, but they code for transmembrane entities that have a receptor and a tyrosine kinase that can activate the *MAPK* pathway leading to proliferation, differentiation, and survival. These genes are typically activated in lung cancers by fusion with *RFWD2*, *CD74*, or *MPRIP*, and can be treated with tyrosine kinase inhibitors like cabozantinib [83].

6.6. *Early detection of lung cancer*

Now that some of the important signposts of the cancer landscape have been identified, the key to success for personalized/precision therapy will lie in the capacity for early lung cancer and bronchogenic cancer detection [84]. However, that is not to say that all of the possible therapeutic targets have been defined. Despite the fact that much work remains to be done, some promising candidate targets are already available.

Early studies of lung cancer screening in the 1970s and 1980s used analog chest x-ray technology. This approach was too insensitive to succeed, and despite the participation of more than 35,000 subjects with as much as 20 years follow-up, no significant mortality benefit was observed. More recently, success for early detection of lung cancer was obtained through the use of low-dose helical computed tomography (LDCT) [85]. This study enrolled 53,454 individuals at high cancer risk (defined as age 55–74 years, 30 pack years tobacco smoking, former smokers who had quit within the past 15 years). From the total participants, 26,722 were randomly assigned to screening with LDCT, and 26,732 were screened with chest x-rays. Compliance with screening tests was higher than 90%, and the rate of positive tests was 24% with LDCT but only 7% by x-ray. One problem with this screening approach is the high rate of false positives, because some individuals are subjected to invasive intervention, which could have been avoided. However, the final number of lung cancers found by LDCT was 1060, and by x-ray 941 were found, for a rate ratio of 1.13 (95%CI = 1.03–1.23). There were 247 lung cancer deaths in the LDCT group and 309 lung cancer deaths in the x-ray group, which equates to a relative reduction in mortality from lung cancer in the LDCT group of 20.0% (95%CI = 6.8–26.7; $p = .004$). The overall mortality was calculated to be almost 7% lower in the LDCT group than in the radiography group ($p = .02$) [85]. Multiple strategies for how to proceed with either an individual with a nodule detected by LDCT, or high-risk individual that was nodule negative by LDCT, have been published [84]. These strategies and improvements in LDCT will no doubt improve the chances of precision therapies to succeed.

7. Conclusions

This chapter describes the multistage process of human chemical carcinogenesis through the example of lung cancer. Major exposures in the carcinogenic process leading to human lung cancer include PAHs, TSNs, DEEF, and others. Evidence of exposure to many classes of chemical carcinogens can be found by evaluating DNA adduct formation in normal lung tissue, mutational spectra in lung tumor DNA, and changes in gene expression specific to lung tumor pathology. Knowledge of the dysregulated biological pathways involved in carcinogenesis, including proliferative mechanisms, tyrosine kinase upregulation, motility, differentiation, angiogenesis, and metastasis, have led to development of individualized or precision therapies.

Acknowledgment

Graphics produced by Kimberly Clough-Thomas of NIOSH-CDC, Morgantown, WV.

Abbreviations

4-ABP	4-aminobiphenyl
APC	adenomatous polyposis coli
ATM	ataxia-telangiectasia mutated
BP	benzo[*a*]pyrene
BPDE	7β,8α-dihydroxy-9α,10α-epoxy-7,8,9,10-tetra hydrobenzo[*a*]-pyrene
BPdG	r7,t8,t9-trihydroxy-c-10-(N^2-deoxyguanosyl)-7,8,9,10-tetrahydrobenzo[*a*]pyrene
CI	confidence interval
CYPs	cytochrome P450 mixed function oxidases

5-meC	5-methylcytosine
DEEF	diesel engine exhaust fumes
DDR	DNA damage recognition
DRZ	diagonal radiation zone
DG-C8-ABP	N-(deoxyguanosin-8-yl)-2-aminobiphenyl
EGFR	epidermal growth factor receptor
ELISA	enzyme-linked immunosorbent assay
ENDS	electronic nicotine delivery systems
GSTM1	glutatione-S-transferase M1
8-OH-dG	8-hydroxy-deoxyguanosine
8-oxo-dG	8-oxo-deoxyguanosine
IARC	International Agency for Research on Cancer
LOH	loss of heterozygosity
LDCT	low-dose helical computed tomography
MAPK	mitogen activated protein kinase
MSH/MLH	mismatch repair genes
NNN	N-NitrosoNorNicotine
NNK	4-(methylnitrosoamino)-1-(3-pyridyl)-1-butanone
NDMA	N-nitrosodimethylamine
N7-me-dG	N7-methyl-deoxyguanosine
O^6-me-dG	O^6-methyl-deoxyguanosine
OR	odds ratio
PAHs	polycyclic aromatic hydrocarbons
PBL	peripheral blood lymphocytes
p53	TP53: transforming protein 53 (see Table 4)
RR	relative risk
TSN	tobacco-specific nitrosamines
WBC	white blood cells

References

1. Who Fact Sheet 297. http://www.who.int/mediacentre/factsheets/fs297/en/
2. SEER. http://seer.cancer.gov/statfacts/html/lungb.html, 2016

3. Goulart BHL, Ramsey SD. Moving beyond the National Lung Screening Trial: discussing strategies for implementation of lung cancer screening programs. *Oncologist* 2013; 18: 941–946.
4. DHHS (NIOSH). Promoting health and preventing disease and injury through workplace tobacco policies. Current Intelligence Bulletin 67, Publication No. 2015–113, pp. 7–17.
5. Weston A, Harris CC. Chemical carcinogenesis. In *Cancer Medicine*, Holland JF, Frei E, III, Bast R, eds., 8th edition, Chap. 15, B.C. Decker, Hamilton, 2008, pp. 225–236.
6. Weston A, Poirier MC. Carcinogen DNA-adducts and DNA repair. In *Encyclopedia of Toxicology*, Wexler P, ed., 3rd edition. Elsevier, Oxford, 2013.
7. DHHS (CDC). How tobacco smoke causes disease. The biology and behavioral basis for smoking-attributal disease. A report of the Surgeon General, 2010; 221–644.
8. Vogelstein B, Papadopoulos N, Velculescu VE, Zhou S, Diaz LA, Kinzler KW. Cancer genome landscapes. *Science* 2013; 339: 1546–1558.
9. Hofseth LJ, Weston A, Harris CC. Chemical carcinogenesis (Chapter 23). In *Cancer Medicine,* Bast RC, Croce CM, Hait WN, Hong WK, Kufe DW, Piccart-Gebhart M, *et al.*, eds., 9th edition. John Wiley & Sons, 287–306, 2016.
10. Kinzler KW, Vogelstein B. Gatekeepers and caretakers. *Nature* 1997; 386: 761–763.
11. Levine AJ. P53, the cellular gatekeeper for growth and division. *Cell* 1997; 88: 323–331.
12. Loeb LA, Springgate CF, Battula N. Errors in DNA replication as a basis of malignant changes. *Cancer Res* 1974; 34: 2311–2312.
13. Leob LA. Human cancers express Mutator phenotype: origin, consequences and targeting. *Nat Rev Cancer* 2011; 11: 450–457.
14. van der Hel OL, Bueno de Mesquita HB, Sandkuijl L, *et al.* Rapid N-acetyltransferase 2 imputed phenotype and smoking may increase risk of colon cancer in women. *Cancer Causes Control* 2003; 14: 293–298.
15. Attfield MD, Schleiff PL, Lubin JH, Blair A, Stewart PA, Vermulen R, *et al.* The diesel exhaust in miners study: a cohort mortality study with emphasis on lung cancer. *J Natl Cancer Inst* 2012; 104: 869–883, 2012.
16. Silverman DT, Samanic CM, Lubin JH, Blair AE, Stewart PA, Vermulen R, *et al.* The diesel exhaust in miners study: a nested case-control study of lung cancer and diesel exhaust. *J Natl Cancer Inst* 2012; 104: 855–868.
17. International Agency for Research on Cancer. Monograph 105, pp. 33–35, 147–260, 2013.
18. International Agency for Research on Cancer. Monograph 46, pp. 321–358, 1989.

19. Beland FA, Heflich RH, Howard PC, Fu PP. The *in vitro* metabolic activation of nitro polycyclic aromatic hydrocarbons. In *Polycyclic Hydrocarbons and Carcinogenesis*, Harvey R, ed., American Chemical Society, Washington, DC, 1985, pp. 371–395.
20. Hecht SS. Tobacco smoke carcinogens and lung cancer. *J Natl Cancer Inst* 1999; 91: 1194–1210.
21. Steiner S, Bisig C, Petri-Fink A, Rothen-Rutishauser B. Diesel-exhaust: current knowledge of adverse effects and underlying cellular mechanisms. *Arch Toxicol* 2016; 90: 1541–1553.
22. Kucab JE, Phillips DH, Arlt VM. Metabolic activation of diesel exhaust carcinogens in primary and immortalized human p53 knock-in (Hupki) mouse embryo fibroblasts. *Environ Mol Mutagen* 2012; 53: 201–217.
23. Von Brocke J, Krais A, Whibley C, Hollstein MC, Schmeiser HH. The carcinogenic air pollutant 3-nitrobenzanthrone induces GC to TA transversion mutations in human p53 sequences. *Mutagenesis* 2009; 24: 17–23.
24. Prasad SB, Vidyullatha P, Vani GT, Devi RPV, Rani UP, Peddy PP, *et al.* Association of gene polymorphism in detoxification enzymes and urinary 8-OHdG levels in traffic policemen exposed to vehicular exhaust. *Inhal Toxicol* 2013; 25: 1–8.
25. Zhang X, Li J, He Z, Duan H, Gao W, Wang H, *et al.* Association between DNA methylation in DNA damage response-related genes and cytokinesis-block micronucleus cytome index in diesel exposed workers. *Arch Toxicol* 2016; 90: 1997–2008.
26. Bin P, Shen M, Li H, Sun X, Niu Y, Meng T, *et al.* Increased levels of urinary biomarkers of lipid peroxidation products among workers occupationally exposed to diesel engine exhaust. *Free Radic Res* 2016; 50: 820–830.
27. Shen M, Bin P, Li H, Zhang X, Sun X, Duan H, *et al.* Increased levels of etheno-DNA adducts and genotoxicity biomarkers of long-term exposure to pure diesel engine exhaust. *Sci Total Environ* 2016; 543(Pt A): 267–273.
28. Duan H, Jai X, Zhai Q, Ma L, Wang S, Huang C, *et al.* Long-term exposure to diesel engine exhaust induces primary DNA damage: a population-based study. *Occup Environ Med* 2016; 73: 83–90.
29. Correa AX, Cotelle S, Millet M, Somensi CA, Wagner TM, Radetski CM. Genotoxicity assessment of particulate matter emitted from heavy-duty diesel-powered vehicles using the *in vivo* Vicia faba L. micronucleus test. *Ecotoxicol Environ Saf* 2016; 127: 199–204.
30. Hamid S, Eckert KA. Effect of DNA polymerase beta loop on discrimination of O6-methyldeoxyguanosine modification present in the nucleotide versus template substrate. *Biochem* 2005; 44: 10378–10387.

31. Crovella S, Bianco AM, Vuch J, Zupin L, Moura RR, Trevisan E, *et al.* Iron signature in asbestos-induced malignant pleural mesothelioma: a population-based autopsy study. *J Toxicol Environ Health A* 2016; 79: 129–141.
32. Joshi GN, Goetjen AM, Knecht DA. Silica particles cause NADPH oxidase-independent ROS generation and transient phagolysosomal leakage. *Mol Biol Cell* 2015; 26: 3150–3164.
33. Schins RP, Knaapen AM, Cakmak GD, Shi T, weishaupt C, Borm PJ. Oxidant-induced DNA damage by quartz in alveolar epithelial cells. *Mutat Res* 2002; 517: 77–86.
34. Perera FP, Poirier MC, Yuspa SH, Nakayama J, Jaretzki A, Curnen MM, *et al.* A pilot project in molecular cancer epidemiology: determination of benzo[*a*]pyrene-DNA adducts in animal and human tissues by immunoassays. *Carcinogenesis* 1982; 12: 1405–1410.
35. Phillips DH, Hewer A, Martin CN, Garner RC, King MM. Correlation of DNA adduct levels in human lung with cigarette smoking. *Nature* 1988; 336: 790–792.
36. Van Schooten FJ, Hillebrand MJN, van Leeuwen FE, Lutgerink JT, van Zandwijk N, Jansen HM, *et al.* Polycyclic aromatic hydrocarbon-DNA adducts in lung tissue from lung cancer patients. *Carcinogenesis* 1990; 11: 1677–1681.
37. Van Schooten FJ, Hillebrand MJN, van Leeuwen FE, van Zandwijk N, Jansen HM, den Engelse L, *et al.* Polycyclic aromatic hydrocarbon-DNA adducts in white blood cells from lung cancer patients: no correlation with adduct levels in lung. *Carcinogenesis* 1992; 13: 987–993.
38. Gyórffy E, Annaq L, Gyóri Z, Segesdi J, Minárovits J, Soltész I, *et al.* DNA adducts in tumour, normal peripheral lung and bronchus, and peripheral blood lymphocytes from smoking and non-smoking lung cancer patients: correlations between tissues and detection by ^{32}P-postlabelling and immunoassay. *Carcinogenesis* 2004; 25: 1201–1209.
39. Ozawa S, Schoket B, McDaniel LP, Tang Y-M, Ambrosone CB, Kostič S, *et al.* Analyses of bronchial bulky DNA adduct levels and *CYP2C9*, *GSTP1* and *NQO1* genotypes in a Hungarian study population with pulmonary diseases. *Carcinogenesis* 1999; 20: 991–995.
40. Phillips DH. Smoking-related DNA and protein adducts in human tissues. *Carcinogenesis* 2002; 23:1979–2004.
41. Tang D, Santella RM, Blackwood AM, Young T-L, Mayer J, Jaretzki A, *et al.* A molecular epidemiological case-control study of lung cancer. *Cancer Epidemiol Biomarkers Prev* 1995; 4: 341–346.

42. Tang D, Rundle A, Warburton D, Santella RM, Tsai W-Y, Chiamprasert S, *et al.* Association between both genetic and environmental biomarkers and lung cancer: evidence of a greater risk of lung cancer in women smokers. *Carcinogenesis* 1998; 19: 1949–1953.
43. Tang D, Phillips DH, Stamfer M, Mooney LA, Hsu Y, Cho S, *et al.* Association between carcinogen-DNA adducts in white blood cells and lung cancer risk in the physicians health study. *Cancer Res* 2001; 61: 6708–6712.
44. Veglia F, Matullo G, Vineis P. Bulky DNA adducts and risk of cancer: a meta-analysis. *Cancer Epidemiol Biomarkers Prev* 2003; 12: 157–160.
45. Veglia F, Loft S, Matullo G, Peluso M, Munnia A, Perera F, *et al.* DNA adducts and cancer risk in prospective studies: a pooled analysis and a meta-analysis. *Carcinogenesis* 2008; 29: 932–936.
46. Miura I, Siegfried JM, Resau J, Keller SM, Zhou JY, Testa JR. Chromosome alterations in 21 non-small cell lung carcinomas. *Genes Chromosomes Cancer* 1990; 2: 328–238.
47. Santos E, Martin-Zanca D, Reddy EP, Pierotti MA, Della Porta G, Barbacid M. Malignant activation of a K-ras oncogene in lung carcinoma but not in normal tissue of the same patient. *Science* 1984; 223: 661–664.
48. Vogelstein B, Kinzler KW. Cancer genes and the pathways they control. *Nat Med* 2004; 10: 789–797.
49. The Cancer Genome Atlas Research Network. Comprehensive molecular profiling of lung adenocarcinomas. *Nature* 2014; 511: 543–550.
50. Marshall CJ, Vousden KH, Phillips DH. Activation of c-Ha-ras-1 proto-oncogene by *in vitro* modification with a chemical carcinogen, benzo(a) pyrene diol-epoxide. *Nature* 1984; 310: 586–589.
51. Reddy EP, Reynolds RK, Santos E, Barbacid M. A point mutation is responsible for the acquisition of transforming properties by the T24 human bladder carcinoma oncogene. *Nature* 1982; 300: 149–152.
52. Fearon ER, Vogelstein B, Feinberg AP. Somatic deletion and duplication of genes on chromosome 11 in Wilms' tumours. *Nature* 1984; 309: 176–178.
53. Cavenee WK. Recessive mutations in the causation of human cancer. *Cancer* 1991; 67: 2431–2435.
54. Hansen MF, Cavenee WK. Retinoblastoma and osteosarcoma the prototypic cancer family. *Acta Paediatr Jpn* 1987; 29: 526–533.
55. Baker SJ, Pressinger AC, Jessup JM, Paraskeva C, Markowitz S, Wilson JK, *et al.* p53 gene mutations occur in combination with p17 allelic deletions as late events in colorectal tumorigenesis. *Cancer Res* 1990; 50: 7717–7722.
56. Weston A, Willey JC, Modali R, Sugimura H, McDowell E, Resau J, *et al.* Differential DNA sequence deletions from chromosomes 3, 11, 13, and 17

in squamous cell carcinoma, large cell carcinoma and adenocarcinoma of the lung. *Proc Natl Acad Sci* 1989; 86: 5099–5103.
57. Nigro JM, Baker SJ, Pressinger AC, Jessup JM, Hostetter R, Cleary K, *et al.* Mutations in the p53 gene occur in diverse human tumour types. *Nature* 1989; 342: 705–708.
58. Harris CC. Structure and function of the p53 tumor suppressor gene: clues for rational cancer therapeutic strategies. *J Natl Cancer Inst* 1996; 88:1442–1455.
59. Ambs S, Hussain SP, Harris CC. Interactive effects of nitric oxide and the p53 tumor suppressor gene in carcinogenesis and tumor progression. *FASEB J* 1997; 11: 443–448.
60. Cho KR, Vogelstein B. Genetic alterations in the adenoma-carcinoma sequence. *Cancer* 1992; 706: 1727–1731.
61. Kwak EL, Bang YJ, Camidge DR, Shaw AT, Solomon B, Maki RG, *et al.* Anaplastic lymphoma kinase inhibition in non-small cell lung cancer. *N Eng J Med* 2010; 363: 1693–1703.
62. Bergethon K, Shaw AT, Ou SH, Katayama R, Lovly CM, McDonald NT, *et al.* Rearrangements define a unique molecular class of lung cancers. *J Clin Oncol* 2012; 30: 863–870.
63. Drilon A, Wang L, Hasanovic A, Suehara Y, Lipson D, Stephens P, *et al.* Response to cabozantinib in patents with RET fusion — positive lung adenocarcinomas. *Cancer Discov* 2013; 3: 630–635.
64. Siegelin MD, Borzuk AC. Epidermal growth factor receptor mutations in lung adenocarcinoma. *Lab Invest* 2014; 94: 129–137.
65. Sharma SV, Bell DW, Settleman J, Haber DA. Epidermal growth factor receptor mutations in lung cancer. *Nat Rev Lung Cancer* 2007; 7: 169–181.
66. Wong DW, Leung EL, So KK, Tam IY, Sihoe AD, Cheng LC, *et al.* The ELM4-ALK fusion gene is involved in various histologic types of lung cancer from nonsmokers with wild-type EGFR and KRAS. *Cancer* 2009; 115: 1723–1733.
67. Koivunen JP, Mermel C, Zejnullahu K, Murpht C, Lifshits E, Holmes AJ, *et al.* ELM4-ALK fusion gene and efficacy of an ALK kinase inhibitor in lung cancer. *Clin Cancer Res* 2008; 14: 4275–4283.
68. Kohno T, Nakaoku T, Tsuta K, Tsuchihara K, Matsumoto S, Yoh K, *et al.* Beyond ALK-RET, ROS1 and other fusion in lung cancer. *Transl Lung Cancer Res* 2015; 4: 156–164.
69. Garrido-Castro AC, Felip E. HER2 driven non-small cell lung cancer (NSCLC): potential therapeutic approaches. *Transl Lung Cancer Res* 2013; 2: 122–127.

70. Li C, Sun Y, Fang R, Han X, Luo X, Wang R, *et al.* Lung adenocarcinomas with HER2-activating mutations are associated with distinct clinical features and HER2/EGFR copy number gains. *J Thoracic Oncol* 2012; 7: 85–89.
71. Li C, Fang R, Sun Y, Han X, Li F, Gao B, *et al.* Spectrum of oncogenic driver mutations in lung adenocarcinomas from East Asian never smokers. *PLOS ONE* 2011; 6: e28204 (1–6).
72. Agus DB, Akita RW, Fox WD, Lewis GD, Higgins B, Pisacane PI, *et al.* Targeting ligand-activated ErbB2 signaling inhibits breast and prostate tumor growth. *Cancer Cell* 2002; 2: 127–137.
73. The Cancer Genome Atlas Research Network. Comprehensive genomic characterization of squamous cell lung cancers. *Nature* 2012; 489: 519–525.
74. Gao W-M, Mady HH, Yu G-Y, Siegfried JM, Luke JD, Melhem MF, *et al.* Comparison of p53 mutations between adenocarcinoma and squamous cell carcinoma of the lung: unique spectra involving G to A transitions and G to T transversions in both histologic types. *Lung Cancer* 2003; 40: 141–150.
75. Bishop JA, Teruya-Feldstein J, Westra WH, Pelosi G, Travis WD, Rekhtman N. p40 deltaNp63 is superior to p63 for the diagnosis of pulmonary squamous cell carcinoma. *Modern Pathol* 2012; 25: 405–435.
76. Bass AJ, Watanabe H, Mermel CH, Yu S, Perner S, Verhaak RG, *et al.* Sox2 is an amplified lineage-survival oncogene in lung and esophageal squamous cell carcinomas. *Nat Genet* 2009; 41: 1238–1242.
77. Hammerman PS, Sos ML, Ramos AH, Xu C, Dutt A, Zhou W, *et al.* Mutations in the DDR2 kinase gene identify a novel therapeutic target in squamous cell lung cancer. *Cancer Discov* 2011; 1: 78–89. doi:10.1158/2159-8274.CD-11-0005.
78. Raponi M, Dossey L, Jatkoe T, Wu X, Chen G, Fan H, *et al.* MicroRNA classifiers for predicting prognosis of squamous cell lung cancer. *Cancer Res* 209; 69: 5776–5782.
79. Hamilton G, Rath B. Smoking, inflammation and small cell lung cancer: recent development. *Wiener Med Wochenschrift* 2015; 165: 379–386.
80. Jiang L, Huang J, Higgs BW, Hu Z, Xiao Z, Yao X, *et al.* Genomic landscape survey identifies SRSF1 as a key oncodriver in small cell lung cancer. *PLOS Genet* 2016; 12: e1005895.
81. Rudin CM, Durinck S, Stawiski EW, Poirier JT, Modrusan Z, Shames DS, *et al.* Comprehensive genomic analysis identifies SOX2 as a frequently amplified gene in small cell lung cancer. *Nat Genet* 2012; 44: 1111–1116.
82. Peifer M, Fernandez-Cuesta L, Sos ML, George J, Seidel GJ, Kasper LH, *et al.* Integrative genome analyses identify key somatic driver mutations of small cell lung cancer. *Nat Genet* 2012; 44: 1104–1110.

83. Amatu A, Sartore-Bianchi A, Siena S. NTRK gene fusion as novel targets of cancer therapy across multiple tumor types. *ESMO Open* 2016; 1: e000023. doi:10.1136/esmoopen-2015-000023.
84. Ruparel M, Quaife SL, Navani N, Wardle J, Janes SM, Baldwin DR. Pulmonary nodules and CT screening: the past present and future. *Thorax* 2016; 71: 367–375.
85. National Lung Screening Trial Research Team, Denise R, Adams AM, Berg CD, Black WC, Clapp JD, *et al.* Reduced lung-cancer mortality with low-dose computed tomography screening. *New Engl J Med* 2011; 365: 395–409.
86. Zhu S-H, Sun JY, Bonnevie E, Cummins SE, Gamst A, Yin L, *et al.* Four hundred and sixty brands of e-cigarettes and counting: implications for product regulation. *Tob Control* 2014; 23: iii3–iii9. doi:10.1136/tobaccocontrol.

Glossary

Aneuploidy: Losses or gains of whole or parts of chromosomes results in aneuploidy, this is when the normal complement of 46 whole chromosome present in a diploid cell is changed by disruptive events in carcinogenesis. Some such events occur during chromosomal segregation resulting in daughter cells with unequal numbers of chromosomes or micronuclei form resulting in loss of genetic material, or aberrant duplication or rearrangement of chromosome fragments.

Apoptosis: Programmed cell death, one of the normal functions of TP53 is to recognize massive unrepairable DNA damage in albeit normal cells and engage a process that removes such a damaged cell by condemning the cell to death.

Caretaker genes: Genes dedicated to DNA repair and that therefore look after the integrity of the genome. Additionally, carcinogen metabolism genes that detoxify and remove carcinogenic chemicals from a cells environment (e.g., some cytochrome p450 genes).

Clastogenic damage: A clastogen is a genotoxic agent that disrupts chromosomes by causing breaks. Such damage may result in deletions, rearrangements, and/or duplications.

DNA adduct: Reactive electrophilic chemical species can form covalent bonds with nucleotides in DNA, thus modifying them chemically. These modified nucleotides are DNA adducts.

Driver mutations: Mutations (often, but not limited to, point mutations) that confer gene signaling that causes a cell to have a survival quality and proliferative advantage.

e-Juices: (Or e-liquids) are chemical mixtures developed ENDS (it should be realized that not all e-juices contain nicotine, which implies that the term ENDS is not always accurate). These sometimes contain extracts of tobacco, but also frequently contain flavorings. There are some 7700 varieties of e-juices [86].

Epigenetic: Any change in gene expression/phenotype that does not involve DNA sequence alterations.

Fusion proteins: Fusion proteins form from rearrangements in DNA. Fusion proteins can drive cell survival, cell proliferation, motility, persistence, metastasis, and other abnormal processes. Examples of fusion proteins include transmembrane chimeroforms that have an extracellular or extravesicular receptor or kinesin motor domain, as well as an intercellular or intervesicular activating domain (e.g., a kinase, or a phosphatase). Examples of fusion proteins can be found in Table 4, such hybrid proteoforms can activate signaling, molecular transportation, survival, cellular and vesicular migration, proliferation, and metastasis.

Gatekeeper genes: Genes that maintain homeostasis with regard to carcinogenesis; such genes involve DNA repair mechanisms and cellular suicide (apoptosis) genes.

Genetic polymorphism (single-nucleotide polymorphism or SNP): In any population this is defined as at least one single nucleotide difference in a single gene. But in the population a defined by the number (or % of human chromosomes) of chromosomes that is two chromosomes per person. So that an SNP is minimally defined as a DNA sequence difference in 1% of human chromosomes [or 1 in every person/50]. Some SNPs are thought to be protective to human disease or in other cases as susceptibility factors in human disease. Studies addressing these questions are still not well formed.

Karyotype: The number and appearance of chromosomes in a cell. Tumors usually are characterized by a karyotype that differs from normal cells from the host of the tumor. These differences represent the genetic changes that took place during carcinogenesis.

Malignant conversion: Evolution of preneoplastic cells into cells that express the malignant phenotype through the accumulation of mutations.

Metastasis: The spreading of cancer cells from the site of the primary cancer to local and distant tissues in the body.

Mutator phenotype: This condition emerges when a genetic or epigenetic event incapacitates a gene critical to arrest of the carcinogenic process. Such critical genes are DNA repair genes or gatekeeper (cell death genes/apoptosis) genes.

Passenger mutations: Mutations in genes that do not endow selective growth advantage to cells.

Personalized medicine: This may be construed to be an attempt to predict an individual's response to administration of a drug to treat a medical condition by analysis of xenobiotic disposition (i.e., assessing inherited cytochrome P450 metabolic propensity). However, it is more logically understanding genetically modified pathways in tumorigenesis and treating the cancer by specifically inhibiting tumor cell survival and proliferation.

Snus: Pulverized, moist tobacco, a variant of snuff. It is ordinarily placed under the top-lip.

Tumor initiation: The initial, irreversible, genetic change occurring in a normal cell that imparts a growth advantage.

Tumor progression: Evolution of a more aggressive phenotype, including metastasis.

Tumor promotion: The epigenetic process whereby an initiated cell is stimulated to grow and form a preneoplastic clone of cells that are at high risk of malignant conversion.

Tumor suppressor gene: Or antioncogene prevents cells from becoming cancerous. An example of a tumor suppressor gene is *TP53*. *TP53* is a tumor suppressor gene, as such it can activate DNA repair proteins upon detection of DNA damage caused by activated xenobiotic molecules (e.g., PAH-diol epoxides, aromatic nitrenium ions, methylating agents, or oxidative stressors). *TP53* can also arrest cell division at the G1/S regulation point allowing time for DNA repair to occur. Or if the DNA damage is too severe p53 can induce apoptosis or programmed cell death. Loss of p53 activity occurs when one copy of the gene is deleted and the other copy suffers a point mutation.

Xenobiotic metabolism: Refers to the metabolic transformation of foreign chemicals by the body.

Chapter 6

Aflatoxin Exposure, Human Liver Cancer Risk, and Chemoprevention

Thomas W. Kensler*,† and John D. Groopman*

**Department of Environmental Health and Engineering, Johns Hopkins Bloomberg School of Public Health, Baltimore, MD 21205, USA*

†Translational Research Program, Fred Hutchinson Cancer Research Center, Seattle, WA, 98109, USA

Abstract

Since its discovery over 50 years ago aflatoxin has been shown to be one of the most potent environmental carcinogens in experimental models and in epidemiologic studies. It is linked to the development of liver cancer and other health problems in people. Exposure to this carcinogen from the diet varies enormously across the world and many human studies have employed aflatoxin specific biomarkers to characterize health risks from these exposures. The development of these biomarkers has been predicated upon mechanistic investigations that has identified and characterized the DNA and protein adducts formed by the metabolism of this compound. In turn these biomarkers have also proven to have efficacy for the measurement of successful primary prevention strategies and chemoprevention technologies. Thus, the translation of mechanistic studies of aflatoxin in experimental models to both epidemiologic investigations in high-risk populations in Asia and Africa and the use of these markers to design public health prevention methods makes aflatoxin a paradigm for other environmental carcinogens.

1. Human Exposure to Aflatoxins

Human exposure to aflatoxins occurs because of the widespread soil distribution of the fungi, *Aspergillus flavus*, which produces aflatoxin B_1 (AFB_1) and aflatoxin B_2 (AFB_2) and *Aspergillus parasiticus*, which produces AFB_1, AFB_2, and aflatoxin G_1 (AFG_1) and aflatoxin G_2 (AFG_2). Since the growth conditions for aflatoxin production are relatively nonspecific, aflatoxins can contaminate a spectrum of crops, including the major staple cereals, edible nuts, and legumes. This contamination can occur either or both prior to harvest when the crops are growing or postharvest during storage [1]. Many experimental studies have explored the relative potency of the aflatoxins leading to acute and chronic toxicities — almost uniformly AFB_1 is the most potent in comparison with the other three forms [1, 2]. Regulatory agencies in many parts of the world have established action levels for the aflatoxins. Some levels are quite restrictive such as those set by the European Commission which are 2 μg/kg for AFB_1 and 4 μg/kg for total aflatoxins for cereals and nuts for direct human consumption [3]. Since exposure is a product of both contamination level and consumption, certain rural communities in economically developing countries can readily exceed these action levels. This is especially true if their diet is composed of high amounts of a single cereal, such as maize, which is a staple in many parts of Africa [4] and historically, in parts of China [5]. While there have been many reports of toxicities in the veterinary literature, acute toxicity in humans has also been documented on several occasions [6–8], most recently with an outbreak of fatal acute aflatoxicosis in Kenya. During the acute aflatoxicosis event in Kenya, total aflatoxin levels in a rural market survey conducted in four districts had a range of total aflatoxin of 1–46,400 μg/kg, with 7% of samples being above 1000 μg/kg [8]. This level greatly exceeds any of the regulatory levels that are in place to assure margins of safety. Of emerging concern is the impact of climate change on the prevalence and magnitude of contamination of foodstuffs by aflatoxins due to combinations of drought, heavy rainfall, and overall stress for plants. Climatic change may thus portend a larger public health problem for aflatoxin over the ensuing decades.

1.1. *Aflatoxin-induced hepatocarcinogenesis*

1.1.1. *Human epidemiology*

Hepatocellular carcinoma (HCC), the most common form of liver cancer, is among the leading causes of cancer death in most parts of the economically developing world. The unequal distribution of this disease is depicted by the map in Figure 1 drawn from the GLOBOCAN database of the International Agency for Research on Cancer (IARC) [9–11]. Since the burden of liver cancer is also coincident with regions where aflatoxin exposure is high, many efforts starting over 50 years ago examined this possible association [12]. These initial studies were hindered by the lack of adequate data on aflatoxin intake, excretion, and metabolism in people, the underlying susceptibility factors such as diet and co-exposure to viruses, as well as by the incomplete statistics on worldwide cancer morbidity and mortality. Despite these deficiencies, early studies did provide data illustrating that increasing HCC rates corresponded to increasing levels of dietary aflatoxin exposure [13]. The use of aflatoxin biomarkers, discussed later in this chapter, has firmed up the association between aflatoxin exposure and risk of developing HCC. Aflatoxin has been classified by IARC as a Group 1, or "known human carcinogen" since 1993.

Collectively liver cancer, including HCC, accounts for 9.1% of all reported cancer deaths and is the second most common cause of cancer mortality worldwide [9]. The incidence of liver cancer varies enormously globally and unfortunately, the burden of this nearly always fatal disease is much greater in the less economically developed countries of Asia and sub-Saharan Africa (Figure 1) [16]. HCC is also the most rapidly rising solid tumor in the United States and Central America, and is overrepresented in minority communities, including African-Americans, Hispanic/Latino-Americans, and Asian-Americans [9, 17–20]. Overall, there are more than 750,000 new cases each year and more than 300,000 deaths annually in the People's Republic of China alone [16]. In contrast with most common cancers in the economically developed world where over 90% of cases are diagnosed after the age of 45, in high-risk regions for liver cancer onset begins to occur in

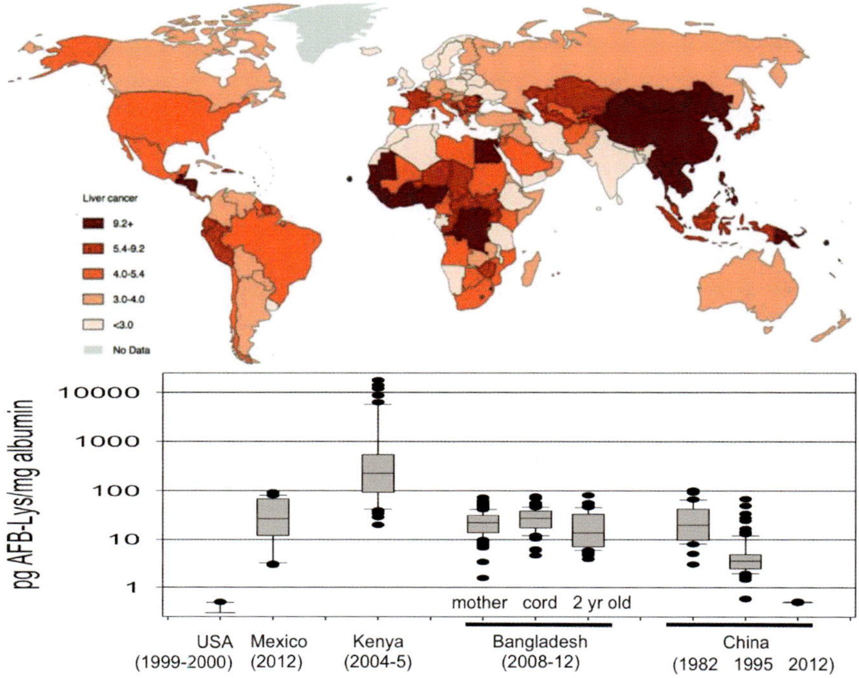

Figure 1. Estimated liver cancer mortality worldwide in 2012 (top): from GLOBOCAN 2012 [11]. Aflatoxin exposure levels vary widely across regions and time periods (bottom). Depicted are medians and 5% and 95% confidence intervals for levels of aflatoxin-albumin adducts measured by mass spectrometry. The US data are from the NHANES [14]; Mexico (unpublished); Kenya [15]; Bangladesh: third trimester, cord blood, and 2 year olds at follow-up [15]; China [5].

both men and women by 20 years of age and peaks between 40 and 49 years of age in men and between 50 and 59 years of age in women [21–23]. This earlier onset of HCC might be attributable to exposures that are both substantial and persistent across the life span. Gender differences in liver cancer incidence have also been well described, and worldwide in 2012, the number of cases among men was 554,000, and among women was 228,000 [10].

1.1.2. *Animal models: carcinogenicity and chemoprevention*

The carcinogenic potency of AFB_1 has been well established in many species of animals, including rodents, nonhuman primates, and fish [24, 25].

The liver is consistently the primary target organ affected and the toxin induces a high incidence of cancer. However, under certain circumstances, depending on animal species and strain, dose, route of administration, and dietary factors, significant numbers of tumors have also been found at other sites, including kidney and colon. Indeed, very few animal species have been found to be resistant to aflatoxin-initiated carcinogenesis. Wide cross-species potency, including sensitivity of primates, provided justification for suspecting that this agent could contribute to human cancer.

Much of the published information on AFB_1 carcinogenicity has been obtained from studies in rats, which are highly susceptible to the toxin. Such experiments have often examined dose-response characteristics, as well as the influences of such parameters as route of administration, dose–response and frequency of dose, and the sex, age, and strain of the test animal. Comparable to observations in humans, male rats have an earlier onset and higher incidence of cancer compared to female animals [26]. A seminal study in which aflatoxin was chronically fed to rats at levels of 1, 5, 15, 50, and 100 ppb (μg/kg) in the diet-induced liver tumors at incidences of 9%, 4.5%, 19%, 80%, and 100%, respectively [27]. Therefore, even μg per day doses of aflatoxin could induce liver cancer, a finding that provided a quantitative context for human cancer investigations exploring the linkage between aflatoxin and HCC. In recent years, there has been an increasing literature pertaining to the carcinogenic responses of the rainbow trout (*Oncorhynchus mykiss*), an even more sensitive species than the rat, and the monkey, possibly a more appropriate model for human risk estimation [28–34].

Early studies found that rhesus monkeys were susceptible to AFB_1 carcinogenicity. In three reports on single animals, two cases of HCC [30–33] and one of cholangiocarcinoma [34] were observed in animals treated for up to 6 years with an oral or a combined oral and intramuscular dosing regimen. Further data on 47 monkeys, representing three species (rhesus, cynomolgus, and African green) that had received AFB_1 by i.p. and/or oral routes for periods greater than 2 months, have been published [30]. Primary liver tumor incidence was 19% (5/26) in animals surviving for longer than 6 months, and total tumor incidence in these animals was 50% (13/26). While out of necessity, these investigations involve small numbers of animals given their cost and long-term nature, the findings support species extrapolation used in risk assessments for humans.

One of the major controversies in risk assessment is the linearity of the dose–cancer response curve. The high sensitivity of the rainbow trout to aflatoxin-induced HCC, along with low spontaneous tumor incidence and cost, made possible design of a study to define an effective tumor-inducing dose in 1% of animals (ED_{01}) that would be impractical in rodents. In this study involving over 42,000 trout, AFB_1 elicited a linear dose–response for liver cancer [28]. These findings complement the earlier observations of a linear dose–response for DNA adduct formation by AFB_1 over an 8-log range [35]. However, recent studies have provided an opposing perspective. Chemopreventive agents, discussed in more detail later, have been used for risk reduction in animal models and in clinical trials in exposed populations. An especially potent agent, a synthetic triterpenoid named CDDO-Im, was used in a lifetime cancer bioassay undertaken in F344 rats dosed daily with AFB_1 (200 μg/kg rat) for 4 weeks and receiving either vehicle or CDDO-Im (three times weekly), 1 week prior to and throughout the exposure period. CDDO-Im completely protected (0/20) against AFB_1-induced liver cancer at 2 years of age compared to a 96% incidence (22/23) observed in the AFB_1 group. With CDDO-Im treatment, the integrated level of aflatoxin-DNA adducts excreted into urine was reduced, but only by 66%. Thus, despite a substantial burden of DNA damage products in the animals treated repeatedly with aflatoxin, they were protected from the development of liver cancer. Consequently, the absence of cancer in this setting requires a concept of a threshold for DNA damage for cancer development [36]. Translating this concept to an actionable threshold for aflatoxin exposure in humans is not possible, however, with current knowledge of exposure, metabolism, and cancer risk.

1.2. *Aflatoxin metabolism: application to biomarkers*

Molecular biomarkers are typically used as indicators of exposure, effect, or susceptibility for both individuals and populations. A biomarker of exposure refers to measurement of the specific compound of interest, its metabolite(s), or its specific interactive products in a body compartment or fluid, which indicates the presence and magnitude of current and past exposure. A biomarker of effect indicates the presence and magnitude of a biological response from exposure to an environmental agent. A biomarker of susceptibility is an indicator or a metric of an inherent or acquired ability of an individual to respond to the challenge of exposure to a specific toxicant. Measures of these biomarkers through molecular epidemiology studies have great utility in addressing the relationships between exposure to environmental agents and development of clinical diseases, and in identifying those individuals at high risk for the disease. Unraveling the aflatoxin/liver cancer link has served as an exemplar of this strategy [12].

Metabolism plays a critical role in the biological activity and disposition of aflatoxin [24, 37, 38]. In general, the enzymes catalyzing aflatoxin metabolism are the same across species, but the principal isoforms of cytochrome P450s and detoxication enzymes within the enzyme families are often species specific. To cause DNA damage, AFB_1 undergoes an initial two-electron oxidation by cytochrome P450-family members (CYP1A2 and CYP3A4 in humans), yielding AFB_1-8,9-oxide (Figure 2). This epoxide reacts with the N^7 atom of guanine to form a promutagenic DNA adduct (aflatoxin–N^7-guanine). This aflatoxin–DNA adduct is chemically unstable within DNA and undergoes depurination, which is followed by export from the tissue and excretion in urine. The aflatoxin N^7-guanine can also undergo ring opening to form formamido adducts that are quite stable. Evidence suggests these "FAPY" adducts contribute importantly to the genotoxic actions of aflatoxin [39]. AFB_1-8,9-oxide is also a substrate for several isoforms of glutathione *S*-transferases (GSTs), which yield a stable, nontoxic, polar product that is excreted in bile. The aflatoxin–glutathione product also undergoes sequential metabolism in the liver and kidneys to be excreted as a stable, water-soluble mercapturic acid (aflatoxin–*N*-acetylcysteine) in the urine. AFB_1 also undergoes extensive oxidation, which is catalyzed by cytochrome P450s. In addition to formation of the 8,9-oxide, oxidation by CYP1A2 yields a stable metabolite, aflatoxin M_1, excreted in milk and urine. Aflatoxin M_1 is less carcinogenic and mutagenic than AFB_1, but equally toxic. Collectively, these end products of aflatoxin biotransformation are biomarkers of exposure to aflatoxin and risk of liver cancer and can be measured in biofluids collected from humans.

Recent studies in genetically engineered mice and rats have provided insights into the critical pathways involved in the detoxication of aflatoxins. Adult mice are exceedingly resistant to the hepatoxicity and carcinogenicity of AFB_1, a phenotype presumed to be related to their enhanced capacity to detoxify aflatoxin. Indeed, mice in which the *Gst a3* gene has been deleted are substantively more sensitive to toxicity by aflatoxins [40]. Collateral genetic amplification of a master regulator (the transcription factor Nrf2) of multiple detoxication genes did not restore a protective function, highlighting the unilateral importance of the *Gst a3* isoform in protection in mice [41]. Interestingly, disruption of the *Nrf2* gene in rats greatly enhanced their sensitivity to aflatoxin hepatotoxicity, as measured by hepatic DNA adduct burden, liver function tests, and survival [42]. The hepatoprotective actions of agents such as CDDO-Im observed in wild-type rats were also abrogated in these knockout rats. Protein adducts have also been thought to contribute to the hepatotoxicity of aflatoxins, as has been shown clearly with other hepatotoxicants such as acetaminophen. However, creation of transgenic lines of rats overexpressing the aldo-keto reductase 7A, which greatly increased the rate of detoxication and urinary excretion of the putative protein-modifying metabolite aflatoxin dialdehyde, had no effect on markers of aflatoxin toxicity or tumorigenicity [43]. The extent to which these rodent studies can be extrapolated to humans remains controversial [44].

In the case of AFB_1, the measurement of the DNA adducts is of major interest because they are direct products of damage to a critical cellular macromolecular target. The finding that the major aflatoxin-nucleic acid adduct, aflatoxin-N^7-guanine, was excreted exclusively in urine of exposed rats spurred interest in using this metabolite as a biomarker of both exposure and risk. This adduct, however, has a short half-life in the body (~8 h) [45]. The serum aflatoxin-albumin adducts, derived from the same initial epoxide metabolite (see Figure 2), have been examined also as a biomarker since the longer half-life of circulating albumin (~3 weeks) integrates exposures over longer time periods. Importantly, aflatoxin-albumin adducts, in human sera were found to be stable for at least 25 years when stored at –20°C [46], furthering their utility in longitudinal studies examining links between exposures and cancer risk in humans. Studies in experimental models found that the formation of aflatoxin-DNA adducts in liver, excretion of the urinary

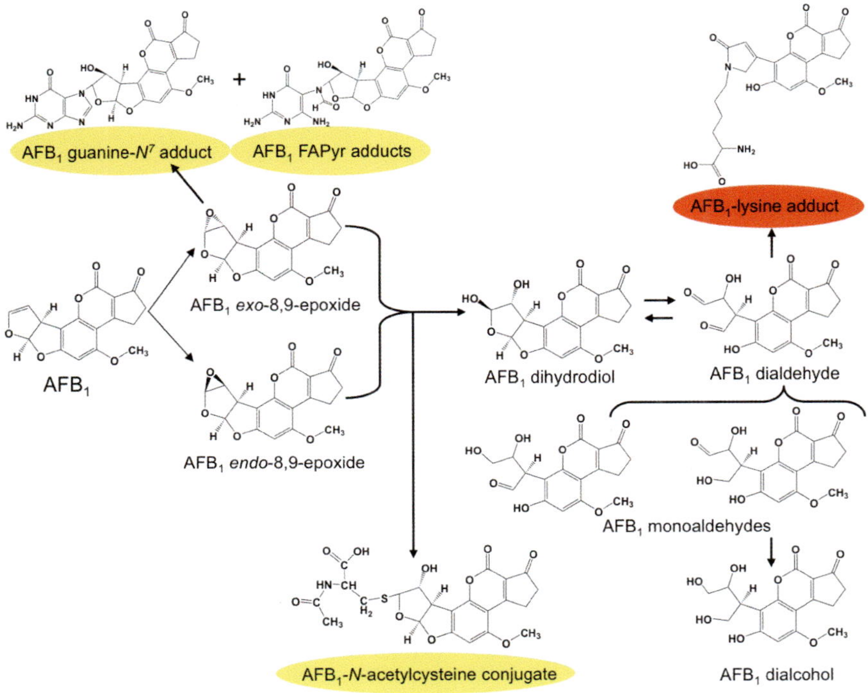

Figure 2. Products measured in biofluids for use as biomarkers in epidemiological and intervention studies are highlighted in yellow (urine) and red (serum).

aflatoxin-nucleic acid adducts, and formation of the serum albumin adduct were highly correlated [47]. More limited studies in humans are consistent with this view [48].

Many different analytical methods are available for quantitation of chemical adducts in biological samples [49–51]. Each methodology has unique specificity and sensitivity. For example, to measure a single aflatoxin metabolite, a chromatographic method can resolve mixtures of aflatoxins into individual compounds, providing that the extraction procedure does not introduce large amounts of interfering chemicals. Antibody-based methods often are more sensitive than chromatography, but immunoassays are less selective because the antibody may cross-react with multiple metabolites. Current studies using isotope-dilution mass spectrometry with liquid chromatography separation have demonstrated an increase in sensitivity of at least 1000-fold over technologies used for the detection of aflatoxin biomarkers 15 years ago [52–54] and are presently the "gold standard" for analysis.

1.3. *Aflatoxin biomarkers in human studies*

Extensive cross-sectional epidemiologic studies have been conducted in high-risk groups for HCC. Initial studies in the Philippines [55] demonstrated that an oxidative metabolite of aflatoxin could be measured in urine and thus had potential to serve as an internal dose marker. Subsequent work conducted in the People's Republic of China and The Gambia, West Africa, determined that the levels of urinary aflatoxin biomarkers showed dose-dependent relationships with aflatoxin intake. Gan *et al.* [56] and Wild *et al.* [57] also monitored levels of aflatoxin-serum albumin adducts and observed a highly significant association between intake of aflatoxin and level of adduct.

Many published case-control studies have examined the relation of aflatoxin exposure and HCC. Compared with cohort studies, case-control studies are both cost- and time-effective. Unfortunately, case-control studies are often initiated long after exposure has occurred and it cannot be assumed that the exposure has not appreciably changed overtime. Also, such studies involve assumptions in the selection of controls, including that the disease state does not alter metabolism of aflatoxin. Thus, matching of cases and controls in a specific chemical biomarker study is much more difficult than in a case-control study involving genetic markers.

Data obtained from cohort studies have the greatest power to determine a true relationship between an exposure and disease outcome because one starts with a healthy cohort, obtains biomarker samples, and then follows the cohort until significant numbers of cases are obtained. A nested study within the cohort can then be designed to match cases and controls. Two major cohort studies with aflatoxin biomarkers have demonstrated the important role of this carcinogen in the etiology of liver cancer. The first study, comprising more than 18,000 men in Shanghai, examined the interaction of hepatitis B virus (HBV) and aflatoxin biomarkers as independent and interactive risk factors for liver cancer. The nested case-control data revealed a statistically significant increase in the adjusted relative risk of 3.4 for those liver cancer cases where urinary aflatoxin biomarkers were detected. For hepatitis B virus surface antigen (HBsAg)-positive people (a marker of infection with HBV), the relative risk was 7, but for individuals with both urinary aflatoxins and positive HBsAg status, the relative risk was 59 [58, 59]. These results strongly support a causal relationship between the presence of the chemical and viral-specific biomarkers and the risk of HCC. However, the underlying mechanisms driving this multiplicative interaction have not been resolved fully.

Subsequent cohort studies in Taiwan have substantially confirmed the results from the Shanghai investigation. Wang *et al.* [60] examined HCC cases and controls nested within a cohort and found that in HBV-infected people there was an adjusted odds ratio of 2.8 for detectable compared with nondetectable aflatoxin-albumin adducts and 5.5 for high compared with low levels of aflatoxin metabolites in urine. In a follow-up study, there was a dose–response relationship between urinary AFM_1 levels and risk of liver cancer in chronic HBV carriers. Similar to the Shanghai study, the liver cancer risk associated with AFB_1 exposure was more striking among the HBV carriers with detectable aflatoxin-N^7-guanine in urine.

1.4. *Intervention trials using aflatoxin biomarkers*

Clinical trials and other interventions are designed to translate findings from human and experimental investigations to public health prevention. Both primary (to reduce exposure) and secondary (to alter metabolism and deposition) interventions can use specific biomarkers as short-term modifiable endpoints of efficacy. Such biomarkers can be applied also to the preselection of exposed individuals for study cohorts, thereby reducing study size requirements [61]. In a primary prevention trial, the goal is to reduce exposure to aflatoxins through the diet. Interventions can range from attempting to lower mold growth in harvested crops to using trapping agents that block the uptake of ingested aflatoxins. In secondary prevention trials, one goal is to modulate the metabolism of ingested aflatoxin to enhance detoxification processes, thereby reducing formation of DNA adducts and enhancing elimination.

1.4.1. *Primary prevention*

The use of aflatoxin biomarkers as efficacy endpoints in primary prevention trials in West Africa has been reported [62]. A package of postharvest measures (hand sorting, drying on mats, sun drying, storage in natural fiber bags and on wooden pallets, and use of insecticides) was used to restrict aflatoxin contamination of groundnut crops. Six hundred people were monitored, and in control (nonintervention) villages mean aflatoxin-albumin concentration increased from 5.5 pg/mg immediately after harvest to 18.7 pg/mg 5 months later. By contrast, mean aflatoxin-albumin concentration in intervention villages after 5 months of groundnut storage was much the same as that immediately postharvest (7.2 pg/mg *vs.* 8.0 pg/mg). Thus, primary prevention maybe an effective means to reduce liver cancer burden, especially in areas where single foodstuffs such a groundnuts are major components of the diet. Devising strategies for sustained implementation and large-scale dissemination of the educational and technical skills required for population-scale benefit remain huge challenges in this arena of primary prevention.

Encouraging dietary diversity offers another route for primary prevention. In rural Africa and Latin America, a high percentage of calories come from maize, which is commonly contaminated by aflatoxin. Another common source of aflatoxin in much of the world is groundnuts [63, 64]. Access to a greater number of types of foods will lower the risk of exposure by lowering the intake of foods contaminated with mycotoxins [65]. Access to foods with better "health value" would include replacing foods with high mycotoxin risk with those of lower risk.

Utilizing the availability of serum samples collected over a 20-year period, changing aflatoxin exposure patterns have been documented (Figure 1). In China, major agricultural reforms in the 1980s led to diminished maize consumption. The population-based cancer registry in Qidong, China has documented a more than 50% reduction in age-standardized mortality rates from primary liver cancer occurring across birth cohorts from the 1960s to the 1980s for Qidongese less than 35 years of age, although all were born before universal vaccination of newborns against HBV [5]. Median levels of the aflatoxin biomarker decreased from 19.3 pg/mg albumin in 1989 to undetectable (< 0.5 pg/mg) by 2009. A population attributable benefit of 65% for reduced liver cancer mortality was estimated from a government-facilitated switch of dietary staple from maize to rice; 83% of this benefit was in those infected with HBV. Food policy reforms in China thus resulted in a dramatic decrease in aflatoxin exposure, which, independent of HBV vaccination, reduced liver cancer risk. The extensive HBV vaccine coverage now in place augurs even greater risk reductions in the future [5].

As described above, economic growth is leading to market basket diversity that will help to reduce exposure to aflatoxin from a single source susceptible to high levels of contamination. A recent monograph from IARC outlines a number of these economic changes along with the development fungal resistant strains of seeds and the potential use of bio-control techniques to limit aflatoxin exposure in regions where economies are in transition [4]. Collectively, all of these primary prevention strategies will help to mitigate aflatoxin toxicities in humans.

1.4.2. *Secondary prevention: human chemoprevention trials*

Chemoprevention is a strategy for the secondary prevention of cancer. This approach entails the use of drugs, dietary supplements, or functional foods to retard, block, or reverse the carcinogenic process. Such interventions can serve to alter cell fate, either by preventing cells from acquiring carcinogen-mediated damage to the genome and epigenome or by impeding proliferation of preneoplastic cells or, alternatively, accelerating their apoptosis. One successful strategy for cancer chemoprevention is modulation of carcinogen-metabolizing enzymes, leading to diminished DNA damage and enhanced elimination. Inducers of conjugating enzymes such as dithiolethiones and sulforaphane inhibit tumorigenesis of environmental carcinogens in various animal models [66, 67]. Increasing lines of evidence show that the Nrf2 signaling pathway is a key molecular target of these chemopreventive enzyme inducers. As already mentioned, genetic disruption of the *Nrf2* gene in mice leads to enhanced sensitivity to carcinogens and the loss of chemopreventive efficacy by inducers [68, 69].

Oltipraz, a substituted 1,2-dithiole-3-thione, was originally developed by the pharmaceutical industry as a possible treatment for schistosomiasis and was extensively evaluated in clinical trials in the early 1980s. In extensive preclinical evaluation by the National Cancer Institute and others, oltipraz was found to be effective as an anticarcinogen in nearly a score of animal models [70]. Therefore, aflatoxin biomarkers (Figure 2) were used as intermediate endpoints in a chemoprevention trial of oltipraz in Qidong, China [71, 72]. This was a placebo-controlled, double-blind study in which participants were randomized to receive placebo or 125-mg oltipraz daily or 500-mg oltipraz weekly. Urinary aflatoxin M_1 levels were reduced by 51% compared with the placebo group in persons receiving the 500-mg weekly dose. No significant differences were seen in urinary aflatoxin M_1 levels in the 125-mg group compared with placebo. This effect was thought to be due to inhibition of cytochrome P450 1A2 activity. Median levels of aflatoxin-mercapturic acid (a glutathione conjugate derivative) were elevated six-fold in the 125-mg group, but were unchanged in the 500-mg group. Increased formation of aflatoxin-mercapturic acid reflects induction of aflatoxin conjugation through the actions of GSTs. The apparent lack of induction in the 500-mg group was thought to reflect masking caused by diminished aflatoxin-8,9-epoxide formation for conjugation through the inhibition of CYPlA2 seen in this group. This study demonstrated for the first time that aflatoxin biomarkers could be modulated in humans in a manner that would predict decreased disease risk.

Although the oltipraz clinical trial demonstrated the proof of principle for elevating flux through pathways leading to aflatoxin detoxication in humans, the practicality of using a drug-based method for prevention in the economically developing world is limited. Not only is there a potential for adverse health effects from any long-term exposure to a drug, but also the expense of this type of intervention may make the intervention cost prohibitive for these populations. There may also be culture-based aversion to the use of drugs. Thus, there was a need for other innovative approaches.

For many years, sodium calcium aluminosilicate marketed as NovaSil clay, a common anticaking agent in animal feeds, has worked to adsorb aflatoxins in the gastrointestinal tract of animals and diminish the bioavailability and adverse effects of these toxins [73] leading to the exploration of clay-based enterosorption for intervention studies in human populations at high risk for aflatoxicosis [74]. A 2-week randomized, double-blinded clinical trial was conducted with 50 volunteers divided into two groups: the low-dose group received nine capsules containing 1.5-g NovaSil/day and the high-dose group received nine capsules containing 3.0 g/day. At the end of the study, there were no statistically significant differences between the two groups for adverse health complaints nor were there differences in any of the clinical chemistries or any other health parameters measured [75]. Subsequently, a 3-month intervention trial was conducted in Ghana. NovaSil was delivered before meals *via* capsules. Levels of aflatoxin-albumin adducts in serum samples collected at baseline and at 1 month were similar among the placebo, low-dose (1.5-g NovaSil/day), and high-dose (3.0-g NovaSil/day) groups. However, the levels of aflatoxin-albumin adducts at 3 months were significantly decreased in both intervention groups compared with levels in the placebo group. A decrease in median levels of aflatoxin M_1 in samples collected at 3 months was found in the high-dose group when compared with the median level in the placebo group. It should be noted that clay is a dietary staple in central Africa and is common in the marketplace. However, the forms of clay local to that region do not absorb aflatoxins.

Many foods have high levels of enzyme inducers that activate Nrf2 signaling [76, 77]. The anticarcinogenic properties of chlorophyllin, a water-soluble derivative of chlorophyll, have been demonstrated in a number of animal models [78]. The initial characterization of chlorophyllin as an anticarcinogen arose from inhibition of HCC development in aflatoxin-treated trout [78, 79]. Although the primary mode of action is thought to be the sequestration of aflatoxin by chlorophyllin through formation of a 1:1 complex [78, 80], experimental data have characterized enzyme-inducing properties that may also contribute to its mechanism of action [81]. Chlorophyll was shown to be protective against aflatoxin-induced cancer in the rat; moreover, the aflatoxin biomarkers tracked with the protective efficacy [82]. In a randomized, double-blind, placebo-controlled chemoprevention trial also conducted in Qidong, chlorophyllin was determined to alter the disposition of aflatoxin [83]. One hundred and eighty healthy adults were randomly assigned to ingest 100 mg of chlorophyllin or a placebo three times a day prior to each meal for 4 months. The primary endpoint was modulation of levels of urinary aflatoxin-N^7-guanine adducts collected 3 months into the intervention. Adherence to the study protocol was outstanding, and no adverse events were reported. Aflatoxin-N^7-guanine could be detected in 105 of 169 available samples. Chlorophyllin consumption at each meal led to an overall 55% reduction in median urinary levels of this aflatoxin biomarker compared with those subjects taking placebo.

In a more recent chemoprevention trial in humans, a beverage formed from hot water infusions of 3-day old broccoli sprouts, containing defined concentrations of glucoraphanin (a stable precursor of the anticarcinogen sulforaphane), was evaluated for its ability to alter the metabolic disposition of aflatoxin. In this study, 200 healthy adults drank infusions containing either 400- or <3-μmole glucoraphanin nightly for 2 weeks. Urinary levels of aflatoxin N^7-guanine (AFB-N^7-Gua) were similar in individuals in the two intervention arms. However, measurement of urinary levels of sulforaphane metabolites indicated striking interindividual differences in bioavailability. This outcome may reflect individual differences in the rates of hydrolysis of glucoraphanin to sulforaphane by the intestinal microflora of the study participants. Accounting for this variability, a significant inverse association was observed for uptake and excretion of sulforaphane and AFB-N^7-Gua adducts in individuals receiving broccoli sprout beverage [84]. This preliminary study illustrates the potential use of an inexpensive, easily implemented, food-based method for secondary prevention in a population at high risk for aflatoxin exposures. Follow-up studies have employed this intervention strategy in China with good success toward enhancing the detoxication of air pollutants over a several month time frame [85]. This strategy of "frugal" or "green" chemoprevention assumes broad relevance as environmental contamination becomes an emerging public health crisis in many areas of the developing world.

Many studies have demonstrated that green tea polyphenols (GTPs) inhibit various chemically induced cancers in experimental animals and epidemiological studies also point to the potential benefit of these compounds [86,87]. GTP in the drinking water for 2 or 4 weeks protected against the development of AFB_1-induced hepatocarcinogenesis in the rat [88]. Aflatoxin-DNA binding in the liver was significantly (20%–30%) inhibited in animals pretreated with green tea and the burden of preneoplastic lesions was also inhibited significantly by 60%–70%. The experimental data on GTP provided the impetus to translate this strategy to human clinical trials. In an initial study in an aflatoxin-exposed high-risk group in Guangxi, People's Republic of China, the effects of GTP (500- or 1000-mg capsules) were assessed by analysis of blood and urine samples collected during a randomized, double-blinded, placebo-controlled chemoprevention trial [89]. Reductions in aflatoxin-albumin adduct levels were observed in both groups receiving GTPs over the 3-month intervention period. An analysis using a mixed-effects model indicated that the reduction in aflatoxin-albumin adduct levels overtime was dose- and time-dependent. Reductions in median aflatoxin M_1 levels, as compared with the placebo, were found in both GTP groups at 3 months of the intervention, while significant elevations in median aflatoxin-mercapturic acid levels were observed in both GTP groups compared with the placebo group at 1 and 3 months of intervention. These results indicate that GTPs effectively modulate aflatoxin metabolism and metabolic activation, as had been previously observed with oltipraz in Qidong.

2. Summary

It is instructive to review certain aspects of the history of the science and policy evolving from the aflatoxins to identify lessons that are applicable to other emerging environmental health issues. A time line highlighting the key milestones in the discovery, toxicological characterization, and regulation of aflatoxins is presented in Figure 3. Since their discovery in 1960,

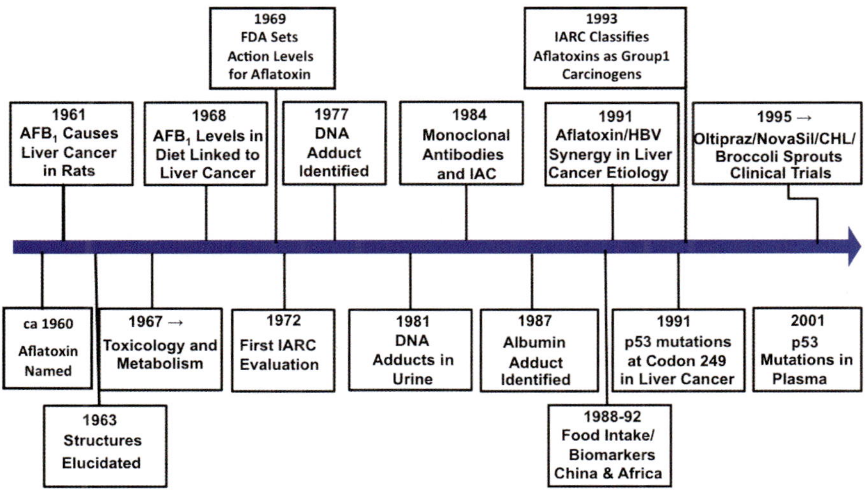

Figure 3. Time line for key events in the discovery, toxicological evaluation, molecular epidemiology, and regulation of aflatoxins. FDA, Food and Drug Administration; IARC, International Agency for Research on Cancer; IAC, immunoaffinity chromatography; CHL, chlorophyllin. Adapted from Kensler *et al.* [12].

elucidation of the chemistry, metabolism, and mode of action has accelerated appreciation of the risks to health associated with acute and chronic exposures to this mycotoxin. It took several decades to affirm the carcinogenic risk to human health and to initiate rigorous, mechanism-based interventions with objective outcomes. In the upcoming decades, much of the population growth in the world is going to occur in countries that have intrinsically high aflatoxin exposure, in addition to exposures to many other chemical and biological agents. Currently, over 5 billion people worldwide experience uncontrolled exposure to aflatoxin [90]. More recent observations that aflatoxin exposure may affect child growth and susceptibility to infection [91], as well HCC risk, serves to emphasize further the public health need for development and implementation of interventions in these countries. It is now well established that aflatoxin exposure begins *in utero* (see Figure 1) and can increase markedly as a child is weaned onto solid family foods [92]. This early exposure, at the time the child is also first exposed to numerous infections including HBV, suggests that there may be particular merit in intervening to reduce exposure during these critical windows. Aflatoxin adducts to DNA and protein will serve as guideposts to assess the efficacy of such intervention strategies. Indeed, the comprehensive studies from "molecules to man" of aflatoxin are templates for providing a map to investigate other environmental carcinogens. While the "hallmarks of cancer" [93] have been well articulated through the tremendous advances in genetic analysis and DNA sequencing, it is still remarkable how little we know about the agents that are suspected to be responsible for initiating these genetic changes. For many of the major human cancers such as pancreas, colon, and prostate, there is a limited knowledge base on chemical agents that contribute to these diseases. Hopefully, a legacy of the aflatoxin investigations outlined in this chapter will be able to accelerate this knowledge base for other cancers so that effective prevention strategies can be more rapidly implemented.

Acknowledgments

The authors acknowledge research support by grants R01 CA190610, R35 CA197222, P30 CA047904, and P30 CA006973 from the USPHS.

Abbreviations

AFB_1	aflatoxin B_1
AFG_1	aflatoxin G_1
AFB_2	aflatoxin B_2
AFG_2	aflatoxin G_2
AFB-N^7-Gua	aflatoxin N^7-guanine
AFB-FAPyr	8,9-dihydro-8-(2,6-diamino-4-oxo-3,4-dihydropyrimid-5-yl formamido)-9-hydroxy-aflatoxin B_1
HCC	hepatocellular carcinoma
HBV	hepatitis B virus
HBsAg	hepatitis B virus surface antigen
GST	glutathione *S*-transferase
GTPs	green tea polyphenols
IARC	International Agency for Research on Cancer
GLOBOCAN	http://globocan.iarc.fr

References

1. Wu F, Groopman JD, Pestka JJ. Public health impacts of foodborne mycotoxins. *Annu Rev Food Sci Technol* 2014; 5: 351–372.
2. Groopman JD, Kensler TW, Wu F. Food safety: mycotoxins—occurrence and toxic effects. In *Encyclopedia of Human Nutrition Edition*, Elsevier, Pittsburgh, PA, 2013, pp. 337–341.
3. Wu F, Guclu H. Aflatoxin regulations in a network of global maize trade. *PLoS One* 2012; 7: e45151.
4. Wild CP, Miller JD, Groopman JD. *Mycotoxin Control in Low- and Middle-Income Countries*. IARC Working Group Reports, No. 9, IARC, Lyon, 2015.
5. Chen JG, Egner PA, Ng D, Jacobson LP, Munoz A, Zhu YR, *et al*. Reduced aflatoxin exposure presages decline in liver cancer mortality in an endemic region of China. *Cancer Prev Res* 2013; 6: 1038–1045.
6. Krishnamachari KA, Bhat VR, Nagarajan V, Tilak TB, Tulpule PG. The problem of aflatoxic human disease in parts of India-epidemiological and ecological aspects. *Ann Nutr Aliment* 1977; 31: 991–996.

7. Chao TC, Maxwell SM, Wong SY. An outbreak of aflatoxicosis and boric acid poisoning in Malaysia: a clinicopathological study. *J Pathol* 1991; 164: 225–233.
8. Lewis L, Onsongo M, Njapau H, Schurz-Rogers H, Luber G, Kieszak S, *et al.* Aflatoxin contamination of commercial maize products during an outbreak of acute aflatoxicosis in eastern and central Kenya. *Environ Health Perspect* 2005; 113: 1763–1767.
9. Ferlay J, Soerjomataram I, Dikshit R, Eser S, Mathers C, Rebelo M, *et al.* Cancer incidence and mortality worldwide: sources, methods and major patterns in GLOBOCAN 2012. *Int J Cancer* 2015; 136: E359–E386. doi:10.1002/ijc.29210. Epub 2014 Oct 9.
10. Torre LA, Bray F, Siegel RL, Ferlay J, Lortet-Tieulent J, Jemal A. Global cancer statistics, 2012. *CA Cancer J Clin* 2015; 65: 87–108. doi:10.3322/caac.21262. Epub 2015 Feb 4.
11. Ferlay J, Soerjomataram I, Ervik M, Dikshit R, Eser S, Mathers C, *et al.* *GLOBOCAN 2012 v1.0, Cancer Incidence and Mortality Worldwide: IARC CancerBase No. 11 [Internet].* International Agency for Research on Cancer, Lyon, 2013.
12. Kensler TW, Roebuck BD, Wogan GN, Groopman JD. Aflatoxin: a 50-year odyssey of mechanistic and translational toxicology. *Toxicol Sci.* 2011; 120 Suppl 1: S28–S48.
13. Bosch FX, Munoz N. Review. Prospects for epidemiological studies on hepatocellular cancer as a model for assessing viral and chemical interactions. *IARC Sci Publ* 1988; 89: 427–438.
14. Schleicher RL, McCoy LF, Powers CD, Sternberg MR, Pfeiffer CM. Serum concentrations of an aflatoxin-albumin adduct in the National Health and Nutrition Examination Survey (NHANES) 1999–2000. *Clin Chim Acta* 2013; 423: 46–50.
15. Groopman JD, Egner PA, Schulze KJ, Wu LS, Merrill R, Mehra S, *et al.* Aflatoxin exposure during the first 1000 days of life in rural South Asia assessed by aflatoxin B1-lysine albumin biomarkers. *Food Chem Toxicol* 2014; 74: 184–189.
16. World Cancer Report 2014: International Agency for Research on Cancer, 2014.
17. El-Serag HB, Kanwal F. Epidemiology of hepatocellular carcinoma in the United States: where are we? *Hepatology* 2014 Nov; 60(5): 1767–1775.
18. Petrick JL, Kelly SP, Altekruse SF, McGlynn KA, Rosenberg PS. Future of hepatocellular carcinoma incidence in the United States forecast through 2030. *J Clin Oncol* 2016; 34(15): 1787–1794.

19. McGlynn KA, Petrick JL, London WT. Global epidemiology of hepatocellular carcinoma: an emphasis on demographic and regional variability. *Clin Liver Dis* 2015; 19: 223–238. doi:10.1016/j.cld.2015.01.001. Epub Feb 26.
20. Bambha K, Belt P, Abraham M, Wilson LA, Pabst M, Ferrell L, *et al.* Ethnicity and nonalcoholic fatty liver disease. *Hepatology* 2012; 55: 769–780. doi:10.1002/hep.24726.
21. Chen JG, Zhu J, Parkin DM, Zhang YH, Lu JH, Zhu YR, *et al.* Trends in the incidence of cancer in Qidong, China, 1978–2002. *Int J Cancer* 2006; 119: 1447–1454.
22. Vatanasapt V, Martin N, Sriplung H, Chindavijak K, Sontipong S, Sriamporn S, *et al.* Cancer incidence in Thailand, 1988–1991. *Cancer Epidemiol Biomarkers Prev.* 1995; 4: 475–483.
23. Parkin DM, Bray F, Ferlay J, Pisani P. Global cancer statistics, 2002. *CA Cancer J Clin* 2005; 55: 74–108.
24. Busby WF, Wogan GN. *Aflatoxins*. American Chemical Society, Washington, DC, 1984.
25. Eaton DL, Groopman JD. *The Toxicology of Aflatoxins: Human Health, Veterinary, and Agricultural Significance*. Academic Press, San Diego, 1994.
26. Wogan GN, Newberne PM. Dose-response characteristics of aflatoxin B1 carcinogenesis in the rat. *Cancer Res* 1967; 27: 2370–2376.
27. Wogan GN, Paglialunga S, Newberne PM. Carcinogenic effects of low dietary levels of aflatoxin B1 in rats. *Food Cosmet Toxicol* 1974; 12: 681–685.
28. Williams DE, Orner G, Willard KD, Tilton S, Hendricks JD, Pereira C, *et al.* Rainbow trout (*Oncorhynchus mykiss*) and ultra-low dose cancer studies. *Comp Biochem Physiol C Toxicol Pharmacol* 2009; 149: 175–181.
29. Bailey GS, Loveland PM, Pereira C, Pierce D, Hendricks JD, Groopman JD. Quantitative carcinogenesis and dosimetry in rainbow trout for aflatoxin B1 and aflatoxicol, two aflatoxins that form the same DNA adduct. *Mutat Res* 1994; 313: 25–38.
30. Thorgeirsson UP, Dalgard DW, Reeves J, Adamson RH. Tumor incidence in a chemical carcinogenesis study of nonhuman primates. *Regul Toxicol Pharmacol* 1994; 19: 130–151.
31. Sieber SM, Correa P, Dalgard DW, Adamson RH. Induction of osteogenic sarcomas and tumors of the hepatobiliary system in nonhuman primates with aflatoxin B1. *Cancer Res* 1979; 39: 4545–4554.
32. Adamson RH, Correa P, Sieber SM, McIntire KR, Dalgard DW. Carcinogenicity of aflatoxin B1 in rhesus monkeys: two additional cases of primary liver cancer. *J Natl Cancer Inst* 1976; 57: 67–78.

33. Gopalan C, Tulpule PG, Krishnamurthi D. Induction of hepatic carcinoma with aflatoxin in the rhesus monkey. *Food Cosmet Toxicol* 1972; 10: 519–521.
34. Tilak TB. Induction of cholangiocarcinoma following treatment of a rhesus monkey with aflatoxin. *Food Cosmet Toxicol* 1975; 13: 247–249.
35. Lutz WK. Quantitative evaluation of DNA-binding data *in vivo* for low-dose extrapolations. *Arch Toxicol* 1987; Suppl 11: 66–74.
36. Johnson NM, Egner PA, Baxter VK, Sporn MB, Wible RS, Sutter TR, *et al.* Complete protection against aflatoxin B1-induced liver cancer with triterpenoid: DNA adduct dosimetry, molecular signature and genotoxicity threshold. *Cancer Prev Res* 2014; 7(7): 658–665.
37. Essigmann JM, Croy RG, Nadzan AM, Busby WF, Jr., Reinhold VN, Buchi G, *et al.* Structural identification of the major DNA adduct formed by aflatoxin B1 *in vitro*. *Proc Natl Acad Sci USA* 1977; 74: 1870–1874.
38. Sabbioni G, Skipper PL, Buchi G, Tannenbaum SR. Isolation and characterization of the major serum albumin adduct formed by aflatoxin B1 *in vivo* in rats. *Carcinogenesis* 1987; 8: 819–824.
39. Smela ME, Hamm ML, Henderson PT, Harris CM, Harris TM, Essigmann JM. The aflatoxin B(1) formamidopyrimidine adduct plays a major role in causing the types of mutations observed in human hepatocellular carcinoma. *Proc Natl Acad Sci USA* 2002; 99: 6655–6660.
40. Ilic Z, Crawford D, Vakharia D, Egner PA, Sell S. Glutathione-S-transferase A3 knockout mice are sensitive to acute cytotoxic and genotoxic effects of aflatoxin B1. *Toxicol Appl Pharmacol* 2010; 242: 241–246.
41. Kensler KH, Slocum SL, Chartoumpekis DV, Dolan PM, Johnson NM, Ilic Z, *et al.* Genetic or pharmacologic activation of Nrf2 signaling fails to protect against aflatoxin genotoxicity in hypersensitive GSTA3 knockout mice. *Toxicol Sci* 2014; 139: 293–300.
42. Taguchi K, Takaku M, Egner PA, Morita M, Kaneko T, Mashimo T, *et al.* Generation of a new model rat: Nrf2 knockout rats are sensitive to aflatoxin B1 toxicity. *Toxicol Sci* 2016; 152: 40–52.
43. Roebuck BD, Johnson DN, Sutter CH, Egner PA, Scholl PF, Friesen MD, *et al.* Transgenic expression of aflatoxin aldehyde reductase (AKR7A1) modulates aflatoxin B1 metabolism but not hepatic carcinogenesis in the rat. *Toxicol Sci* 2009; 109: 41–49.
44. Eaton DL, Schaupp CM. Of mice, rats, and men: could Nrf2 activation protect against aflatoxin heptocarcinogenesis in humans? *Cancer Prev Res* 2014; 7: 653–657.

45. Bennett RA, Essigmann JM, Wogan GN. Excretion of an aflatoxin-guanine adduct in the urine of aflatoxin B1-treated rats. *Cancer Res* 1981; 41: 650–654.
46. Scholl PF, Groopman JD. Long-term stability of human aflatoxin B1 albumin adducts assessed by isotope dilution mass spectrometry and high-performance liquid chromatography-fluorescence. *Cancer Epidemiol Biomarkers Prev* 2008; 17: 1436–1439.
47. Groopman JD, DeMatos P, Egner PA, Love-Hunt A, Kensler TW. Molecular dosimetry of urinary aflatoxin-N7-guanine and serum aflatoxin-albumin adducts predicts chemoprotection by 1,2-dithiole-3-thione in rats. *Carcinogenesis* 1992; 13: 101–106.
48. Groopman JD, Wild CP, Hasler J, Junshi C, Wogan GN, Kensler TW. Molecular epidemiology of aflatoxin exposures: validation of aflatoxin-N7-guanine levels in urine as a biomarker in experimental rat models and humans. *Environ Health Perspect* 1993; 99: 107–113.
49. Wang JS, Groopman JD. *Biomarkers for Carcinogen Exposure: Tumor Initiation*. Taylor & Francis, Washington, DC, 1998.
50. Santella RM. Immunological methods for detection of carcinogen-DNA damage in humans. *Cancer Epidemiol Biomarkers Prev* 1999; 8: 733–739.
51. Poirier MC, Santella RM, Weston A. Carcinogen macromolecular adducts and their measurement. *Carcinogenesis* 2000; 21: 353–359.
52. Scholl PF, McCoy L, Kensler TW, Groopman JD. Quantitative analysis and chronic dosimetry of the aflatoxin B1 plasma albumin adduct Lys-AFB1 in rats by isotope dilution mass spectrometry. *Chem Res Toxicol* 2006; 19: 44–49.
53. Scholl PF, Turner PC, Sutcliffe AE, Sylla A, Diallo MS, Friesen MD, *et al.* Quantitative comparison of aflatoxin B1 serum albumin adducts in humans by isotope dilution mass spectrometry and ELISA. *Cancer Epidemiol Biomarkers Prev* 2006; 15: 823–826.
54. Egner PA, Groopman JD, Wang JS, Kensler TW, Friesen MD. Quantification of aflatoxin-B1-N7-Guanine in human urine by high-performance liquid chromatography and isotope dilution tandem mass spectrometry. *Chem Res Toxicol* 2006; 19: 1191–1195.
55. Campbell TC, Caedo JP, Jr., Bulatao-Jayme J, Salamat L, Engel RW. Aflatoxin M1 in human urine. *Nature* 1970; 227: 403–404.
56. Gan LS, Skipper PL, Peng XC, Groopman JD, Chen JS, Wogan GN, *et al.* Serum albumin adducts in the molecular epidemiology of aflatoxin

carcinogenesis: correlation with aflatoxin B1 intake and urinary excretion of aflatoxin M1. *Carcinogenesis* 1988; 9: 1323–1325.

57. Wild CP, Hudson GJ, Sabbioni G, Chapot B, Hall AJ, Wogan GN, *et al.* Dietary intake of aflatoxins and the level of albumin-bound aflatoxin in peripheral blood in The Gambia, West Africa. *Cancer Epidemiol Biomarkers Prev* 1992; 1: 229–234.
58. Ross RK, Yuan JM, Yu MC, Wogan GN, Qian GS, Tu JT, *et al.* Urinary aflatoxin biomarkers and risk of hepatocellular carcinoma. *Lancet* 1992; 339: 943–946.
59. Qian GS, Ross RK, Yu MC, Yuan JM, Gao YT, Henderson BE, *et al.* A follow-up-study of urinary markers of aflatoxin exposure and liver-cancer risk in Shanghai, Peoples-Republic-of-China. *Cancer Epidemiol Biomarkers Prev* 1994; 3: 3–10.
60. Wang LY, Hatch M, Chen CJ, Levin B, You SL, Lu SN, *et al.* Aflatoxin exposure and risk of hepatocellular carcinoma in Taiwan. *Int J Cancer* 1996; 67: 620–625.
61. Kensler TW, Groopman JD, Wogan GN. Use of carcinogen-DNA and carcinogen-protein adduct biomarkers for cohort selection and as modifiable end points in chemoprevention trials. *IARC Sci Publ* 1996: 237–248.
62. Turner PC, Sylla A, Gong YY, Sutcliffe AE, Hall AJ, Wild CP. Reduction in exposure to carcinogenic aflatoxin by postharvest intervention measures in west Africa: a community-based intervention study. *Lancet* 2005; 365: 1950–1956.
63. Wild CP, Gong YY. Mycotoxins and human disease: a largely ignored global health issue. *Carcinogenesis* 2010; 31: 71–82.
64. Liu Y, Wu F. Global burden of aflatoxin-induced hepatocellular carcinoma: a risk assessment. *Environ Health Perspect* 2010; 118: 818–824.
65. Groopman JD, Kensler TW, Wild CP. Protective interventions to prevent aflatoxin-induced carcinogenesis in developing countries. *Annu Rev Public Health*. 2008; 29: 187–203.
66. Roebuck BD, Liu YL, Rogers AE, Groopman JD, Kensler TW. Protection against aflatoxin B1-induced hepatocarcinogenesis in F344 rats by 5-(2-pyrazinyl)-4-methyl-1,2-dithiole-3-thione (oltipraz): predictive role for short-term molecular dosimetry. *Cancer Res*. 1991; 51: 5501–5506.
67. Zhang Y, Talalay P, Cho CG, Posner GH. A major inducer of anticarcinogenic protective enzymes from broccoli: isolation and elucidation of structure. *Proc Natl Acad Sci USA* 1992; 89: 2399–2403.
68. Ramos-Gomez M, Kwak MK, Dolan PM, Itoh K, Yamamoto M, Talalay P, *et al.* Sensitivity to carcinogenesis is increased and chemoprotective efficacy

of enzyme inducers is lost in nrf2 transcription factor-deficient mice. *Proc Natl Acad Sci USA* 2001; 98: 3410–3415.

69. Fahey JW, Haristoy X, Dolan PM, Kensler TW, Scholtus I, Stephenson KK, *et al.* Sulforaphane inhibits extracellular, intracellular, and antibiotic-resistant strains of *Helicobacter pylori* and prevents benzo[a]pyrene-induced stomach tumors. *Proc Natl Acad Sci USA* 2002; 99: 7610–7615.
70. Kensler TW, Groopman JD, Sutter TR, Curphey TJ, Roebuck BD. Development of cancer chemopreventive agents: oltipraz as a paradigm. *Chem Res Toxicol* 1999; 12: 113–126.
71. Kensler TW, He X, Otieno M, Egner PA, Jacobson LP, Chen B, *et al.* Oltipraz chemoprevention trial in Qidong, People's Republic of China: modulation of serum aflatoxin albumin adduct biomarkers. *Cancer Epidemiol Biomarkers Prev* 1998; 7: 127–134.
72. Wang JS, Shen X, He X, Zhu Yr, Zhang BC, Wang JB, *et al.* Protective alterations in phase 1 and 2 metabolism of aflatoxin B1 by oltipraz in residents of Qidong, People's Republic of China. *J Natl Cancer Inst* 1999; 91: 347–354.
73. Phillips TD, Lemke SL, Grant PG. Characterization of clay-based enterosorbents for the prevention of aflatoxicosis. *Adv Exp Med Biol* 2002; 504: 157–171.
74. Afriyie-Gyawu E, Mackie J, Dash B, Wiles M, Taylor J, Huebner H, *et al.* Chronic toxicological evaluation of dietary NovaSil clay in Sprague-Dawley rats. *Food Addit Contam* 2005; 22: 259–269.
75. Wang JS, Luo H, Billam M, Wang Z, Guan H, Tang L, *et al.* Short-term safety evaluation of processed calcium montmorillonite clay (NovaSil) in humans. *Food Addit Contam* 2005; 22: 270–279.
76. Talalay P, Fahey JW. Phytochemicals from cruciferous plants protect against cancer by modulating carcinogen metabolism. *J Nutr* 2001; 131: 3027S–3033S.
77. Fahey JW, Kensler TW. Role of dietary supplements/nutraceuticals in chemoprevention through induction of cytoprotective enzymes. *Chem Res Toxicol* 2007; 20: 572–576.
78. Breinholt V, Hendricks J, Pereira C, Arbogast D, Bailey G. Dietary chlorophyllin is a potent inhibitor of aflatoxin B1 hepatocarcinogenesis in rainbow trout. *Cancer Res* 1995; 55: 57–62.
79. Dashwood R, Negishi T, Hayatsu H, Breinholt V, Hendricks J, Bailey G. Chemopreventive properties of chlorophylls towards aflatoxin B1: a review of the antimutagenicity and anticarcinogenicity data in rainbow trout. *Mutat Res* 1998; 399: 245–253.

80. Breinholt V, Schimerlik M, Dashwood R, Bailey G. Mechanisms of chlorophyllin anticarcinogenesis against aflatoxin B1: complex formation with the carcinogen. *Chem Res Toxicol* 1995; 8: 506–514.
81. Fahey JW, Stephenson KK, Dinkova-Kostova AT, Egner PA, Kensler TW, Talalay P. Chlorophyll, chlorophyllin and related tetrapyrroles are significant inducers of mammalian phase 2 cytoprotective genes. *Carcinogenesis* 2005; 26: 1247–1255.
82. Simonich MT, Egner P, Roebuck BD, Orner G, Jubert C, Pereira C, *et al.* Natural chlorophyll inhibits aflatoxin B1 induced multi-organ carcinogenesis in the rat. *Carcinogenesis* 2007; 28(6): 1294–1302.
83. Egner PA, Wang JB, Zhu YR, Zhang BC, Wu Y, Zhang QN, *et al.* Chlorophyllin intervention reduces aflatoxin-DNA adducts in individuals at high risk for liver cancer. *Proc Natl Acad Sci USA* 2001; 98: 1460–1466.
84. Kensler TW, Chen JG, Egner PA, Fahey JW, Jacobson LP, Stephenson KK, *et al.* Effects of glucosinolate-rich broccoli sprouts on urinary levels of aflatoxin-DNA adducts and phenanthrene tetraols in a randomized clinical trial in He Zuo township, Qidong, People's Republic of China. *Cancer Epidemiol Biomarkers Prev* 2005; 14: 2605–2613.
85. Egner PA, Chen JG, Zarth AT, Ng DK, Wang JB, Kensler KH, *et al.* Rapid and sustainable detoxication of airborne pollutants by broccoli sprout beverage: results of a randomized clinical trial in china. *Cancer Prev Res* 2014; 7: 813–823.
86. Moyers SB, Kumar NB. Green tea polyphenols and cancer chemoprevention: multiple mechanisms and endpoints for phase II trials. *Nutr Rev* 2004; 62: 204–211.
87. Yang CS, Lambert JD, Hou Z, Ju J, Lu G, Hao X. Molecular targets for the cancer preventive activity of tea polyphenols. *Mol Carcinog* 2006; 45: 431–435.
88. Qin G, Gopalan-Kriczky P, Su J, Ning Y, Lotlikar PD. Inhibition of aflatoxin B1-induced initiation of hepatocarcinogenesis in the rat by green tea. *Cancer Lett* 1997; 112: 149–154.
89. Luo H, Tang L, Tang M, Billam M, Huang T, Yu J, *et al.* Phase IIa chemoprevention trial of green tea polyphenols in high-risk individuals of liver cancer: modulation of urinary excretion of green tea polyphenols and 8-hydroxydeoxyguanosine. *Carcinogenesis* 2006; 27: 262–268.
90. Strosnider H, Azziz-Baumgartner E, Banziger M, Bhat RV, Breiman R, Brune MN, *et al.* Workgroup report: public health strategies for reducing

aflatoxin exposure in developing countries. *Environ Health Perspect* 2006; 114: 1898–1903.

91. Gong YY, Cardwell K, Hounsa A, Egal S, Turner PC, Hall AJ, *et al*. Dietary aflatoxin exposure and impaired growth in young children from Benin and Togo: cross sectional study. *BMJ* 2002; 325: 20–21.
92. Wild CP, Turner PC. The toxicology of aflatoxins as a basis for public health decisions. *Mutagenesis* 2002; 17: 471–481.
93. Hanahan D, Weinberg RA. Hallmarks of cancer: the next generation. *Cell* 2011; 144: 646–674.

Chapter 7

Metal Carcinogenesis and DNA Damage: A Case Study Using Hexavalent Chromium

Sandra S. Wise and John Pierce Wise

Department of Pharmacology and Toxicology, School of Medicine, University of Louisville, Louisville, Kentucky 40292, USA

Abstract

Many metal compounds are classified by many agencies around the world as known or probable human carcinogens. Occupational and environmental exposure to metals are risk factors for both workers and the general population. However, mechanisms for metal induced carcinogenesis are poorly understood. Metals can act as mimics for essential elements, increase free radicals or disrupt protein structure and folding. The physical and chemical properties of metals can also affect their potency and bioavailability. Here we use chromium as a model metal carcinogen to address the potential mechanisms of its carcinogenicity. We discuss the cellular uptake and intracellular reduction of hexavalent chromium, its effect on DNA repair, its effect on cell cycle and mitosis, and epigenetic alterations including methylation, histone modification and microRNAs.

1. Overview of Metal Carcinogenesis

Metal ions have unique chemical properties that allow for a variety of roles in biotransformation. Some metals are classified as essential for proper biological functions, while others have no biological function and can pose a toxic threat to the health of an organism. Many metal compounds, including arsenic, beryllium, cadmium, chromium, cobalt, lead, nickel, plutonium, radium, and thorium, are classified by many agencies around the world as known or probable human carcinogens. For example, the International Agency for Research on Cancer (IARC) classifies arsenic, beryllium, cadmium, chromium, nickel, plutonium, radium, and thorium as Group 1 carcinogens, which means that they are carcinogenic to humans. Lead is classified as a Group 2A carcinogen, or a probable human carcinogen. Cobalt is classified as a Group 2B carcinogen, or a possible human carcinogen. For IARC analyses, the strength of the data at the time of evaluation dictates the categorization chosen. While the IARC Group 1 carcinogens are widely accepted around the world as known human carcinogens, there are some variations in the rankings of the others. For example, in 2016, the US Department of Health and Human Services evaluated the data for cobalt and declared it to be a reasonably anticipated human carcinogen.

Metal exposures, which can occur as a result of both occupational and environmental insult, are considered risk factors for both workers and the general population. The complete mechanism by which any particular metal causes a normal cell to become a cancer cell is poorly understood, with no consistent theme found applicable to all the metals. Some metals mimic essential elements to cause toxic outcomes. For example, lead can mimic calcium, but also block calcium channels interfering with cell signaling. Arsenic can interfere with protein zinc fingers and disrupt DNA repair. Many toxic metals increase free radicals, such as reactive oxygen species (ROS) and reactive nitrogen species (RNS), resulting in increased oxidative stress, as well as damage to DNA, proteins, and lipids. All of these events are considered to be linked to tumor induction. The carcinogenic potential of a particular metal may be dependent on several physicochemical properties such as oxidation state, solubility, and complexation with other chemicals. Physicochemical properties may influence cellular uptake, intracellular transport and distribution, and bioavailability of metals. Furthermore, once inside the cell, the metal may cause a myriad of intracellular effects, which also contribute to carcinogenesis.

This chapter illustrates the various contributions that metal exposure may make toward carcinogenesis, using chromium as a case study. Chromium is a known human carcinogen and considered to act through a genotoxic mechanism. This chapter progresses from considering physicochemical properties that are key to chromium carcinogenesis, such as metal ion uptake and intracellular metabolism, to DNA damage and repair, dysregulation of cell division, and alterations in epigenetic homeostasis.

2. Metal Ion Uptake and Metabolism

Uptake and metabolism are important factors in the carcinogenicity of metals. For the metals that cause cancer, entering the cell is a necessary step. Some metals, like lead and cadmium, enter cells as ions. For others, such as nickel and cobalt, the more potent form is as a particle, which enters the cell in a vacuole and dissolves inside the cell. Some metals, including lead and nickel, undergo little or no intracellular metabolism to cause carcinogenic outcomes. Others, including arsenic and chromium, are extensively metabolized and the metabolic products play major roles in the carcinogenic process.

Chromium (Cr) provides an interesting example of the importance of uptake and intracellular metabolism (Figure 1). It has two stable valence states in biological systems: hexavalent chromium [Cr(VI)] and trivalent chromium [Cr(III)]. However, while Cr(VI) is a human carcinogen, Cr(III) is not considered a carcinogen [1]. This difference can be explained primarily by differences in Cr cellular uptake. Specifically, Cr(VI) forms a chromate (CrO_4^{2-}) anion, which structurally resembles phosphate (PO_4^{3-}) and sulfate (SO_4^{2-}), and consequently is rapidly transported into cells by facilitated diffusion [2]. In contrast, Cr(III) binds to multiple ligands (up to 8) forming a bulky molecule that enters cells very poorly and slowly by passive diffusion [3].

After uptake, metabolism plays a key role in chromium carcinogenesis (Figure 1). Once inside the cell, the Cr(VI)-containing chromate anion

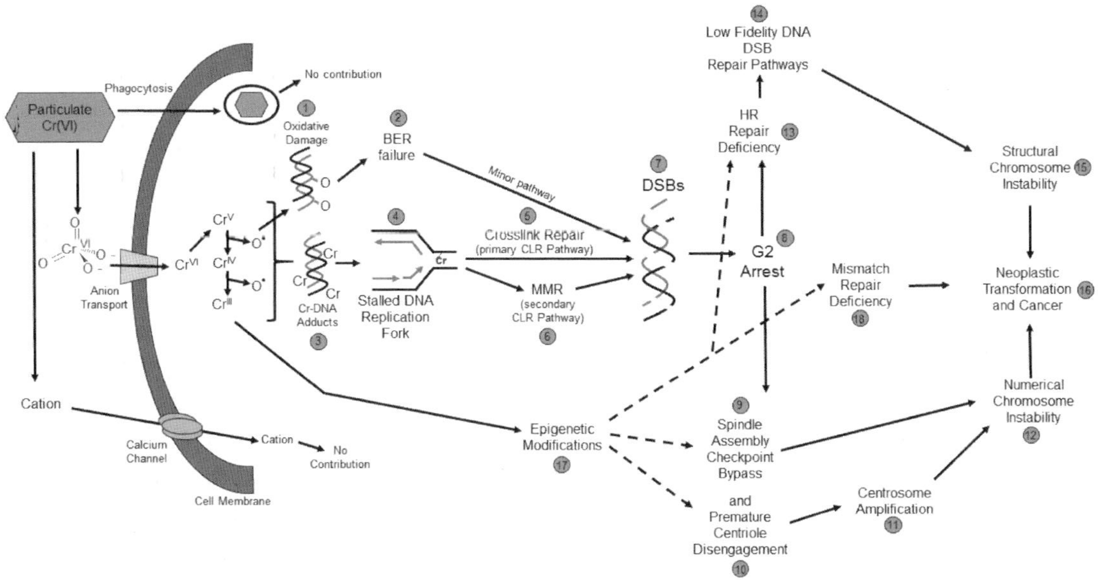

Figure 1. Cr(VI)-induced DNA damage and related effects within the cell.

Particulate chromate, the most carcinogenic form of hexavalent chromium, dissolves outside of the cell into the chromate anion and a cation (e.g., zinc, lead, calcium). The cation can enter the cell *via* various ion channels. The Cr(VI) anion enters the cell by phagocytosis, facilitated diffusion using an anion transporter. Once particulate Cr(VI) is inside the cell, the chromate anion is reduced intracellularly, producing ROS causing DNA adducts. The oxidative damage (1) can lead to the formation of DSBs if BER fails (2). In addition, the Cr(VI)-reduction species Cr^{5+}, Cr^{4+}, and Cr^{3+} can form DNA adducts (3), which lead to stalled replication forks (4). The Cr-DNA adducts are repaired by cross-link repair (5) and/or MMR (6). These proficient repair pathways as well as the failed BER lead to DNA DSBs (7). Formation of the breaks induces a G2 arrest (8). Prolonged Cr(VI) exposure results in spindle assembly checkpoint bypass (9) premature centriole disengagement (10), centrosome amplification (11), and ultimately numerical chromosome instability (12). Prolonged Cr(VI) exposure also leads to failure of HR repair of DNA DSBs (13), forcing the use of low-fidelity DNA DSB repair pathways (14), and structural chromosome instability (15). Both structural and numerical chromosome instability lead to neoplastic transformation and cancer (16). Epigenetic modifications may contribute to structural and numerical chromosome instability (17); including methylation of the hMLH1 promoter may inhibit MMR late in this mechanism inducing microsatellite instability (18), further contributing to neoplastic transformation (16).

is rapidly reduced to Cr(III). This metabolism is carried out by naturally occurring intracellular antioxidants such as glutathione, cysteine, ascorbic acid, cytochrome (CYP) P450s, mitochondria, nicotinamide adenine dinucleotide phosphate (NADPH), and proteins such as hemoglobin and glutathione reductase. In the process of reducing Cr(VI) to Cr(III), several unstable and highly reactive intermediates such as Cr(V), Cr(IV), and various ROS, are produced [2, 4, 5]. Along with Cr(III), the highly reactive Cr(IV) and CR(V) intermediates can react with a variety of intracellular ligands functionally trapping Cr and producing high intracellular Cr levels. Technological limitations do not allow identification of the ultimate carcinogenic metabolite responsible for the carcinogenicity of Cr(VI). However, it is clear that the intracellular reduction of Cr(VI) is crucial to tumor induction, as Cr(VI) alone cannot react with DNA or other intracellular molecules [6].

Another major factor in the uptake and metabolism of Cr(VI) is water solubility. Cr(VI) compounds can range in solubility from highly water soluble to poorly water soluble, and it is the insoluble forms (i.e., the particulate forms) that are the more potent carcinogens [1, 7, 8]. The explanation for this difference relies partially on the route of Cr administration into the body. Cr(VI) is primarily considered an inhalation carcinogen. The Cr(VI) particles are inhaled and impact at bronchiole bifurcation sites in the lung where the particles accumulate and persist. Soluble Cr(VI) compounds, by contrast, would be expected to be more rapidly cleared by the body. Indeed, human pathology studies of workers with Cr(VI)-induced lung cancer show both Cr accumulation and tumor formation at bronchiole bifurcation sites [9]. Studies in cultured human lung cells showed that the Cr-containing particles do enter the cells by phagocytosis, but in short-term studies this does not appear to have an effect. Rather, the particles appear to dissolve outside the cell releasing the chromate anion, which then enters the cell and is reduced inside the cell to Cr(III) [10, 11].

It is important to note, recent data suggest that Cr(VI) may also be a drinking water carcinogen and not just an inhalation carcinogen. Previously, this outcome was thought to be unlikely because the digestive tract can reduce Cr(VI) to Cr(III) and prevent its uptake into cells. However, recent data from rodent studies indicate tumors can form in rodents exposed to Cr(VI) in the drinking water [12]. These results are controversial as the doses used were quite high, and while proponents argue that the data show Cr(VI) is a drinking water carcinogen, opponents argue that the natural reductive capacity of the digestive system was overwhelmed by the unusually high levels of Cr(VI). The authors of this chapter will not draw any conclusions about this particular controversy, but it is mentioned to further illustrate the importance of metabolism in the carcinogenicity of Cr(VI), and to acknowledge the current debate over whether or not Cr(VI) is only an inhalation carcinogen.

3. Metals and DNA Damage

DNA damage is a central concept and contributing factor in the process of carcinogenesis (see Chapter 1 of this volume). Metals have been extensively studied for their ability to induce DNA damage. Metals can cause a variety of types of DNA damage including oxidative DNA base lesions, DNA adducts and cross-links, DNA single-strand breaks, and DNA double-strand breaks (DSB). Some metals are capable of directly interacting with DNA while others have more indirect effects such as bonding either covalently or electrostatically with intracellular molecules, such as amino acids, ascorbate, cysteine, histidine, and glutathione, which then lead to adducts with DNA. Cr(VI) has long been established as a potent genotoxic agent. Studies have shown that Cr(VI) has the ability to induce multiple types of DNA damage (discussed in detail below), including oxidative lesions in DNA bases, DNA adducts and cross-links, DNA single-strand breaks, and DNA DSBs. All of these events may produce chromosome instability, and ultimately contribute to carcinogenesis.

Oxidative lesions: Metals are well known for their ability to create ROS. Metal ions can react with superoxide anions and hydrogen peroxide to produce hydroxyl radicals, which are highly reactive with DNA. Of the four DNA bases, guanine oxidation products, such as 7,8-dihydro-8-oxo-guanine (8-oxo-dG), are the most common because they have the lowest reduction potential [13]. Oxidized DNA bases produce base and sugar lesions, strand breaks, DNA–protein cross-links, and abasic sites. If these types of damage are left unrepaired, they can lead to mutations and neoplastic transformation.

Chromium causes damage to both purines and pyrimidines [14–17]. As with other metals, the major lesion on DNA nucleobases is 8-oxo-dG, which produces G→T transversion mutations during transcription. This type of damage can be further oxidized by Cr to form another lesion, spiroiminodihydantoin (Sp) [17]. In addition to guanine, recent studies have shown that 2′-deoxyadenosine of the p53 gene can also be oxidized by Cr treatment [18].

DNA adducts and cross-links: A DNA adduct is a covalently bonded addition product between a chemical and DNA. DNA cross-links occur when chemicals react with two different positions on DNA or between DNA and protein. DNA adducts and cross-links both lead to blocked DNA replication, cell cycle arrest, mutagenesis, and potentially cell death if left unrepaired. Metals are generally able to induce DNA adducts and cross-links because of their reactive nature.

Focusing specifically on Cr(VI), there are Cr-DNA adducts, Cr-DNA cross-links and DNA–protein Cr cross-links which can all lead to subsequent cellular dysfunction, including DNA polymerase arrest, mutagenesis, and changes in gene expression [19]. The repair processes required for these lesions often produce additional DNA damage, including single-strand breaks and DSBs. One major challenge has been the isolation and structural identification of a Cr-DNA-specific adduct. Initial strategies to detect this adduct have included simply isolating DNA and counting the radiolabeled Cr associated with the DNA [20]. However, using this approach, it is unclear if the Cr is covalently bound to DNA bases, or if it is electrostatically associated with the DNA backbone. The fact that Cr is interacting with DNA in any fashion is biologically important, as any interaction may lead to more dangerous consequences, including DNA strand breaks and aneuploidy.

Single-strand breaks: Single-strand breaks are distinct breaks in one strand of the DNA double helix, often accompanied by the loss of a single nucleotide at the breakage site. Single-strand breaks are typically found as part of the repair process for other types of DNA damage, and form during base excision repair (BER), nucleotide excision repair (NER), and mismatch repair (MMR). If not repaired accurately, single-strand breaks can lead to genomic instability and threaten cell survival, so cells have developed rapid and efficient repair processes.

Several cell culture studies have shown that Cr(VI) induces DNA single-strand breaks [21–23]. Single-strand breaks have also been demonstrated in whole animal studies, with damage occurring in lymphocytes and somatic cells of the liver, kidney, spleen, lung, and brain following Cr(VI) exposure [24, 25]. These breaks can be worsened by coexposures with cadmium, cobalt, and arsenic [26–28], but it is unclear whether Cr(VI)-induced single-strand breaks occur as result of direct damage or a result of excision repair. Cr can induce breaks in cell-free systems, but the method of detection in these studies requires high pH, thus suggesting that Cr-DNA adducts indirectly cause single-strand breaks [29]. There is likely a contribution from both direct and indirect breaks although it is currently not possible to determine which is the major cause of these effects.

Double-strand breaks: These lesions can occur in both strands of the DNA double helix, and are the most dangerous type of DNA damage because, if unrepaired they can cause cell death. If improperly repaired, they can lead to deletions, translocations, and improper DNA fusions.

All forms of Cr(VI) have been shown to cause DNA DSBs in cultured human cells and in animals [30–36]. Cr(VI)-induced DSBs were measured by the production of γ-H2A.X foci, and by increased migration of broken DNA in an agarose gel electrophoresis assay (comet assay). The DSBs are thought to be formed as a consequence of excision repair, or collapsed replication forks, as their formation is limited to the late S- and G2-phases of the cell cycle, when DNA replication is occurring or complete [30, 35].

4. Metals and DNA Repair

The ability to repair DNA damage is also an important consideration in carcinogenesis. Clearly, to induce a permanent change in the DNA, that damage would have to be unrepaired in some manner. Thus, chemical carcinogens must overcome DNA repair to cause cancer. Normal healthy cells have a great capacity for DNA repair, and can employ a variety of repair strategies depending on the type of damage incurred. However, metals can not only damage DNA but also interfere with DNA repair processes, thus enhancing the carcinogenic potential of metals. This interference can occur by disrupting DNA–protein interactions, displacement of essential metal ions by toxic metal ions, or by altering the expression or structure of DNA repair proteins. Interactions between metals and the major types of DNA repair will be discussed below, and Table 1 shows a

Table 1. DNA repair gene families and their functions.

BER	
DNA glycosylase	Catalyzes the first step in BER, removes the damaged nitrogenous base leaving the sugar–phosphate backbone intact
APE1	Removes the damaged base and generates a nick in the backbone with a 5′-dRP moiety
Pol β	Removes the 5′-dRP and incorporates a single nucleotide into the DNA
FEN1	Removes approximately 2–7 base pairs
Pol ε/δ	Resynthesizes the gap produced by FEN1
Lig3 (short patch) Lig1 (long patch)	Seals the nick to restore the DNA strand
NER	
XPC/RAD23B	Recognizes the lesion in the DNA, opens the DNA around the lesion slightly. Required for the recruitment of other NER factors
TFIIH	10 subunit complex: the core (XPB, p52, p8, p62, p34, p44) and CAK CDK7, cyclin H, and MAT1, with XPD acting as a protein bridge. Key enzyme involved in dual excision
XPB and XPD	Catalytic subunits of TFIIH, unwind the DNA around the lesion
RPA, XPA, and XPG	RPA binds and stabilizes ssDNA. XPA confirms lesion presence. Together they anchor excision nuclease. 3′ Incision by XPG
XPF-ERCC1	Cuts at the 5′ end of the damaged DNA strand
Pol δ/RFC/PCNA Pol k/PCNA/XRCC1 Pol ε/RFC(Ctf18)	Resynthesizes the gap and ligation
SSBR	
PARP1	Synthesize and transfer ADP-ribose polymers onto glutamate, aspartate or lysine residues of nuclear proteins to modify their functional properties
XRCC1	Orchestrates processing of DNA ends. No enzymatic activity itself but acts as a recruiter and scaffold for other enzymatic proteins (APE1, Pol β, PNKP, APTX, and TDP1, depending on the type of damaged termini)
Pol β, Pol δ and/ or Pol ε (Pol δε) and DNA ligases	Gap filling and ligation

(*Continued*)

Table 1. (*Continued*)

MMR	
MSH2/MSH6	MSH2 stabilizes MSH6. MSH6 has a nuclear localization sequence that delivers the dimer to the nucleus. MSH6 interacts with DNA recognizing mismatches or small insertion and deletion loops
MSH2/MSH3	Recognizes large DNA insertions and deletions which intrinsically bend the DNA double helix. It can separate the two DNA strands.
MLH1/PMS2	Discriminates the new strand from the parental strand. PMS2 is an endonuclease and introduces nicks into a discontinuous DNA strand
Exo1	An exonuclease, excises the incorrect DNA strand
Pol δ	Fills in the gap
NHEJ	
Ku70/Ku80	Binds to DNA DSB ends, recruits DNA-PKcs
DNA-PKcs	Catalytic subunit of a nuclear DNA-dependent serine/threonine protein kinase called DNA-PK, induces a conformational change that allows end processing enzymes to access the ends of the double-strand break
Artemis, MRN, PNK, WRN	End processing enzymes with nuclease activity
Pol λ or μ	Fills in the gaps from the partially complementary overhangs
Lig IV/XRCC4	Seals the nick
HR	
γ-H2A.X	Phosphorylated histone protein that signals a DSB, recruits the MRN complex
MRN complex	MRE11 component binds the DNA ends; RAD50 interacts within MRN complex on the opposing side of the DSB holding the ends in proximity to one another; Nbs1 activates autophosphorylation signals on ATM dimers
pATM	Transduces the signal to the HR repair pathway

(*Continued*)

Table 1. (*Continued*)

Rad52 and Rad51 paralogs (Rad51B, Rad51C, Rad51D, XRCC2, and XRCC3)	Promote binding of RAD51 to the single-stranded DNA and stabilize the nucleoprotein filament which is essential for the homology search and strand invasion
Rad 54	Facilitates homology search and stabilizes the D-loop
SLX4, MUS81/GEN1, or BLM	Holliday junction resolvase

summary of what gene families are involved with the different types of DNA repair.

Base Excision Repair (BER): The primary pathway used to repair DNA alkylation, oxidative lesions, and single-strand breaks, BER consists of two subpathways, short patch and long patch. Both of these pathways use either a bifunctional DNA glycosylase or a monofunctional DNA glycosylase, along with the apurinic/apyrimidinic endonuclease (APE1), to remove the damaged base and generate a single-strand break with a 5′deoxy-ribose-phosphate (5′-dRP) moiety. Short patch BER, which is dominant in mammalian systems, utilizes DNA polymerase β to remove the 5′-dRP and incorporates a single nucleotide into the DNA. Long patch removes approximately two to seven base pairs by utilizing flap structure-specific endonuclease 1 (FEN1), and the gap is resynthesized by DNA polymerase ε/δ (Pol ε/δ). Finally, ligase 3, in combination with X-ray repair cross-complementing protein (XRCC1), seals the nick to restore the DNA strand [37, 38].

Chromium can cause oxidative damage in DNA bases, the primary manifestation of which is the 8-oxo-dG lesion. Proof that BER is involved in this repair is the fact that yeast knockouts of BER genes (apn1, ntg1, and ntg2) sustain more cytotoxicity as a result of Cr(VI) treatment, compared to normal controls [39]. Individual human blood samples treated with Cr(VI) for 1 hour showed no difference in levels of DNA single-strand breaks, compared to controls, but individuals with the Cys326 8-oxo-guanine DNA glycosylase (OGG1) polymorphism had more formamidopyrimidine glycosylase (FPG)-dependent oxidative damage with Cr(VI) exposure, compared to those with no exposure [22, 40]. Cr(VI)-exposed rats were shown to have increased lung levels of 8-oxo-dG after 1 week of exposure; however, in rats exposed 2 weeks or longer there was complete recovery by 3 weeks [41]. Cr caused a concentration-dependent decrease in levels of the OGG1, but no effect on APE1, in A549 lung carcinoma cells [42]. Finally, generation of Sp was 20 times greater than 8-oxo-dG with chromium-treated oligonucleotides, and endonuclease 8-like 1 (Neil1) responded to this lesion [17]. Using methoxyamine hydrochloride to inhibit repair in CHO cells, Brooks *et al.* found reduced mutagenesis with Cr(VI) as well as methyl methane sulfonate (MMS), but increased mutagenesis following ultraviolet (UV) exposure suggesting different agents may have different BER mechanisms [43].

Single-strand break repair: Single-strand break repair is often the final step in the repair of other types of lesions, as repair processes (BER, NER, and MMR) may produce breaks in the DNA when the damaged base or portion is removed. Single-strand breaks are generally detected by poly [ADPribose] polymerase 1 (PARP1), and this is followed by processing of the DNA ends, events which are orchestrated by XRCC1. XRCC1 has no enzymatic activity itself but acts as a recruiter and scaffold for other enzymatic proteins, which can include APE1, DNA polymerase (Pol) β, polynucleotide kinase 3′-phosphatase (PNKP), tyrosyl-DNA phosphodiesterase 1 (TDP1), and aprataxin (APTX). Finally, gap filling and ligation occurs through the action of Pol β, Pol δ, and/or Pol ε followed by DNA ligases [44].

Chinese hamster ovary (CHO) cell lines deficient in XRCC1 were found to be more sensitive to the cytotoxic effects of Cr(VI) than wild-type cells [45]. In addition, these repair-deficient cells showed Cr(VI) concentration-dependent increases in chromosome damage, with more chromatid lesions than wild-type cells. Therefore, XRCC1 appears to be involved in the repair of Cr(VI)-induced lesions, and that failed repair leads to more complex forms of damage [46, 47]. Additionally, an epidemiological study, in which genetic polymorphisms in DNA repair genes of Cr(VI) workers were examined, found that a polymorphism of Arg399Gln in XRCC1 was associated with high levels of DNA damage in lymphocytes of chromium-exposed workers [48].

Nucleotide Excision Repair (NER): Chemically modified bases and helical distortions in the DNA are repaired by NER. With these types of DNA lesions and intrastrand cross-links, the XPC/RAD23B recognizes the lesion in the DNA, opens the DNA around the lesion slightly, and then recruits transcription factor II H (TFIIH) to the lesion. TFIIH contains helicases, XPB, and XPD, which unwind the DNA around the lesion. Once one of the helicases encounters the DNA adduct, it stalls and the other helicase continues to unwind the DNA for another 20 base pairs. XPC/RAD23B then dissociates from the complex and replication protein A (RPA), XPA, and XPG bind to the DNA. XPF-ERCC1 is then recruited to the NER complex and cuts at the 5′ end of the damaged DNA strand. This induces a conformational change and activates XPG to cut at the 3′ end of the segment. The 24–32 base pair segment containing the adduct is excised and DNA polymerases and their partners (Pol δ/RFC/PCNA, Pol k/PCNA/XRCC1, or Pol ε/RFC(Ctf18)) fill in the gap [49].

Cr is known to cause nucleotide damage in the form of Cr-DNA adducts and cysteine-Cr-DNA adducts [20]. Yeast knockouts of rad1 were more sensitive to Cr(VI) treatment than normal controls [37]. CHO cells deficient in NER (XPD and ERCC1) showed increased sensitivity to Cr and were unable to remove Cr-DNA adducts. Repair of Cr-DNA was observed in normal and complemented cells, but not in NER-deficient cells [45, 50]. When CHO cells deficient in XPD and XPF were exposed to Cr, there was less Cr-induced mutagenesis and chromosome damage [43] compared to wild-type cells. Cells pretreated with Cr for 48 hour showed an enhanced benzo[*a*]pyrene-induced cell killing and mutation frequency, effects which were not observed in ERCC2-deficient cells [51]. Studies in NER-proficient human cells showed efficient removal of Cr-DNA adducts, but cells deficient in XPA, XPC, and XPF were severely compromised and showed increased mutagenicity of binary Cr(III)-DNA and ternary cysteine-Cr(III)-DNA adducts. The XPA knockdowns had increased cytotoxicity and apoptosis [52]. Repair of protein-Cr-DNA cross-links by NER was found to be negligible in human cells, but NER did act to protect against larger lesions [53]. Finally, studies of human patients with Cr-induced lung cancer showed increased polymorphisms in the XPD gene but no association with changes in XRCC1 and OGG1 [54].

Mismatch Repair (MMR): During normal DNA replication, often there are erroneous insertions of nucleotides, accidental deletions, or misincorporated nucleotides collectively referred to as mismatches. MMR functions to recognize and repair these types of errors. The heterodimer MSH2/MSH6 is responsible for recognizing mismatches or small insertion and deletion loops. The heterodimer MSH2/MSH3 recognizes larger insertion and deletion loops. Once recognized, the MLH1/PMS2 heterodimer is recruited to the mismatch and works together with the recognizing heterodimer pair to search for a break in the strand discriminating the new strand from the parental strand. Then Exo1 excises the incorrect strand and Pol δ fills the gap [55].

Studies of Cr-induced lung cancer patients have shown higher rates of replication errors than lung cancer patients without Cr exposure. They also showed that MLH1 and MSH2 genes were repressed at a higher rate compared to non-Cr-induced lung tumors and that inactivation of MLH1 expression correlated with increased microsatellite instability in the Cr(VI)-exposed tumors [56]. A study considering gene polymorphisms in Cr(VI)-exposed cancer patients found that alterations in MLH1 were associated with increased risk of lung cancer development [57]. Studies using cells deficient in the MMR genes MLH1, PMS2, MSH2, and MSH6 showed increased clonogenic survival and deficient activation of apoptosis after Cr(VI) treatment suggesting that MMR proteins may potentiate the genotoxicity of Cr-DNA adducts by causing replication fork collapse leading to DNA DSBs [58]. In addition, these MMR repair-deficient cells showed less γ-H2A.X foci formation suggesting fewer DNA DSBs. Inhibition of MLH1 or MSH2 decreased the number of DNA DSBs and micronuclei and increased survival [35, 59]. MMR proteins were able to recognize ternary Cr-DNA adducts at a comparable level to their natural substrate, G–T mismatch, but did not recognize binary Cr-DNA adducts; in addition the generation of DSBs and chromosomal breaks was dependent on the ternary adducts which were mistakenly processed by MMR proteins as mismatches and subsequently processed into DNA breaks [60].

These data indicate that MMR is active after Cr(VI) exposure. The types of lesions the MMR genes may be responding to is unclear in part because the two cell culture studies employed a shuttle vector construct to consider repair and such a system has meaningful shortcomings for repair studies. The specific repair mechanism is uncertain, but the data do indicate that MLH1 and MSH2 are both candidate genes to be involved. Interestingly, as was observed for NER; proficient MMR seems to be more genotoxic than deficient repair which is interesting as the Cr-induced tumors appear MMR deficient. This suggests that acquiring a state of MMR deficiency is likely a later event in the overall carcinogenic process and that mismatch damage is likely acquired after other carcinogenic changes have occurred.

Double-strand break repair: There are two major pathways for repairing DNA DSBs. Nonhomologous end joining (NHEJ) is more error prone and is primarily used in G1 and early S when a sister chromatid is not available. Homologous recombination (HR) repair is primarily used in late S- and G2-phases when a sister chromatid is available to use as a template for efficient and error-free repair.

In NHEJ, the damage is first recognized by the Ku70/Ku80 heterodimer and then recruits DNA-dependent protein kinase, catalytic subunit (DNA-PKcs) to form a holoenzyme that facilitates binding of short complementary sequences between two ends of the DSB. This creates displaced noncomplementary strands which can be processed by Artemis, the Mre11/Rad50/Nbs1 (MRN) complex, phosphorylated polynucleotide kinase (PNK), or the Werner syndrome protein (WRN) all of which have nuclease activity. DNA polymerase λ or μ is recruited and fills in the gaps from the partially complementary overhangs. Finally, DNA ligase IV/XRCC4 seals the nick in both strands of the DNA [61].

In HR, damage is initially sensed by phosphorylated γ-H2A.X [62–64] which then signals to the MRN complex, the MRE11 component binds the DNA ends and RAD50 interacts within the MRN complex on the other side of the DSB, functioning to hold the two ends of the broken DNA in proximity to one another [65]. Ataxia telangiectasia mutated (ATM) dimers are recruited to the break and a conformational change in Nbs1 activates autophosphorylation signals on ATM dimers [66]. pATM then transduces the signal to the HR repair pathway. DNA DSBs are then converted to 3′ single-stranded DNA tails, possibly by Mre11. Rad52, along with Rad51 paralogs (Rad51B, Rad51C, Rad51D, XRCC2, and XRCC3), promote binding of RAD51 to the single-stranded DNA and stabilize the nucleoprotein filament [67]. The nucleoprotein filament invades a homologous molecule in a reaction stimulated by RAD54, which searches for sequence homology most likely on the sister chromatid. This invasion leads to Holliday junction formation, which is resolved by resolvase complexes that contain SLX4, MUS81 and GEN1, or Bloom's syndrome protein (BLM) [68–72].

For Cr(VI) and NHEJ, Chinese hamster cell lines deficient in DNA-PKcs were more sensitive to the cytotoxic effects of Cr(VI) than their normal counterparts [44]. Likewise, Chinese hamster cell lines deficient in Ku80 were more sensitive to Cr(VI) than Ku80-proficient cells. However, Ku80 played no role in protecting cells from genotoxicity, as the amounts of chromosome damage seen in Ku80-deficient cells and normal and complemented cells were the same [73].

HR repair has been observed in cells exposed to Cr(VI), and many of the genes involved have been investigated. Cr(VI) is known to affect the ATM-repair pathway, which includes Chk2, Mre11, RAD51C, BRCA2, and XRCC3. HR-deficient yeast were hypersensitive to Cr lethality, and RAD52 mutants exhibited terminal arrest in G2/M suggesting HR is essential for their survival [36]. Cr(VI) induces Rad51 repair foci in CHO cells, indicating that HR is activated by Cr(VI)-induced damage [45]. CHO cells deficient in XRCC3 or RAD51C had increases in complex chromosome damage compared to their repair-proficient counterparts after Cr(VI) exposure, showing that HR-specific proteins are essential to maintain normal chromosome integrity [74]. Human cells deficient in Mre11 exhibited loss of contact inhibition and anchorage-dependent growth when treated with Cr(VI), both of these are indicative of oncogenic transformation and do not occur in repair-proficient cells [75]. In repair-proficient human cells, Cr(VI) induces DNA DSBs, indicated by phosphorylated H2A.X; Mre11 is induced which then activates ATM to phosphorylate p53 and Chk2, all crucial steps for facilitating efficient recovery from Cr-induced damage and cell cycle arrest [30, 33, 75, 76].

Additional studies have elucidated a role for the WRN protein specifically in response to MMR activity. As mentioned in the previous section, MMR can process ternary DNA adducts resulting in DNA DSBs [60]. Cells lacking the WRN protein exhibit more persistent Cr-induced DSBs [77, 78]. In addition, WRN was required for the formation of RAD51 foci indicating activation of HR repair [77]. DNA DSBs were also observed to occur preferentially in euchromatin increasing the probability of deleting tumor suppressors and inducing oncogenic translocations [79].

More recent studies have shown that Cr(VI) induces DNA DSBs at both acute and prolonged exposure times but it also impairs HR with prolonged exposures measured as a reduction in expression of Rad51 as well as reduced Rad51 foci formation in the nucleus [80]. Interestingly, this study also showed a change in the subcellular localization of Rad51, as it was retained in the cytoplasm with longer exposure time. Browning *et al.* [81] also showed a reduction in nucleoprotein filament formation with prolonged Cr(VI) exposure, and that the Rad51 nuclear transport protein, Rad51C, was also compromised as measured by reduced focus formation. With longer exposures, the absence of Rad51 in the nuclei forces the cells to use a more error-prone pathway, such as alternative NHEJ or single-strand annealing, which are activated only when HR repair is aborted.

5. How Metals Impact Cell Division and Mitosis

Normal cell division is critical for maintaining the integrity of DNA and protecting cells against carcinogenesis. Cell division is the process by which a parent cell divides into two daughter cells, the main goal of which is to properly replicate DNA. Normal cell division is a tightly regulated process, which entails several important steps designed to maintain both the structure and quantity of DNA. Multiple checkpoints employed throughout the cell cycle ensure that the steps, which involve a delicate balance of effectors, cyclins, and cyclin-dependent kinases, as well as cyclin-dependent kinase inhibitors (e.g., p27Kip1, p21Waf1/Cip1, p15INK4B, p16INK4A) are accomplished appropriately. In addition, once mitosis ensues, there are factors that control the proper segregation of chromosomes into the mother and daughter cells. In this section, we consider the functioning of the cell cycle, including centrosome regulation, spindle assembly checkpoints, kinetochores, and chromosome cohesion, and how they can be affected by metals.

Metals and cell cycle disruption: The cell cycle comprises events that occur in order for a cell to replicate its DNA. G0 is the resting phase in which cells are considered quiescent. It is followed by G1, in which the cell increases in size, increases protein supply and organelles, and prepares for DNA synthesis. The G1 checkpoint, therefore, ensures that there are sufficient materials available before the cell can proceed to DNA synthesis, or S-phase. In S-phase, DNA replication occurs, followed by G2 which involves additional protein synthesis to prepare the cell for mitosis or M-phase. At the end of G2, the G2 checkpoint ensures that all the DNA is properly replicated and the cell is ready to enter M-phase. M-phase is discussed separately below because it has additional important features.

As discussed earlier, Cr(VI) is known to induce a variety of types of DNA damage, and is therefore able to affect many parts of the cell cycle. Most normal healthy cells are able to repair this damage; however, most DNA repair requires a delay in the cell cycle to ensure that replication will not proceed on a damaged DNA template. In human lung carcinoma cells, Cr(VI) exposure was shown to dysregulate key cell cycle proteins, including INK4p19 and p34 CDC2 [82]. Rats treated with Cr(VI) showed significant alterations of cyclin D1-3 in their lungs [83]. In HLF cells, Cr(VI) upregulated key cell cycle inhibitors p21 and p15, through decreased cyclin A gene expression and increased GADD45 expression [84]. Interestingly, higher concentrations of Cr(VI) induced an initial G1 arrest followed by delayed S-phase, but lower concentrations induced a G2/M arrest [32, 39, 85–87]. Protein tyrosine phosphorylation is necessary for cell survival, and conversely protein tyrosine phosphatases (PTP) reduce cell survival. Cr(VI) exposure was shown to inhibit PTP function, leading to an increase in human lung cell survival despite the concurrent induction of DNA damage [88]. Furthermore, PTP inhibition was shown to bypass the G2/M cell cycle checkpoint [89].

Metals and mitosis: Events that impact mitosis are of growing importance in carcinogenesis, as disruption of mitosis can have profound effects on chromosome number and aneugenesis. Mitosis, which consists of nuclear division or karyokinesis, involves arrest of cell growth, condensation of DNA into duplicate packaged chromosomes, and segregation of these chromosomes into two separate daughter cells. Mitosis can be considered to occur in phases (Figure 2): prophase or chromosome condensation, metaphase or alignment of chromosome at the center of the cell, anaphase, where chromosome move to two opposite poles of the cell guided by two centrosomes, and telophase, where daughter nuclei form and cytokinesis begins. At the completion of metaphase, there is a spindle assembly checkpoint, which ensures proper spindle attachment and alignment prior to proceeding to anaphase.

The spindle assembly checkpoint (Figure 2) functions in mitosis at the transition of metaphase to anaphase. It operates to ensure that all the chromosomes are aligned along the metaphase plate, and all kinetochores are attached to the proper number of microtubules with tension applied equally to each side of the chromosome. Unlike other cell cycle checkpoints, the spindle assembly checkpoint is active upon entry into metaphase and produces a "stop" signal only when the spindle assembly checkpoint is satisfied and a "go" signal is transduced allowing progression to anaphase.

Metals induce cell cycle arrest, but it varies by metal. For example, arsenic induces a profound M-phase arrest, while chromium arrests cells in G2-phase. Emerging data indicate metals can impact various aspects of the spindle assembly checkpoint function. Lung cancers in general are known to exhibit alterations in the spindle assembly checkpoint proteins, but it is unclear whether these changes are observed only in Cr-induced lung cancers. Cell culture studies have investigated the effects of Cr on spindle assembly checkpoint proteins. Chronic exposures to particulate chromate show increases in bypass of the spindle assembly checkpoint as indicated by increases in centromere spreading, premature anaphase, and premature centromere division [87, 90]. Mitotic analysis after Cr(VI) exposure show inappropriate increases of cells in anaphase [90], suggesting a bypass of the spindle assembly checkpoint. The underlying molecular mechanisms are uncertain, but data do show that Mad2 expression is decreased [87].

Metals and centrosome amplification: Centrosomes also play an important role in mitosis. Centrosome amplification is associated with aneuploidy and is a common phenotype in many cancers. Centrosomes are the microtubule organizing centers within a cell, and are duplicated

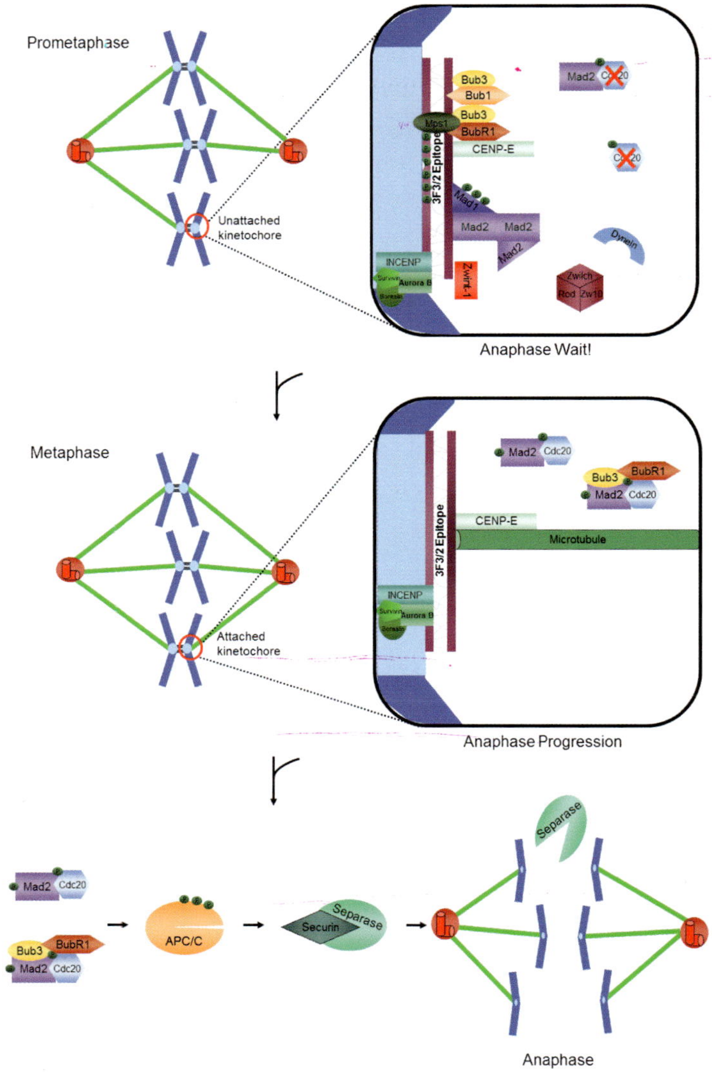

Figure 2. Spindle assembly checkpoint. *Prophase*: sister chromatids are held together by cohesin until prophase when cohesin on the chromatids is phosphorylated by Plk1. Cohesin remains at the kinetochore until the SAC is satisfied and anaphase can progress. *Prometaphase*: proteins assemble at the kinetochore to form the mitotic checkpoint complex (MCC). *Metaphase*: The MCC is maintained until the last kinetochore is attached to a microtubule and all of the chromosomes are properly aligned, it is then dissolved and Mad2 is phosphorylated activating Cdc20 allowing it to bind the APC/C which in turn ubiquitinates securin for degradation allowing separase to be active. Separase then cleaves cohesin at the kinetochore and *anaphase* ensues.

in cell division because they comprise the point of organization of chromosomal segregation. Improper chromosome segregation is associated with the development of aneuploidy, a common characteristic of cancer cells.

Interestingly, because there is a significant amount of cross talk involved in many cell replication mechanisms, metals that perturb one or more proteins may affect multiple outcomes. For example, cells depleted in homologous repair proteins XRCC2, XRCC3, Rad51D, and Rad51B exhibit centrosome fragmentation [91–93]. The spindle assembly checkpoint and the centrosome duplication cycle also overlap with the role of Nek2. In the centrosome cycle, Nek2 is involved in the centrosome separation step in G2 [94], and in the spindle assembly checkpoint it is required for localizing Mad2 to the kinetochore and for proper microtubule attachment [95, 96]. In addition, cohesin, Shugoshin 1 (Sgo1), and separase are involved in both centrosome maintenance and separation as well as sister chromatid cohesion and are involved in the final steps in chromosome separation.

Emerging data indicate metals can cause centrosome amplification. Prolonged Cr(VI) exposure induces centrosome amplification in human lung cells and correlates with the induction of aneuploidy [89, 97]. Centrosome amplification occurs through multiple mechanisms including but not limited to centrosome overduplication, failure of cytokinesis, centriole splitting, and acentriolar centrosome formation. Cr(VI) induces centrosome overduplication and is associated with a persistent G2 arrest [89]. Failure of cytokinesis leads to tetraploid cells with four centrosomes which are observed after prolonged Cr(VI) exposure; however, cytokinesis failure typically results in binucleated cells which are not seen with Cr(VI) treatment [88, 96]. Cr(VI) does induce centriole splitting [89], which can generate multiple centrosomes with only one centriole and may be related to Cr(VI)-induced DNA damage present during mitosis [98]. Acentriolar centrosomes are also observed after prolonged Cr(VI) exposures [89]. In addition, cells that were neoplastically transformed by Cr(VI) had amplified centrosomes suggesting it is a permanent characteristic that can contribute to chromosome instability in tumors [99].

6. Epigenetic Changes

Epigenetic changes do not deal with direct damage or disruption of the nucleotide sequence of DNA, but rather change the way that it is interpreted by the cell and thus the organism. Since epigenetic events regulate gene activity they can dictate the normal process of cell differentiation and development but have also been demonstrated to affect a variety of disease states including cancer [100]. Epigenetic alterations include methylation at the five position of cytosine to form 5-methyl-cytosine (5-meC), post-translational histone modification (methylation, acetylation, phosphorylation), noncoding RNAs, micro-RNAs (miRNA), nonhistone chromatin binding proteins, as well as the higher ordering of chromatin structure (open or closed). The classic types of epigenetic alterations are formation of 5-meC and histone modifications, although the role of miRNAs is a rapidly expanding field of study and will likely become exceedingly important in the future.

Epigenetic processes regulate the organizational structure of DNA, examples of which include condensation of chromatin, such that the gene cannot be transcribed, or enabling decondensation of chromatin so that genes may become available for transcription. Studies performed in the last 20 years have revealed that environmental factors have a profound effect on epigenetic modifications, and metals are no exception. Several recent reviews highlight the effects of arsenic, cadmium, chromium, copper, lead, mercury, and nickel, with special focus on changes in the epigenome [101–104].

Methylation of cytosine in DNA: In mammalian cells, 5-meC formation occurs almost exclusively on cysteine residues that are immediately followed by a guanine base. This combination, commonly referred to as CpG islands, is often found in the promoter regions of genes. In normal cells, these regions tend to be free of methylation, while CpG repeats found outside of promotor regions are most often methylated. DNA methyltransferases (DNMTs) are responsible for maintaining methyl status from generation to generation. Aberrant methylation contributes to the development of cancer and can occur as a result of environment, diet, and pharmaceuticals. Gene-specific hypermethylation of tumor suppressor genes can lead to inactivation and global hypomethylation can lead to activation of genes involved in cancer development [105, 106]. There are also several proteins identified with the ability to reverse methylation, for example, ten-eleven translocation (TET) and activation-induced cytidine deaminase (AID) [107]. Thus, methylation changes are inducible as well as reversible making them likely targets for intervention in the prevention and treatment of disease [108, 109]. Metals are known to alter methylation patterns.

For Cr(VI), methylation of the promoter region of the p16 tumor suppressor gene as well as suppression of the p16 protein, were observed in chromate-exposed lung cancer patients [110]. Two additional tumor suppressor genes, adenomatosis polyposis coli (APC) and hMLH1, were found to be hypermethylated in chromate-induced cancers [56, 111]. Another study, which included individuals with current exposure to Cr(VI) who had Cr accumulation in their blood, showed that these individuals had global DNA hypomethylation correlating with low levels of folate, as well as urinary 8-hydroxy-2′-deoxyguanine and DNA strand breaks [112]. In cell culture studies, Cr(VI)-treated cells were transfected with a *gpt* reporter gene and found that chromate-induced aberrant DNA methylation in the *gpt* gene and silenced gene expression [113].

Histone modification: Histones are responsible for the packaging and order of DNA into nucleosomal structures. The end terminal tail of the histones is subject to a variety of modifications that can alter the structure of chromatin making it more or less available for transcription. Histone modification occurs posttranslationally and includes lysine acetylation, arginine and lysine methylation and demethylation, serine phosphorylation, lysine ubiquitylation, sumoylation, and several others [114, 115]. The type of modification determines the effect on gene expression; for example, lysine acetylation is associated with gene activation whereas lysine methylation leads to signals that are dependent on the specific residue [116, 117]. Modification of histones can thus impact both gene and protein expression and have been implicated in a variety of disease states including cancer. Metals can induce histone modifications.

For example, Cr(VI) interacts with arginine and lysine residues of histones *in vitro* [118]. Cells exposed to both Cr(VI) and benozo[a]pyrene, a procarcinogen, resulted in Cr cross-linking of histone deacetylase 1-DNA methyltransferase 1 of the CYp1a1 promoter which led to inhibition of histone and acetylation marks [119]. Chromate also induced increased global levels of H3K9me2, HSK9me3, and H3K4me3; in addition, the H3K9 dimethylation correlated with decreased MLH1 expression, a protein required in DNA repair [120, 121]. However, these studies were performed in a lung carcinoma cell line (A549) and in mouse hepatoma cells (Hepa-1c1c7), and may not reflect induced events leading to carcinogenesis but rather events that occur after transformation has already occurred. In nontransformed human bronchial cells, Cr inhibited biotinidase expression through decreasing histone acetylation as well as more globally decreased acetylation of H3 and H4 and increased levels of histone deacetylases HDAC2 and HDAC3 [122, 123]. More recently, Cr was shown to increase the level of Nupr1, a small stress response protein implicated in lung cancer, and that this was regulated by acetylation of H4K16 [124].

Micro-RNA: miRNA are small noncoding RNAs, typically 20–24 nucleotides, which function to silence RNA. MiRNAs induce posttranscriptional gene silencing by pairing complementary sequence motifs in target mRNAs preventing translation. Aberrant expression of miRNAs is associated in a variety of human diseases including cancer [125]. In addition, circulating miRNAs have been correlated with the presence of tumors and thus may serve as potential biomarkers.

Metals can induce miRNAs. Workers exposed to metal particulates, including chromate, had increased levels of miR-222 and miR-21 which are associated with oxidative stress and inflammation [126]. MiRNA-143 levels were reduced 35-fold in Cr-transformed cells and were associated with upregulation of IGF-IR and IRS1; increased extracellular regulated kinase (ERK), HIP-1alpha, and NF-kB signaling; and upregulation of IL-8 leading to increased angiogenesis [127]. Microarray analysis of miRNAs in *Drosophila melanogaster* treated with chromate showed increased expression of 13 miRNAs and decreased expression of six miRNAs which were targets involved in DNA repair, oxidation reduction, and stress-activated MAPK [128].

7. Summary: Metals and Carcinogenesis Paradigms

Metal carcinogenesis is a complex process involving DNA damage, effects on cell division, escape from DNA repair, mitotic interference, and epigenetic changes. There are two major carcinogenesis paradigms to put these outcomes into an understandable context. One paradigm, the multistage model of carcinogenesis, suggests that the driving events are initiating mutations in oncogenes and tumor suppressor genes, which are selected and promoted to grow and expand and ultimately progress into tumors. The second paradigm, the genomic instability model of carcinogenesis, suggests the driving events may comprise the acquisition of a mutator phenotype, which occurs *via* the loss of DNA repair and/or the acquisition of aneuploidy.

While there are few generalizations to place metals in these categories, one common outcome is the observation that metals are weak mutagens. While they can induce base mutations if the doses are high enough, they typically do not induce these changes at low doses or in robust systems. Thus, the multistage model does not apply well to metals. By contrast, metals do commonly inhibit DNA repair, albeit often by very different mechanisms. Moreover, most metals will impact chromosome structure or number if not both. Consequently, the genomic instability model of carcinogenesis appears to fit metal carcinogenesis much better.

For Cr(VI), the available literature discussed earlier supports a mechanism where the particles are inhaled and impact at bronchial bifurcations sites. The particles dissolve over time outside the cell releasing the chromate anion. Chromate enters the cell by anion transport and once inside the cell, the Cr(VI) is reduced to Cr(III) *via* multiple reactive intermediates. One or more of the products of Cr(VI) reduction interact with DNA to induce repair systems such as mismatch and excision repair, pathways which produce DNA DSBs. Simultaneously, these metabolic products induce epigenetic changes that inhibit DNA DSB repair *via* the HR repair pathway. This outcome forces the cell to switch to more error-prone repair mechanisms resulting in damage to chromosome structure. Thus, breaks are produced that cannot initially be repaired. The effort to repair the breaks results in G2 cell cycle arrest which becomes prolonged due to the delays in repair. The prolonged repair results in slippage through the cell cycle decoupling it from the centrosome duplication cycle and interfering with the spindle assembly checkpoint. These outcomes result in changes in chromosome number. Thus, a prolonged exposure to Cr(VI) results in changes in chromosome number and structure that ultimately progress into a tumor.

Abbreviations

5′-dRP	5′-deoxy-ribose-phosphate
5meC	5-methylcytosine
8-oxo-dG	8-hydroxy-deoxyguanine
AID	activation-induced cytidine deaminase
APC	adenomatosis polyposis coli
APE 1	apurinic/apyrimidinic endonuclease 1
APTX	aprataxin
ATM	ataxia telangiectasia mutated
BER	base excision repair
BLM	Bloom's syndrome protein
BRCA	breast cancer
CAK	cyclin-activated kinase
Cdc	cell division cycle
Cdk	cyclin-dependent kinase
CHK	checkpoint kinase
CHO	Chinese hamster ovary
Cr	chromium
CYP	cytochrome P-450
DNA-Pkcs	DNA-dependent protein kinase, catalytic subunit
DNMT	DNA methyltransferase
DSB	double-strand break
Exo	exonuclease
ERCC	excision repair cross-complementing
ERK	extracellular regulated kinase
FEN 1	flap structure-specific endonuclease 1
FPG	formamidopyrimidine glycosylase
GADD	growth arrest and DNA damage
ϒ-H2A.X	phosphorylated H2A histone family, member X
HDAC	histone deacetylase
HIP1	Huntingtin interacting protein 1
HR	homologous recombination
IARC	International Agency for Research on Cancer
Ku70/80	"Ku" is derived from the name of the Japanese patient in which it was discovered

MAD	mitotic arrest deficient
miRNA	micro-RNA
MLH	MutL protein homolog
MMR	mismatch repair
MRN	complex of Mre11, Rad50, and Nbs1 proteins
MSH	MutS protein homolog
MMS	methyl methane sulfonate
NADPH	nicotinamide adenine dinucleotide phosphate
Nek	NIMA-related kinase
NER	nucleotide excision repair
Neil1	endonuclease 8-like 1
NF-kB	nuclear factor kappa-light-chain-enhancer of activated B cells
NHEJ	nonhomologous end joining
Nupr1	nuclear protein 1
OGG1	8-oxo-guanine DNA glycosylase
PARP1	poly [ADP ribose] polymerase 1
Pol	DNA polymerase
PMS	postmeiotic segregation increased
PNKP	polynucleotide kinase 3′-phosphatase
PTP	protein tyrosine phosphatase
RAD	family of proteins identified in yeast as radiation sensitive
ROS	reactive oxygen species
RNS	reactive nitrogen species
RPA	replication protein A
Sgo1	Shugoshin 1
Sp	spiroiminodihydantoin
SSBR	single strand break repair
TDP1	tyrosyl-DNA phosphodiesterase 1
TET	ten-eleven translocation
TFIIH	transcription factor IIH
UV	ultraviolet
WRN	Werners syndrome protein
XP	xeroderma pigmentosum, an autosomal recessive DNA repair deficiency gene of patients having mutations in proteins: XPA, XPB, XPC, XPD, XPF, XPG
XRCC	X-ray repair cross-complementing protein

References

1. IARC. Chromium, nickel and welding. *IARC Monogr Eval Carcinog Risks Hum* 1990; 49: 1–648.
2. De Flora S, Wetterhahn KE. Mechanisms of chromium metabolism and genotoxicity. *Life Chem Rep* 1989; 7: 169–244.
3. Zhitkovich A. Chromium in drinking water: sources, metabolism, and cancer risks. *Chem Res Toxicol* 2011; 24(10): 1617–1629.
4. Shi X, Chiu A, Chen CT, Halliwell B, Castronova V, Valluathan A. Reduction of chromium VI and its relationship to carcinogenesis. *J Toxicol Environ Health* 1999; 2(1): 87–104.
5. Liu KJ, Shi X, Jiang JJ, Goda F, Dalal N, Swartz HM. Chromate-induced chromium(V) formation in live mice and its control by cellular antioxidants: an L-band electron paramagnetic resonance study. *Arch Biochem Biophys* 1995; 323(1): 33–39.
6. Fornace AJ, Seres DS, Lechner JF, Harris CC. DNA-protein crosslinking by chromium salts. *Chem Biol Interact* 1981; 36: 345–354.
7. Patierno SR, Banh D, Landolph JR. Transformation of C3H/10T1/2 mouse embryo cells by insoluble lead chromate but not soluble calcium chromate: relationship to mutagenesis and internalization of lead chromate particles. *Cancer Res* 1988; 47: 3815–3823.
8. Elias Z, Poirot O, Pezerat H, Suquet H, Schneider O, Danière MC, *et al.* Cytotoxic and neoplastic transforming effects of industrial hexavalent chromium pigments in Syrian hamster embryo cells. *Carcinogenesis* 1989; 10(11): 2043–2052.
9. Ishikawa Y, Nakagawa K, Satoh Y, Kitagawa T, Sugano H, Hirano T, *et al.* Characteristics of chromate workers' cancers, chromium lung deposition and precancerous bronchial lesions: an autopsy study. *Br J Cancer* 1994; 70(1):160–166.
10. Wise SS, Holmes AL, Ketterer ME, Hartsock WJ, Fomchenko E, Katsfis SP, *et al.* Chromium is the proximate clastogenic species for lead chromate-induced clastogenicity in human bronchial cells. *Mutat Res* 2004; 560: 79–89.
11. Xie H, Holmes AL, Wise SS, Gordon N, Wise JP, Sr. Lead chromate-induced chromosome damage requires extracellular dissolution to liberate chromium ions but does not require particle internalization or intracellular dissolution. *Chem Res Toxicol* 2004; 17(10): 1362–1367.
12. Stout MD, Herbert RA, Kissling GE, Collins BJ, Travlos GS, Witt KL, *et al.* Hexavalent chromium is carcinogenic to F344/N rats and B6C3F1

mice after chronic oral exposure. *Environ Health Perspect* 2009; 117(5): 716–722.
13. Crespo-Hernandez CE, Close DM, Gorb L, Leszczynski J. Determination of redox potentials for the Watson-Crick base pairs, DNA nucleosides, and relevant nucleoside analogues. *J Phys Chem B* 2007; 111: 5386–5395.
14. Blasiak J, Kowalik J. A comparison of the *in vitro* genotoxicity of tri- and hexavalent chromium. *Mutat Res* 2000; 469: 135–145.
15. Qi W, Reiter RJ, Tan DX, Garcia JJ, Manchester LC, Karbownik M, *et al.* Chromium (III)-induced 8-hydroxydeoxyguanosine in DNA and its reduction by antioxidants: comparative effects of melatonin, ascorbate and vitamin E. *Environ Health Perspect* 2000; 108(5): 399–402.
16. Sugden KD, Martin BD. Guanine and 7,8-dihydro-8-Oxo-guanine-specific oxidation in DNA by chromium (V). *Environ Health Perspect* 2002; 110(5): 725–728.
17. Slade PG, Hailer MK, Martin BD, Sugden KD. Guanine-specific oxidation of double stranded DNA by Cr(VI) and ascorbic acid forms spiroiminohydantoin and 8-Oxo-2′-deoxyguanosine. *Chem Res Toxicol* 2005; 18(7): 1140–1149.
18. Arakawa H, Weng MW, Chen WC, Tang MS. Chromium (VI) induces both bulky DNA adducts and oxidative DNA damage at adenines and guanines in the p53 gene of human lung cells. *Carcinogenesis* 2012; 33(10): 1993–2000.
19. Singh J, Carlisle DL, Pritchard DE, Patierno SR. Chromium induced genotoxicity and apoptosis: relationship to chromium carcinogenesis. *Oncol Rep* 1998; 5: 1307–1318.
20. Wise SS, Holmes AL, Wise JP, Sr. Hexavalent chromium-induced DNA damage and repair mechanisms. *Rev Environ Health* 2008; 23: 39–57.
21. Sugiyama M, Wang XW, Costa M. Comparison of DNA lesions and cytotoxicity induced by calcium chromate in human, mouse and hamster cell lines. *Cancer Res* 1986; 46: 4547–4551.
22. Hodges NJ, Adam B, Lee AJ, Cross HJ, Chipman JK. Induction of DNA-strand breaks in human peripheral blood lymphocytes and A549 lung cells by sodium dichromate: association with 8-oxo-2-deoxyguanosine formation and inter-individual variability. *Mutagenesis* 2001; 16: 467–474.
23. Yang PH, Gao HY, Cai J, Chiu JF, Sun H, He QY. The stepwise process of chromium-induced DNA breakage: characterization by electrochemistry, atomic force microscopy and DNA electrophoresis. *Chem Res Toxicol* 2005; 18: 1563–1566.

24. Ueno S, Kashimoto T, Susa N, Furukawa Y, Ishii M, Yokoi K, *et al.* Detection of dichromate (VI)-induced DNA stand breaks and formation of paramagnetic chromium in multiple mouse organs. *Toxicol Appl Pharmacol* 2001; 170: 56–62.
25. Kleinsasser NH, Gamarra F, Bergner A, Wallner BC, Harreus UA, Juchhoff J, *et al.* Genotoxicity of nitroso compounds and sodium dichromate in a model combining organ cultures of human nasal epithelia and the Comet Assay. *ORL J Otorhinolaryngol Relat Spec* 2001; 63: 141–147.
26. Bagchi D, Joshi SS, Bagchi M, Balmoori J, Benner EJ, Kuszynski CA, *et al.* Cadmium- and chromium-induced oxidative stress, DNA damage, and apoptotic cell death in cultured human chronic myelogenous leukemia K562 cells, promyelocytic leukemic HL-60 cells and normal peripheral blood mononuclear cells. *J Biochem Mol Toxicol* 2000; 14(1): 33–41.
27. De Boeck M, Lombaert N, De Backer S, Finsy R, Lison D, Kirsch-Volders M. *In vitro* Genotoxic effects of different combinations of cobalt and metallic carbide particles. *Mutagenesis* 2003; 18(2): 177–186.
28. Sugden KD, Rigby KM, Martin BD. Oxidative activation of the human carcinogen chromate by arsenite: a model for synergistic metal activation leading to oxidative damage. *Toxicol In Vitro* 2004; 18: 741–748.
29. Messer J, Reynolds M, Stoddard L, Zhitkovich A. Causes of DNA single-strand breaks during reduction of chromate by glutathione *in vitro* and in cells. *Free Radic Biol Med* 2006; 40: 1981–1992.
30. Ha L, Ceryak S, Patierno SR. Chromium (VI) activates ataxia telangiectasia mutated (ATM) protein. *J Biol Chem* 2003; 278(20): 17885–17894.
31. Xie H, Wise SS, Holmes AL, Xu B, Wakeman T, Pelsue SC, *et al.* Carcinogenic lead chromate induces DNA double-strand breaks and activates ATM kinase in human lung cells. *Mutat Res* 2005; 586(2): 160–172.
32. Xie H, Holmes AL, Young JL, Qin Q, Joyce K, Pelsue SC, *et al.* Zinc chromate induces chromosome instability and DNA double strand breaks in human lung cells. *Toxicol Appl Pharmacol* 2009; 234(3): 293–299.
33. Wakeman TP, Kim WJ, Callens S, Chiu A, Brown KD, Xu B. The ATM-SMC1 pathway is essential for activation of the chromium[VI]-induced S-phase checkpoint. *Mutat Res* 2004; 554: 241–251.
34. Wang XF, Xing ML, Shen Y, Zhu X, Xu LH. Oral administration of Cr(VI) induced oxidative stress, DNA damage and apoptotic cell death in mice. *Toxicology* 2006; 228: 16–23.

35. Reynolds M, Stoddard L, Bespalov I, Zhitkovich A. Ascorbate acts as a highly potent inducer of chromate mutagenesis and clastogenesis: linkage to DNA breaks in G2 phase by mismatch repair. *Nucleic Acids Res* 2007; 35: 465–476.
36. Wise SS, Holmes AL, Qin Q, Xie H, Katsifis SP, Thompson WD, *et al.* Comparative genotoxicity and cytotoxicity of four hexavalent chromium compounds in human bronchial cells. *Chem Res Toxicol* 2010; 23: 365–372.
37. Thompson LH, West MG. XRCCI keeps DNA from getting stranded. *Mutat Res* 2000; 459: 1–18.
38. Sung JS, Demple B. Roles of base excision repair subpathways in correcting oxidized abasic sites in DNA. *FEBS J* 2006; 273: 1620–1629.
39. O'Brien TJ, Fornsaglio JL, Ceryak S, Patierno SR. Effects of hexavalent chromium on the survival and cell cycle distribution of DNA repair-deficient *S. cerevisiae*. *DNA Repair* 2002; 1: 617–627.
40. Lee AJ, Hodges NJ, Chipman JK. Interindividual variability in response to sodium dichromate-induced oxidative DNA damage: role of the Ser 326 Cys polymorphism in the DNA-repair of 8-Oxo-7,8-dihydro-2′-Deoxguanosine DNA glycosylase. *Cancer Epidemiol Biomarkers Prev* 2005; 14(2): 497–505.
41. Maeng SH, Chung HW, Yu IJ, Kim HY, Lim CH, Kim KJ, *et al.* Changes of 8-OH-dG levels in DNA and its base excision repair activity in rat lungs after inhalation exposure to hexavalent chromium. *Mutat Res* 2003; 539(1–2): 109–116.
42. Hodges NJ, Chipman JK. Down regulation of the DNA-repair endonuclease 8-oxo-guanine DNA glycosylase 1 (hOGG1) by sodium dichromate in cultured human A549 lung carcinoma cells. *Carcinogenesis* 2002; 23(1): 55–60.
43. Brooks B, O'Brien TJ, Ceryak S, Wise JP, Sr., Wise SS, Wise JP, Jr., *et al.* Excision repair is required for genotoxin-induced mutagenesis in mammalian cells. *Carcinogenesis* 2008; 29(5): 1064–1069.
44. Caldecott KW. Single-strand break repair and genetic disease. *Nat Rev Genet* 2008; 9(8): 619–631.
45. Bryant HE, Ying S, Helleday T. Homologous recombination is involved in repair of chromium-induced DNA damage in mammalian cells. *Mutat Res* 2006; 99: 116–123.
46. Grlickova-Duzevik EG, Wise SS, Munroe RC, Thompson WD, Wise JP, Sr. XRCC1 protects cells against particulate chromate-induced chromosome damage and cytotoxicity in Chinese hamster ovary cells. *Toxicol Sci* 2006; 92(2): 409–415.

47. Grlickova-Duzevik E, Wise SS, Munroe RC, Thompson WD, Wise JP, Sr. XRCC1 protects cells from chromate-induced chromosome damage, but does not affect cytotoxicity. *Mutat Res* 2006; 610(1–2): 31–37.
48. Zhang X, Zhang X, Zhang L, Chen Q, Yang Z, Yu J, *et al.* XRCC1 Arg399Gln was associated with repair capacity for DNA damage induced by occupational chromium exposure. *BMC Res Notes* 2012; 5: 263.
49. Schärer OD. Nucleotide excision repair in eukaryotes. *Cold Spring Harb Perspect Biol* 2013; 5(10): a012609.
50. O'Brien TJ, Brooks BR, Patierno SR. Nucleotide excision repair functional in the removal of chromium-induced DNA damage in mammalian cells. *Mol Cell Biochem* 2005; 279: 86–95.
51. Hu W, Feng Z, Tang MS. Chromium (VI) enhances (±)-anti-7β, 8α-dihydroxy-9α, 10α-epoxy-7,8,9,10-tetrahydrobenzo[a]pyrene-induced cytotoxicity and mutagenicity in mammalian cells through its inhibitory effects on nucleotide excision repair. *Biochemistry* 2004; 43: 14282–14289.
52. Reynolds M, Peterson E, Quievryn G, Zhitkovich A. Human nucleotide excision repair efficiently removes chromium-DNA phosphate adducts and protects cells against chromate toxicity. *J Biol Chem* 2004; 279(29): 30419–30424.
53. Zecevic A, Hagan E, Reynolds M, Poage G, Johnston T, Zhitkovich A. XPA impacts formation but not proteasome-sensitive repair of DNA-protein crosslinks induced by chromate. *Mutagenesis* 2010; 25(4): 381–388.
54. Sarlinova M, Majerova L, Matakova T, Musak L, Slovakova P, Škereňová M, *et al.* Polymorphisms of DNA repair genes and lung cancer in chromium exposure. *Adv Exp Med Biol* 2015; 833: 1–8.
55. Jiricny J. The multifaceted mismatch-repair system. *Nat Rev Mol Cell Biol* 2006; 7: 335–346.
56. Takahashi Y, Kondo K, Hirose T, Nakagawa H, Tsuyuguchi M, Hashimoto M, *et al.* Microsatellite instability and protein expression of the DNA mismatch repair gene, hMLH1, of lung cancer in chromate-exposed workers. *Mol Carcinog* 2005; 42: 15.
57. Halasova, E, Matakova T, Skerenova M, Krutakova M, Slovakova P, Dzian A, *et al.* Polymorphisms of selected DNA repair genes and lung cancer in chromium exposure. *Adv Exp Med Biol* 2016; 911: 17–22.
58. Peterson-Roth E, Reynolds M, Quievryn G, Zhitcovich A. Mismatch repair proteins are activators of toxic responses to chromium-DNA damage. *Mol Cell Biol* 2005; 25(9): 3596–3607.
59. Reynolds M, Zhitkovich A. Cellular vitamin C increases chromate toxicity via a death program requiring mismatch repair but not p53. *Carcinogenesis* 2007; 28(7): 1613–1620.

60. Reynolds MF, Peterson-Roth EC, Bespalov IA, Johnston T, Gurel VM, Menard HL, *et al.* Rapid DNA double-strand breaks resulting from processing of Cr-DNA cross-links by both MutS dimers. *Cancer Res* 2009; 69(3): 1071–1079.
61. Pastwa E, Blasiak J. Non-homologous DNA end joining. *Acta Biochim Pol* 2003; 50: 891–908.
62. Rogakou EP, Pilch DR, Orr AH, Ivanova VS, Bonner WM. DNA double-stranded breaks induce histone H2AX phosphorylation on serine 139. *J Biol Chem* 1998; 273: 5858–5868.
63. Abraham RT. Checkpoint signaling focusing on 53BP1. *Nat Cell Biol* 2002; 4: E277–E279.
64. Bassing CH, Alt FW. H2AX may function as an anchor to hold broken chromosomal DNA ends in close proximity. *Cell Cycle* 2004; 3: 149–153.
65. Lee J, Paull TT. ATM activation by DNA double-strand breaks through the Mre11-Rad50-Nbs1 complex. *Science* 2005; 308: 551–554.
66. You Z, Chahwan C, Bailis J, Hunter T, Russell P. ATM activation and its recruitment to damaged DNA require binding to the C terminus of Nbs1. *Mol Cell Biol* 2005; 25: 5363–5379.
67. West SC. Molecular views of recombination proteins and their control. *Nat Rev Mol Cell Biol* 2003; 4(6): 435–445.
68. Sung P. Catalysis of ATP-dependent homologous DNA pairing and strand exchange by yeast RAD51 protein. *Science* 1994; 265: 1241–1243.
69. Fleck O, Nielsen O. DNA repair. *J Cell Sci* 2004; 117: 515–517.
70. Garner E, Kim Y, Lach FP, Kottemann MC, Smogorzewska A. Human GEN1 and the SLX4-associated nucleases MUS81 and SLX1 are essential for the resolution of replication-induced Holliday junctions. *Cell Rep* 2013; 5: 207–215.
71. Wyatt HDM, Sarbajna S, Matos J, West SC. Coordinated actions of SLX1-SLX4 and MUS81-EME1 for Holliday junction resolution in human cells. *Mol Cell* 2013; 52: 234–247.
72. Wu L, Hickson ID. The Bloom's syndrome helicase suppresses crossing over during homologous recombination. *Nature* 2003; 426: 870–874.
73. Camyre E, Wise SS, Milligan P, Gordon N, Goodale B, Stackpole M, *et al.* Ku80 deficiency does not affect particulate chromate-induced chromosome damage and cytotoxicity in Chinese hamster ovary cells. *Toxicol Sci* 2007; 97(2): 348–354.
74. Stackpole MM, Wise SS, Goodale BC, Duzevik EG, Munroe RC, Thompson WD, *et al.* Homologous recombination protects against

particulate chromate-induced genomic instability in Chinese hamster cells. *Mutat Res* 2007; 625: 145–154.

75. Xie H, Wise SS, Wise JP, Sr. Deficient repair of particulate chromate-induced DNA double strand breaks leads to neoplastic transformation. *Mutat Res* 2008; 649: 230–238.
76. Ha L, Ceryak S, Patierno SR. Generation of S phase-dependent DNA double strand breaks by Cr(VI) exposure: involvement of ATM in Cr(VI) induction of γH2AX. *Carcinogenesis* 2004; 25(11): 2265–2274.
77. Zecevic A, Menard H, Gurel V, Hagan E, DeCaro R, Zhitkovich A. WRN helicase promotes repair of DNA double-strand breaks caused by aberrant mismatch repair of chromium-DNA adducts. *Cell Cycle* 2009; 8(17): 2769–2778.
78. Liu FJ, Barchowsky A, Opresko P. The Werner syndrome protein suppresses telomere instability caused by chromium (VI) induced DNA replication stress. *PLOS ONE* 2010; 5(6): e11152.
79. DeLoughery Z, Luczak MW, Ortega-Atienza S, Zhitkovich A. DNA double-strand breaks by Cr (VI) are targeted to euchromatin and cause ATR-dependent phosphorylation of histone H2AX and its ubiquitination. *Toxicol Sci* 2014; 143(1): 54–63.
80. Qin Q, Xie H, Wise SS, Browning CL, Thompson KN, Holmes AL, *et al.* Homologous recombination repair signaling in chemical carcinogenesis: prolonged particulate hexavalent chromium exposure suppresses the Rad51 response in human lung cells. *Toxicol Sci* 2014; 142: 117–125.
81. Browning CL, Qin Q, Kelly DF, Prakash R, Vanoli F, Jasin M, *et al.* Prolonged particulate hexavalent chromium exposure suppresses homologous recombination repair in human lung cells. *Toxicol Sci* 2016; 153(1): 70–78.
82. Ye J, Shi X. Gene expression profile in response to chromium-induced cell stress in A549 cells. *Mol Cell Biochem* 2001; 222(1–2): 189–197.
83. Izzotti A, Cartiglia C, Balansky R, D'Agostini F, Longobardi M, De Flora S. Selective induction of gene expression in rat lung by hexavalent chromium. *Mol Carcinog* 2002; 35(2): 75–84.
84. Ceryak S, Zingariello C, O'Brien T, Patierno SR. Induction of pro-apoptotic and cell cycle-inhibiting genes in chromium (VI)-treated human lung fibroblasts: lack of effect of ERK. *Mol Cell Biochem* 2004; 255(1–2): 139–149.
85. Xu J, Bubley GJ, Detrick B, Blankenship LJ, Patierno SR. Chromium (VI) treatment of normal human lung cells results in guanine-specific DNA

polymerase arrest, DNA-DNA cross-links and S-phase blockade of cell cycle. *Carcinogenesis* 1996; 17(7): 1511–1517.

86. Wise JP, Wise SS, Little JE. The cytotoxicity and genotoxicity of particulate and soluble hexavalent chromium in human lung cells. *Mutat Res* 2002; 517: 221–229.
87. Wise SS, Holmes AL, Xie H, Thompson WD, Wise JP, Sr. Chronic exposure to particulate chromate induces spindle assembly checkpoint bypass in human lung cells. *Chem Res Toxicol* 2006; 19(11): 1492–1498.
88. Bae D, Camilli TC, Chun G, Lal M, Wright K, O'Brien TJ, *et al.* Bypass of hexavalent chromium-induced growth arrest by a protein tyrosine phosphatase inhibitor: enhanced survival and mutagenesis. *Mutat Res* 2009; 660: 40–46.
89. Chun G, Bae D, Nickens K, O'Brien TJ, Patierno SR, Ceryak S. Polo-like kinase 1 enhances survival and mutagenesis after genotoxic stress in normal cells through cell cycle checkpoint bypass. *Carcinogenesis* 2010; 31(5): 785–793.
90. Holmes AL, Wise SS, Pelsue SC, Abouissa AM, Lingle W, Salisbury J, *et al.* Chronic exposure to zinc chromate induces centrosome amplification and spindle assembly checkpoint bypass in human lung fibroblasts. *Chem Res Toxicol* 23: 386–395.
91. Griffin, CS, Simpson PJ, Wilson CR, Thacker J. Mammalian recombinant-repair genes XRCC2 and XRCC3 promote correct chromosome segregation. *Nat Cell Biol* 2000; 2: 757–761.
92. Smiraldo PG, Gruver MA, Osborn JC, Pittman DL. Extensive chromosomal instability in Rad51d-deficient mouse cells. *Caner Res* 2005; 65: 2089–2096.
93. Date O, Katsura M, Ishida M, Yoshihara T, Kinomura A, Sueda T, *et al.* Haploinsufficiency of RAD51B causes centrosome fragmentation and aneuploidy in human cells. *Cancer Res* 2006; 66: 6018–6024.
94. Faragher AJ, Fry AM. Nek2A kinase stimulates centrosome disjunction and is required for formation of bipolar mitotic spindles. *Mol Cell Biol* 2003; 14: 2876–2889.
95. Liu Q, Hirohashi Y, Du K, Greene MI, Wang Q. Nek2 targets the mitotic checkpoint proteins Mad2 and Cdc20: a mechanism for aneuploidy in cancer. *Exp Mol Pathol* 2010; 88: 225–233.
96. Du J, Cai X, Yao J, Ding X, Wu Q, Pei S, *et al.* The mitotic checkpoint kinase Nek2A regulates kinetochore microtubule attachment stability. *Oncogene* 2008; 27: 4107–4114.

97. Holmes AL, Wise SS, Sandwick SJ, Lingle WL, Negron VC, Thompson WD, *et al.* Chronic exposure to lead chromate causes centrosome abnormalities and aneuploidy in human lung cells. *Cancer Res* 2006; 66: 4041–4048.
98. Holmes AL, Wise JP, Sr. Mechanisms of metal-induced centrosome amplification. *Biochem Soc Trans* 2012; 38(6): 1687–1690.
99. Xie H, Holmes AH, Wise SS, Huang S, Peng C, Wise JP, Sr. Neoplastic transformation of human bronchial cells by lead chromate particles. *Am J Respir Cell Mol Biol* 2007; 37: 544–552.
100. Sharma S, Kelly TK, Jones PA. Epigenetics in cancer. *Carcinogenesis* 2010; 31(1): 27–36.
101. Arita A, Costa M. Epigenetics in metal carcinogenesis: nickel, arsenic, chromium and cadmium. *Metallomics* 2009; 1: 222–228.
102. Chervenova Y, Arita A, Costa M. Carcinogenic metals and the Epigenome: understanding the effect of nickel, arsenic, and chromium. *Metallomics* 2012; 4(7): 619–627.
103. Martinez-Zamudio R, Ha HC. Environmental epigenetics in metal exposure. *Epigenetics* 2011; 6(7): 820–827.
104. Ryu H-W, Lee DH, Won H-R, Kim KH, Seong YJ, Kwon SH. Influence of toxicologically relevant metals on human epigenetic regulation. *Toxicol Res* 2015; 31: 1–9.
105. Dueñas-González A, Lizano M, Candelaria M, Cetina L, Arce C, Cervera E. Epigenetics of cervical cancer. An overview and therapeutic perspectives. *Mol Cancer* 2005; 4(1): 38.
106. Zhang AH, Bin HH, Pan XL, Xi XG. Analysis of p16 gene mutation, deletion and methylation in patients with arseniasis produced by indoor unventilated-stove coal usage in Guizhou, China. *J Toxicol Environ Health A* 2007; 70(11): 970–975.
107. De Carvalho DD, You JS, Jones PA. DNA methylation and cellular reprogramming. *Trends Cell Biol* 2010; 20: 609–617.
108. Herceg Z. Epigenetics and cancer: towards an evaluation of the impact of environmental and dietary factors. *Mutagenesis* 2007; 22(2): 91–103.
109. Ferguson LR, Karunasinghe N, Philpott M. Epigenetic events and protection from colon cancer in New Zealand. *Environ Mol Mutagen* 2004; 44(1): 36–43.
110. Kondo K, Takahashi Y, Hirose Y, Nagao T, Tsuyuguchi M, Hashimoto M, *et al.* The reduced expression and aberrant methylation of p16(INK4a) in chromate workers with lung cancer. *Lung Cancer* 2006; 53: 295–302.

111. Ali AH, Kondo K, Namura T, Senba Y, Takizawa H, Nakagawa Y, *et al.* Aberrant DNA methylation of some tumor suppressor genes in lung cancers from workers with chromate exposure. *Mol Carcinog* 2011; 50: 89–99.
112. Wang TC, Song YS, Wang H, Zhang J, Yu SF, Gu YE, *et al.* Oxidative DNA damage and global DNA hypomethylation are related to folate deficiency in chromate manufacturing workers. *J Haz Mat* 2012; 213–214: 440–446.
113. Klein C, Su L, Bowser D, Leszczynska J. Chromate induced epimutations in mammalian cells. *Environ Health Perspect* 2002; 110: 735–743.
114. Bannister AJ, Kouzarides T. Regulation of chromatin by histone modifications. *Cell Res* 2011; 21: 381–395.
115. Suganuma T, Workman JL. Signals and combinatorial functions of histone modifications. *Annu Rev Biochem* 2011; 80: 473–499.
116. Hon GC, Hawkins RD, Ren B. Predictive chromatin signatures in the mammalian genome. *Hum Mol Genet* 2009; 18: R195–R201.
117. Hawkins RD, Hon GC, Lee LK, Ngo Q, Lister R, Pelizzola M, *et al.* Distinct epigenomic landscapes of pluripotent and lineage-committed human cells. *Cell Stem Cell* 2010; 6: 479–491.
118. Levina A, Harris HH, Lay PA. Binding of chromium(VI) to histones: implications for chromium(VI)-induced genotoxicity. *J Biol Inorg Chem* 2006; 11: 225–234.
119. Schnekenburger M, Talaska G, Puga A. Chromium cross-links histone deacetylase 1-DNA methyltransferase 1 complexes to chromatin, inhibiting histone-remodeling marks critical for transcriptional activation. *Mol Cell Biol* 2007; 27: 7089–7101.
120. Zhou X, Li Q, Arita A, Sun H, Costa M. Effects of nickel, chromate and arsenite on histone 3 lysine methylation. *Toxicol Appl Pharmacol* 2009; 236: 78–84.
121. Sun H, Zhou X, Chen H, Li Q, Costa M. Modulation of histone methylation and MLH1 gene silencing by hexavalent chromium. *Toxicol Appl Pharmacol* 2009; 237: 258–266.
122. Xia B, Yang LQ, Huang HY, Pang L, Hu GH, Liu QC, *et al.* Chromium (VI) causes down regulation of biotinidase in human bronchial epithelial cells by modifications of histone acetylation. *Toxicol Lett* 2011; 205(2): 140–145.
123. Xia B, Ren XH, Zhuang ZX, Yang LQ, Huang HY, Pang L, *et al.* Effect of hexavalent chromium on histone biotinylation in human bronchial epithelial cells. *Toxicol Lett* 2014; 228(3): 241–247.

124. Chen D, Kluz T, Fang L, Zhang X, Sun H, Jin C, *et al.* Hexavalent Chromium (Cr (VI)) down-regulates acetylation of histone H4 at lysine 16 through induction of stressor protein Nupr1. *PLOS ONE* 2016; 11(6): e0157317.
125. Jansson MD, Lund AH. MicroRNA and cancer. *Mol Oncol* 2012; 6(6): 590–610.
126. Bollati V, Marinelli B, Apostoli P, Bonzini M, Nordio F, Hoxha M, *et al.* Exposure to metal-rich particulate matter modifies the expression of candidate microRNAs in peripheral blood leukocytes. *Environ Health Perspect* 2010; 118: 763–768.
127. He J, Qian X, Carpente R, Xu Q, Wang L, Qi Y, *et al.* Repression of miR-143 mediates Cr (VI)–induced tumor angiogenesis via IGF-IR/IRS1/ERK/IL-8 pathway. *Toxicol Sci* 2013; 134(1): 26–38.
128. Chandra S, Khatoon R, Pandey A, Saini S, Vimal D, Singh P, *et al.* Dme-miR-314-3p modulation in Cr (VI) exposed Drosophila affects DNA damage repair by targeting mus309. *J Hazard Mater* 2016; 304: 360–369.

Chapter 8

Monitoring DNA Damage Induced by Chemotherapeutic Agents as a Predictor of Clinical Outcome

Soterios A. Kyrtopoulos

Institute of Biology, Medicinal Chemistry and Biotechnology, National Hellenic Research Foundation, 48 Vassileos Constantinou Avenue, Athens 11635, Greece

Abstract

A number of important cancer chemotherapeutic drugs, best represented by those belonging to the groups of platinum drugs, nitrogen mustards, nitrosoureas and methylating agents, bring about their therapeutic effects by forming with the DNA of cancer cells covalent complexes (DNA adducts) which block cell replication and thus lead to cell death. The amount of such adducts accumulating in the cells of a given patient represents the biologically relevant dose of drug reaching the patient's tissues and, as such, may constitute a marker susceptibility to the drug's therapeutic effects at the individual level. Based on this anticipation, extensive efforts have been made to develop analytical methodologies for measuring drug-DNA adducts and to apply them in studies aimed at the evaluation of the relationship between DNA adduct accumulation and patient therapeutic response.

The great majority of such studies have focused on the categories of drugs mentioned above, and have utilized a range of analytical methodologies, including immunochemical assays, liquid chromatography with different modes of detection including mass spectrometry, and molecular biology-based methods. Although cancer tissues would be the ideal targets in which adducts should be measured, the great majority of studies have utilized blood leukocytes as surrogate tissues in view of their ease of accessibility. Almost all reported studies have been of relatively small size, a limitation whose consequences are compounded by the complexity imposed by the clinical settings in terms of the variability of treatment protocols (use of a single drug or drug cocktails, different dosages and modes of administration, different types of cancer and sub-types of tumor, etc.). As a result, the degree of success of the reported studies, i.e. the observed correlations between adduct accumulation and patient response to treatment, varies between different drugs and different studies. However, despite these complications, overall clear evidence has accumulated that patients who accumulate higher levels of drug-DNA adducts, even in surrogate tissues like blood cells, have a higher probability of a positive response. Overcoming the difficulties mentioned and achieving further progress in this field would require studies on a much larger scale, which can be achieved through multi-center collaborations, and ideally combining measurements of adduct accumulation with downstream biological responses.

1. Introduction

Carcinogen-DNA adducts are considered to be cancer-initiating lesions, and the levels of carcinogen-specific DNA adducts found in human tissues constitute a measure of the amount of carcinogen that reached DNA, the critical cellular target, thus serving as a dosimeter of biologically significant exposure. Elsewhere in this volume, DNA damage is discussed in relation to its role in the induction of cancer via its ability to cause mutations or other forms of heritable damage in cells. In this chapter, we are considering DNA damage as an event that enhances the efficacy of chemotherapy and improves clinical response. Specific types of DNA damage have been recognized as potential cell-killing lesions because they prevent cells from replicating their DNA. On this basis, efforts have been made over the years to develop DNA-damaging drugs that can preferentially kill cancer cells, the essential characteristic of which is increased proliferation, compared to normal cells.

Some of the earliest cancer chemotherapeutic drugs were derived from mustards, bifunctional alkylating agents that can react covalently with both strands of DNA, and produce interstrand crosslinks (ICLs), which prevent strand separation during cell replication and thus lead to cell death. Over the years, different types of DNA-damaging agents have been developed for use in cancer chemotherapy. Similarly, ionizing radiation, also used in cancer chemotherapy, acts by causing DNA damage. Concomitantly, there have been major efforts to understand the mechanisms by which DNA adducts bring about cell death, and the repertoire of cellular responses to DNA damage. Such investigations are of great importance in drug development, not only because they inform the design of drugs of improved efficacy, but also because they help identify means of enhancing drug efficiency. A discussion of cellular responses to DNA damage caused by cancer chemotherapeutic drugs is beyond the purpose of the present chapter (for reviews see [1, 2]). Here, we will focus primarily on the utility of DNA adducts formed by chemotherapeutic drugs as markers of therapeutic efficacy at the level of the individual patient.

Given the extent to which DNA damage is a major determinant of the therapeutic efficacy of DNA alkylating anticancer drugs, the ability of different patients to respond to treatment would be expected to be dependent upon their ability to accumulate damage in tumor DNA. Possible sources of interindividual variation in DNA damage accumulation could include, among others, factors related to drug tissue distribution, metabolic activation (if required), drug detoxification, and DNA repair. Evaluation, prior to therapeutic treatment, of DNA damage accumulation in a particular patient, might permit the identification of subjects who are the most likely to benefit from a particular drug. Furthermore, monitoring of DNA-damage levels induced in patients during the early treatment phase might guide the subsequent adjustment of treatment protocols, with the aim of optimizing the DNA-damage induced, thus allowing the development of personalized chemotherapy.

Ideally, DNA-damage measurements conducted for the abovementioned purposes should be performed on the tumor itself. Although in principle such specimens may be accessible via biopsy, in practice most related studies have used peripheral blood, which is the most readily accessible surrogate tissue. The use of peripheral blood raises obvious questions regarding the relevance of the resulting measurements to disease response, and these can only be answered post-hoc. In relation to the use of peripheral blood for monitoring human exposure to environmental toxicants and evaluation of correlated disease risks, it is worth noting that the responses observed in blood cells reflect to a surprising degree those occurring in distant target tissues [3].

The DNA-damaging activity of drugs used in cancer chemotherapy has the obvious corollary that such drugs can also act as genotoxic carcinogens. Indeed, the appearance of secondary cancers in patients previously treated with genotoxic chemotherapeutic agents is well-documented [4, 5]. The administration of potentially carcinogenic agents to humans as therapeutic drugs is justified by the urgent need to fight their existing disease, as well as the relatively low probability and long induction latency of a second cancer. Nevertheless, drug-induced DNA damage can in principle serve as a biomarker of carcinogenic exposure, which could inform the risk of a secondary cancer. The study of DNA damage induced by chemotherapeutic drugs provides researchers of chemical carcinogenesis with a unique model for investigating the induction and biological consequences of DNA damage by chemicals in humans, *in vivo* and under highly controlled conditions, making it possible to understand better such phenomena as interindividual variation in toxicity, dose–response relationships, DNA repair, and other events relevant to risk assessment.

This chapter will mainly discuss the use of DNA adducts induced by DNA-reactive drugs as biomarkers in cancer chemotherapy. The drugs to be presented here include the platinum-based therapeutics, nitrogen mustards, chloroethyl nitrosoureas, and methylating agents. Other DNA-reactive drugs include the aziridine-related drugs, of which best known is mitomycin C, the phosphoramide thiotepa, and 8-methoxypsoralen, which is used, in combination with UVA irradiation, for the treatment of skin conditions and cutaneous T-cell lymphoma. Because very limited work on the DNA adducts of these drugs in relation to chemotherapy has been reported, they will not be discussed here. In addition to drugs acting via direct reaction with DNA, cancer chemotherapeutic agents include antimetabolites (analogues of normal molecules involved in DNA replication), and topoisomerase inhibitors (which regulate the coiling status of DNA and prevent the repair of DNA strand breaks formed). For these types of drugs, cell damage consists primarily of DNA strand breaks. The accumulation of DNA strand breaks induced by such drugs has attracted increasing attention during recent years as a biomarker of utility in chemotherapy and will also be briefly discussed.

2. DNA Adduct Formation by Cancer Chemotherapeutic Drugs

The main categories of cancer chemotherapeutic drugs that form covalent adducts with DNA, and for which studies on DNA damage in a clinical context have been reported, include (a) platinum-based agents, (b) nitrogen mustards, (c) chloroethyl nitrosoureas, and (d) methylating agents. The first three of the above categories consist of bifunctional agents capable of reacting with two nucleophilic sites in a DNA molecule and giving rise to DNA inter- and intrastrand crosslinks as their main cytotoxic lesions. The fourth category includes monofunctional alkylating agents, which transfer a methyl group to nucleophilic sites in DNA. Interestingly, the chloroethyl nitrosoureas initially form, via chemistry similar to that of the methylating agents, chloroethylated monoadducts with DNA, which are repaired by a mechanism similar to that acting on methylated DNA adducts. For this reason, the alkylating agents and the chloroethyl nitrosoureas will be discussed together, even though the chloroethylated monoadducts rapidly react with a second site in DNA to give rise to crosslinks. These four drug classes will be discussed in detail in the remainder of this review.

We will not give a detailed discussion of other DNA-reactive drugs, including the aziridine-related drugs (e.g., mitomycin C), the phosphoramide thiotepa, and 8-methoxypsoralene, because there is only very limited data available on their DNA adducts formed in relation to chemotherapy and clinical response.

2.1. *DNA adduct formation by cancer chemotherapeutic drugs: platinum-based agents*

The major types of covalent adducts formed between DNA and the most frequently-used platinum-based chemotherapeutic agents [cis-diamminedichloroplatinum (II) (cisplatin), diamminecyclobutanedicarboxylatoplatinum (II) (carboplatin), [(1R,2R)-cyclohexane-1,2-diamine](ethanedioato-O,O′)platinum(II) (oxaliplatin)] are the intrastrand cross

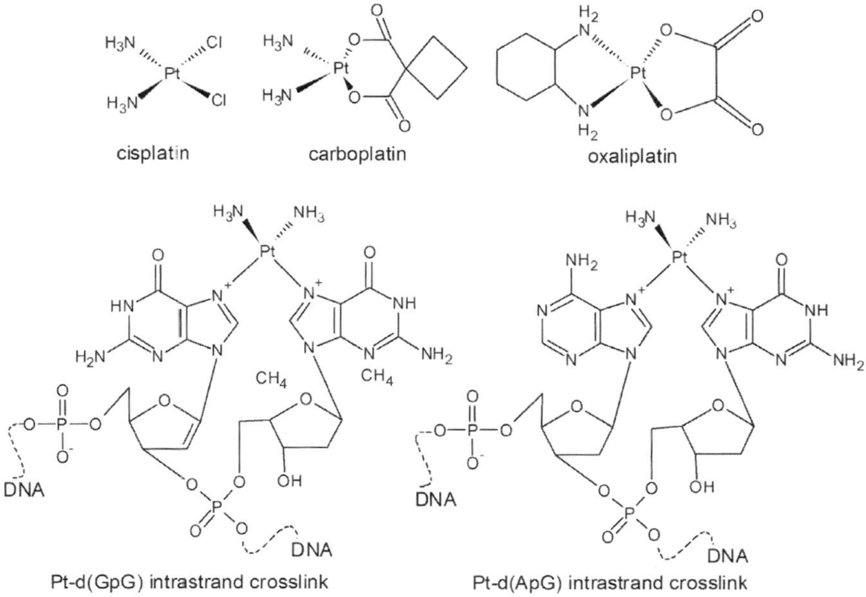

Figure 1. Chemical structures for commonly used platinum-based drugs (cisplatin, carboplatin, oxaliplatin) and their main DNA adducts, the intrastrand crosslinks Pt-d(GpG) and Pt-d(ApG).

links, the cis-diammineplatinum adduct formed on the N7 positions of adjacent deoxyguanines [Pt-d(GpG)], and the cis-diammineplatinum adduct formed on the N7 positions of adjacent deoxyadenines and deoxyguanines [(Pt-d(ApG)] (Figure 1).

Platinum-based anticancer drugs are among the most widely employed DNA-damaging chemotherapeutic drugs and have been the focus of the largest number of studies involving the measurement of drug-DNA adducts in humans. The most important agents in this category, cisplatin and carboplatin, are employed in the treatment of diverse types of solid tumors including testicular, ovarian, bladder, head and neck, cervical, and small-cell lung cancers, while oxaliplatin is mainly used against primary colorectal cancer. The major reaction with DNA involves initially the formation of a covalent monoadduct at position N7 of guanine, with subsequent intrastrand crosslink formation by crosslinking with an adjacent guanine or adenine [Figure 1: Pt-d(GpG) and

Pt-d(ApG)]. Minor amounts of monofunctional adducts and ICLs (not shown) are also formed [6].

2.1.1. *Methods for measuring DNA adducts of platinum-based drugs*

Many different methodologies have been employed for the measurement of DNA adducts formed between platinum-based drugs and DNA [7, 8]. A brief discussion of the relevant methodologies is important because sometimes methods targeted at the same chemical adducts measure only sub-groups of such species. The majority of human studies, going back to the early 1980s, made use of antibody-based methods or atomic absorption spectroscopy (AAS). Other approaches include ^{32}P-postlabeling and cyclic voltammetry, as well as highly sensitive methodologies based on mass spectrometry, including tandem mass spectrometry (MS/MS), inductively coupled plasma mass spectrometry (ICP-MS), and accelerator mass spectrometry (AMS).

A number of antibody-based methods have been used in enzyme-linked immunosorbent assays (ELISAs) and immunohistochemistry (IHC). In 1982, Poirier *et al.* [9] reported the production of a rabbit antiserum against DNA modified approximately 1% by *in vitro* reaction with cisplatin, and the establishment of a competitive ELISA using the same DNA as standard, for the analysis of intact (unhydrolyzed) DNA. The antiserum appeared specific for the bidentate intrastrand Pt-d(GpG) and Pt-d(ApG) adducts, which make up approximately 80% of the total adducts formed by cisplatin, but the early ELISAs measured only 0.2%–33.0% of the total platinum bound to DNA as estimated by AAS [10]. It was later found that the conformation of the DNA modified to 1% with cisplatin (the immunogen) was different than the biological sample DNA modified in the range of 1 adduct/10^8 nucleotides. When ELISA employed a standard curve of DNA modified in the same range as the biological samples, the measured adduct values were much closer to those estimated by AAS [11]. In an alternative immunochemical approach, Fichtinger-Schepman *et al.* [12] raised rabbit antisera by immunizing them with synthetic haptens of specific cisplatin-DNA adducts. ELISAs were established for four types of DNA adducts, including the Pt-d(GpG) and the Pt-d(ApG) intrastrand crosslinks. This approach required DNA samples to be enzymatically hydrolyzed, followed by HPLC to separate the adducts of interest before performing the immunoassays. In a comparative analysis of identical human samples, this method showed 14- to 300-fold higher adduct levels than the early Poirier *et al.* [10] assay, and gave values similar to those obtained by AAS. A third type of antibody-based assay, designed by Terheggen *et al.* [13], employed quantitative immunostaining of buccal cells using a polyclonal antiserum raised against DNA modified *in vitro* with cisplatin. This IHC approach has been employed in a small number of studies [13].

AAS [14–18] and ICP-MS [19, 20] have been used to measure the total amount of platinum bound to DNA. Whereas individual adducts cannot be discriminated by AAS, the majority of platinum-DNA adducts formed are the intrastrand Pt-d(GpG) and Pt-d(ApG) adducts, which are thought to be the main contributors to cytotoxicity and chemotherapeutic efficacy. Newer methods for the measurement of DNA adducts of platinum-based drugs, which combine high-resolution liquid chromatography with detection by sensitive tandem mass spectrometry have yet to be applied in clinical studies [20, 21]. In addition, the ultrasensitive AMS, which requires administration of a radiolabeled compound, measures total platinum bound to DNA in cancer patients (reviewed in ref. [8]), but has also not seen widespread use in the clinic.

Determination of the Pt-d(GpG) and Pt-d(ApG) adducts in DNA by ^{32}P-postlabeling has been reported and validated by comparison with AAS and ELISA [22]. While the high sensitivity of this assay allows for use of small amounts of tissue, including tumor biopsies, the need for radioactive phosphorus makes this assay unsuitable for large-scale clinical studies.

In addition to intrastrand DNA crosslinks, DNA ICLs can be found as a relatively small fraction of the total platinum bound to DNA. Early methods for the measurement of ICL were based on the use of alkaline elution [23], a generic, rather than a chemical-specific method. In addition, for the measurement of platinum drug-induced ICL, a method based on a modification of the single-cell electrophoresis (comet) assay has been developed by Hartley [24]. In this assay, cells are irradiated with X-rays in order to acquire a controlled amount of DNA strand breaks, which is reflected in an increased comet tail moment upon electrophoresis. Cells containing ICL show a smaller increase in tail moment relative to ICL-free controls, and comparison between the two reflects the frequency of ICL.

2.1.2. *Measurement of DNA damage induced by platinum-based chemotherapy and disease response in ovarian and testicular cancer patients*

The methodologies summarized above have been employed for the monitoring of DNA adducts in patients treated with platinum-based drugs, with a focus on the relationship between DNA damage and response to therapy (Table 1). The majority of studies have been performed using tissues from ovarian or testicular cancer patients, while a smaller number of studies have evaluated patients with other kinds of tumors.

The first report on the measurement of platinum drug-related DNA adducts in ovarian and testicular cancer patients came from Reed *et al.* [25]. Using the cisplatin-DNA antiserum and the original ELISA by Poirier *et al.* [9], these workers measured the intrastrand Pt-d(GpG) and Pt-d(ApG) adducts in DNA from white blood cells of 34 ovarian cancer patients and 11 testicular cancer patients. A wide interindividual variation was observed. However, adduct levels increased during multiple, 21–28-day treatment cycles where drug was given for the first 5 days, suggesting that the adducts were not completely repaired during the period between successive treatments. In addition, the study showed a marginally significant (p = .048) positive correlation between the ability of different patients to accumulate measurable levels of adducts and response to therapy. Two subsequent studies of disease response and blood cell platinum-DNA adduct formation from the same group showed similar findings. The first was based on 55 patients treated with therapeutic protocols involving cisplatin as single agent, cisplatin in combination with additional drugs, or carboplatin alone [26]. Peak adduct levels (shown in Figure 2) varied substantially between different patients and a statistically significant correlation between accumulated DNA adduct levels and disease response was observed for the combined groups of patients. This result may have been driven by data from the 25 patients undergoing single-agent therapy. Importantly, the association of DNA adduct levels with response to single-agent therapy, in this group, was stronger than for

Table 1. The major studies evaluating a correlation (positive or negative) between patient response to platinum (Pt) drug-based chemotherapy and the accumulation of Pt-DNA adducts in blood leukocytes after treatment *in vivo* or *in vitro*.

Study	Tumour type (No. of patients)	Pt-based drug	DNA adduct measurement	Conclusion
in vivo				
Reed *et al.* [25]	Ovarian (34), Testicular (11)	cis-Pt	ELISA	Marginally significant correlation
Reed *et al.* [26]	Ovarian (55)	cis-Pt, carbo-Pt	ELISA	Significant correlation
Gupta-Burt *et al.* [15]	Ovarian (27), Breast (25)	cis-Pt, carbo-Pt	ELISA, AAS	Significant correlation for ovarian/ELISA, no correlation for breast/ ELISA; no correlation for AAS
Reed *et al.* [30]	Testicular (17)	cis-Pt	ELISA	Nonsignificant correlation
Motzer *et al.* [18]	Testicular (36)	cis-Pt, carbo-Pt	ELISA, AAS	No correlation
Boffetta *et al.* [32]	Testicular (48)	cis-Pt	GpG adduct by HPLC/ELISA	No correlation
Parker *et al.* [14]	15 types of solid tumour (19)	carbo-Pt	AAS	Significant correlation for pooled tumour group
Reed *et al.* [17]	24 types of solid tumour (49)	cis-Pt, carbo-Pt	AAS	Significant correlation for pooled tumour group
Schellens *et al.* [34]	6 types of solid tumour (35)	cis-Pt	AAS	Significant correlation for pooled tumour group
Blommaert *et al.* [35]	20 types of solid tumour (26)	cis-Pt, carbo-Pt	Quantitative immunostaining of buccal cells	Significant correlation for pooled tumour group, for carbo-Pt
Bonetti *et al.* [19]	Multiple (>7) types of solid tumour (48)	cis-Pt, carbo-Pt	ICP-MS	No correlation
van de Vaart *et al.* [36]	NSLC (27)	cis-Pt	Quantitative immunostaining of buccal cells	Significant correlation
in vitro				
Oshita *et al.* [46]	NSLC (21)	cis-Pt	PCR-stop	Significant correlation
Nel *et al.* [47]	NSLC (11)	cis-Pt	Immunostaining of circulating tumour cells	Significant correlation

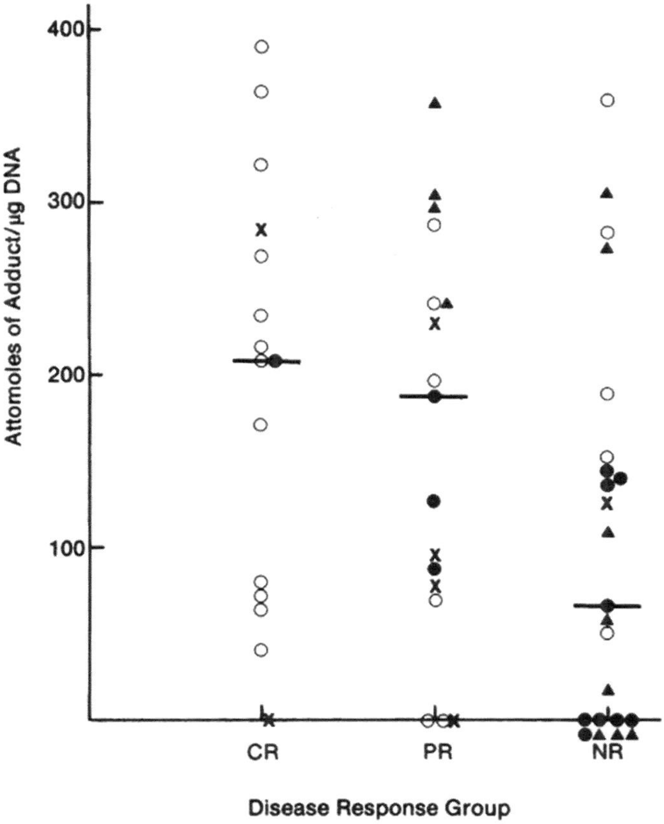

Figure 2. Peak platinum-DNA adduct levels in leukocytes of ovarian cancer patients treated with platinum-based chemotherapy. Patients are groups by disease response. CR, complete response; PR, partial response; NR, no response. Different symbols denote different treatment groups. Each data point represents the peak adduct level for a single patient. The median adduct value for the group is shown by the heavy horizontal bar. Reproduced from Reed *et al.* [26].

several other prognostic factors, underlying the importance of DNA damage in the therapeutic efficacy [27]. Finally, a similar significant correlation of high DNA adduct values with a positive therapeutic response was supported by an additional study involving 27 ovarian cancer patients [15] (Figure 2).

It is worth noting that, in the study by Reed *et al.* [26], levels of Pt-d(GpG) and Pt-d(ApG) adducts in the tumor and other tissues were reported, based on data from a single autopsy. A widespread distribution of DNA adducts among different tissues including the brain, bone marrow, kidney, spleen, and lung as well as in breast tumor tissue, was observed. Furthermore, in an additional study [28] the same group showed Pt-d(GpG) and Pt-d(ApG) intrastrand adducts persistent in the kidney and spleen obtained from a patient who had received the last cisplatin therapy 22 months before autopsy. This study supports the previous findings of accumulation of platinum bound to blood cell DNA in some patients receiving platinum-based therapy [29].

Using the ELISA developed in the Poirier laboratory, two studies showed no support for a positive correlation between DNA adduct accumulation and therapeutic response in testicular cancer patients. One small study of 17 patients showed a marginally nonsignificant excess of adducts in responders [30]. Furthermore, no evidence of a corresponding correlation was observed in a separate study of 36 patients with testicular tumors treated with three different combination chemotherapy protocols involving cisplatin or carboplatin [18]. It is possible that, because almost all testicular cancers are cured with platinum-drug therapies, DNA adduct formation is only one of several factors that impact the chemotherapeutic outcome.

Fichtinger-Schepman *et al.* [31] used antisera raised against four types of specific platinum-nucleoside adducts to study the formation and repair of these adducts in 30 patients with different tumor types, including five patients with ovarian and three with testicular cancer. This study confirmed the wide interindividual variability in DNA adduct levels found in different patients receiving the same cisplatin dose, and also showed that 50% of the adducts initially present were lost in approximately the first 12 hours. However the rate of adduct loss appears to slow over time, resulting in an accumulation of adducts after repeated daily doses. Interestingly, in samples obtained at autopsy detectable adducts were found 20 days after the last drug administration. The small size of the study did not permit an evaluation of the relationship between adduct accumulation and patient therapeutic response. A subsequent larger study, using the same methodology and 48 testicular cancer patients, was also not informative, again likely because of the high rate of therapy efficacy, as only two patients relapsed during follow-up [32].

Two studies using AAS to measure total platinum bound to DNA, in patients mainly with ovarian and testicular tumors, failed to find a correlation with response to chemotherapy [15, 18]. This is in contrast to the results discussed above, as well as some AAS-based studies involving other types of cancers (see below). There is no obvious explanation for this inconsistency, however the procedures used while handling blood samples destined for AAS analysis may have impacted the adduct levels measured [33].

2.1.3. *Measurement of DNA damage induced by platinum-based chemotherapy and disease response in miscellaneous tumor types*

In addition to testicular and ovarian cancers, a relatively large number of tumor types has been examined in studies designed to establish a relationship between platinum drug-based DNA adduct formation and response to chemotherapy. Because for each tumor type there are only small numbers of patients, there is limited statistical power at the level of individual tumor types. However, taken together, the studies support the findings (described above) that individuals who form DNA adducts well respond better than those who accumulate DNA adducts poorly.

There were three studies using AAS to measure adducts. In one, based on 19 patients with 15 types of solid tumors who had received treatment with carboplatin, the mean levels of total platinum bound to DNA were significantly higher in the six responders compared to the 11 nonresponders [14]. Analogous results were reported from two larger studies, one involving 49 patients with 24 cancer types who received chemotherapy consisting of only carboplatin and cisplatin [17], and another involving 35 patients with six types of solid tumors treated with cisplatin alone or in combination with etoposide [34].

Blommaert *et al.* [35] used the quantitative immunostaining of Terheggen *et al.* [13] to examine buccal cells collected from 26 patients who received cisplatin or carboplatin and with 20 different types of solid tumors. Large interindividual differences were observed, but carboplatin-induced nuclear staining was significantly higher in partial responders than in nonresponders, even after correction for the dose level received. In a second study using the same methodology, van de Vaart *et al.* [36] found, in 27 patients with non-small cell lung carcinoma receiving radiotherapy and cisplatin, that higher intensity of DNA adduct staining in buccal cells was significantly associated with longer survival, even after controlling for other prognostic factors such as age and tumor stage.

In three studies, there was no significant association between platinum-drug based DNA adduct levels and disease response. A comparison of adduct levels measured by ELISA in intact DNA in a group of 25 breast cancer patients receiving combination chemotherapy showed no difference in adduct levels in complete or partial responders [15]. Interestingly, the platinum-based drugs are not frequently effective against breast cancer. Bonetti *et al.* [19] employed a sensitive ICP-MS methodology to measure total platinum bound to DNA in patients with different types of solid tumors treated with cisplatin or carboplatin, in most cases in combination with other drugs. In 48 patients, examination of platinum bound to DNA 24 hours following completion of the first drug administration showed no excess of adducts in complete or partial responders relative to nonresponders. Antisera specific for different platinum-DNA adducts were used to measure individual DNA adducts in a pilot-scale study involving seven patients with unspecified types of cancer who received their first round of treatment with cisplatin [37] and had blood samples drawn 3 hours after the start of infusion. Unfortunately, the small number of patients did not permit statistical evaluation of DNA damage and therapeutic response in this study.

2.1.4. *How can platinum drug dosing be optimized?*

In view of evidence suggesting the existence of correlations between positive disease response and high levels of platinum drug bound to patient blood cell DNA, a number of investigators have attempted to establish optimal drug-dosing schedules and predict chemotherapeutic success in individual patients. Thus Schellens *et al.* [38], working with non-small cell lung cancer patients treated with cisplatin, measured total platinum bound to DNA in blood leukocytes after the first round of treatment, and adapted drug dosage in subsequent rounds to achieve adduct levels previously observed in positive responders [34]. Dose escalations of 10%–50% were applied, and the overall response rate was high, however, in the absence of a randomized trial the true efficacy could not be evaluated. The same investigators studied head and neck squamous cell carcinoma patients treated with cisplatin (by intravenous vs. intra-arterial infusion), measuring platinum-DNA adduct levels by ^{32}P-postlabeling in normal blood leukocytes, in buccal cells, and in tumor biopsies [39]. The tumor tissue accumulated substantially higher platinum-DNA adduct levels than the normal tissue, regardless of treatment route, and patients with the highest DNA adduct levels in tumors had a significantly better disease-free survival. Interestingly, blood cell DNA adduct levels did not correlate with adducts in tumors, ether in this study, or in a subsequent study [40].

Ledermann *et al.* [41] used the modified comet assay to measure ICL present in blood cells of ovarian cancer patients treated with a combination of carboplatin and gemcitabine. They found that gemcitabine inhibits the repair of carboplatin-induced ICL, suggesting that this inhibition may be the reason for the enhanced therapeutic efficacy, compared to cisplatin alone.

As previously mentioned, AMS also has the potential to be used for measurement of platinum-related adducts in a clinical setting (reviewed in ref. [8]). Owing to its extremely high sensitivity, this analytical method may permit evaluation of both accumulation and repair of DNA adducts in patients treated with the drugs of interest. If applied in the context of pretherapy testing, the approach may be of use for the optimization of therapeutic dosing on an individual level, as well as for the identification of patients who would be anticipated to be resistant to therapy.

2.1.5. *In vitro DNA adducts of platinum-based drugs, and their repair, as predictive markers of therapeutic response*

The clinical utility of DNA adduct determination in patients after being treated with DNA-damaging drugs, as illustrated by studies described above, lies in their potential to illuminate causes of an individual patient's lack of response. A good example is DNA repair, which was first shown as an inverse factor impacting disease response in women with ovarian cancer. Negative correlations were found between levels of the excision repair cross-complementation group 1 (ERCC1) DNA repair enzyme and both platinum-induced DNA damage and disease response [42]. If DNA adduct formation were to be evaluated prior to treatment, it might be possible to identify patients likely to respond and those likely to be resistant to chemotherapy. A particularly interesting finding of the study of Boffetta *et al.* [32] discussed previously, which extends results from the same group [43], is that the Pt-d(GpG) adducts measured in testicular cancer patients immediately after infusion of cisplatin correlate closely with those found after *in vitro* cisplatin treatment of blood samples from the same patients collected prior to therapy. A similar association was found in cisplatin-treated children by Veal *et al.* [44]; however, only limited supportive data was reported concerning adducts measured following *in vitro* treatment of tumor fragments by Welters *et al.* [45].

The above results imply that measurement of DNA adducts formed after *in vitro* incubation of drug with patient blood collected prior to treatment may reflect DNA adduct accumulation following *in vivo* drug administration, and thereby predict therapeutic response. In further support of this relationship, a positive correlation between *in vitro*-induced DNA damage and response to chemotherapy was reported in a pilot-scale study of non-small cell lung cancer patients by Oshita *et al.* [46], where the extent of DNA damage was estimated via the inhibition of DNA fragment replication by polymerase chain reaction (the "PCR-stop" method). Replication is arrested by Pt-d(GpG) and Pt-d(ApG) adducts, and the method detects most DNA adducts formed by cisplatin. In an analogous approach, Nel *et al.* [47] used immunostaining to measure the formation and repair of Pt-GpG adducts after *in vitro* cisplatin treatment of circulating tumor cells isolated from peripheral blood. In a pilot study based on 11 patients with non-small cell lung cancer, there was a statistically significant correlation between higher adduct accumulation, as reflected by area-under-the-curve for adducts over time, and therapeutic response. Furthermore, the rate of Pt-GpG adduct removal after *in vitro* treatment, and that observed in circulating cancer cells following *in vivo* treatment with the drug, was similar, emphasizing the importance of DNA repair as a determinant of adduct accumulation and tumor response.

The importance of DNA repair was further highlighted by the results of a large study on 591 non-small cell lung carcinoma patients treated with platinum drugs [48]. In this study, the DNA repair capacity of different individuals was measured in peripheral lymphocytes using an *in vitro* host-cell reactivation assay. An inverse association was observed between the DNA repair capacity and patient survival, especially for patients with early-stage disease, supporting the suggestion that DNA repair, even if measured in blood lymphocytes, is an independent predictor of response to therapy.

2.1.6. *Conclusions regarding the relationship between DNA damage and therapeutic response to platinum-based drugs*

- Most patients given platinum drug-based chemotherapy form DNA adducts, the majority of which are intrastrand Pt-d(GpG) and Pt-d(ApG) adducts, in both blood cell DNA and in tumor DNA. There is, however, a high rate of interindividual variability in patients given the same drug dose.
- In some cases, where patients have been examined at autopsy, platinum drug-based DNA damage has been found in multiple organs weeks to months after the last chemotherapy.
- In blood cell DNA of some patients, platinum-DNA damage has been shown to accumulate with total drug dose over a period of multiple 21- or 28-day cycles of therapy, suggesting that there is a dose response for this DNA damage and that DNA repair is not complete over a period of several months.
- There is a great deal of clinical evidence indicating that for some tumors (particularly ovarian cancer) patients with the highest levels of platinum-DNA damage (typically measured in blood cell DNA) are the most efficient clinical responders. Conversely, individuals with efficient DNA repair capacity respond more poorly in the clinic.
- The broad distribution of interindividual platinum-DNA damage values across clinical response groups makes it virtually impossible to identify individual patients likely to respond.
- Furthermore, an obvious weakness of many of the studies reported to date relates to their relatively small size, which limits statistical power.

2.2. DNA adduct formation by cancer chemotherapeutic drugs: nitrogen mustards

Nitrogen mustards (Figure 3) are perhaps the oldest type of alkylating drugs used in cancer chemotherapy. In this category, melphalan (multiple myeloma, ovarian and breast cancer cancer), cyclophosphamide (lymphoma, multiple myeloma, breast cancer, and systemic rheumatic diseases), mechlorethamine (Hodgkin's disease, non-Hodgkin lymphoma, chronic myeloid leukemia, bronchogenic carcinoma), chlorambucil and ifosfamide (pediatric tumors, especially sarcomas) are currently employed widely in the clinic.

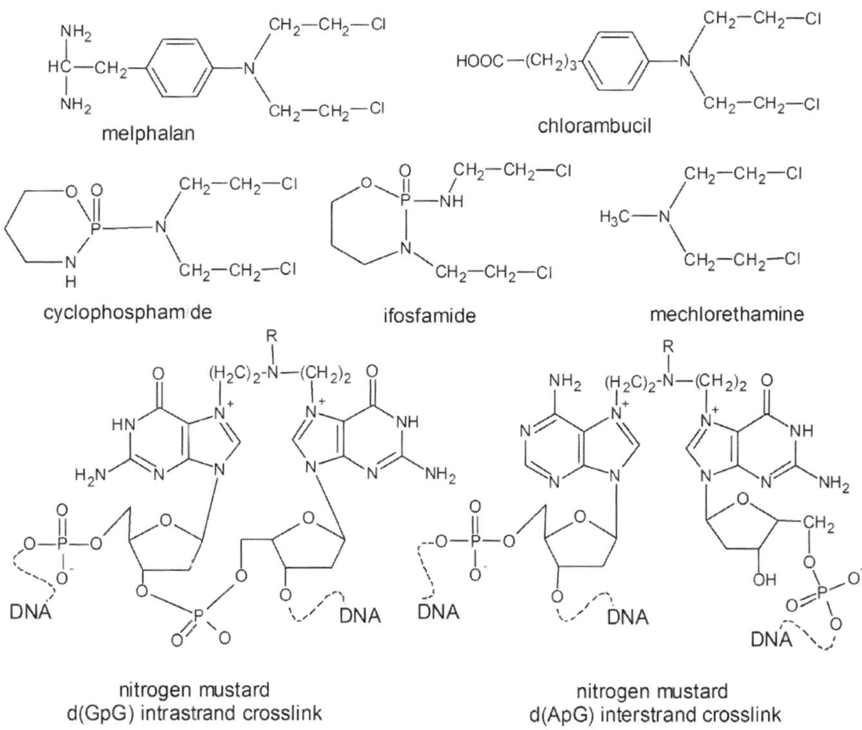

Figure 3. Structural formulas of commonly used nitrogen mustard drugs, (melphalan, chlorambucil, cyclophosphamide, ifosfamide, and mechlorethamine), and their main DNA adducts, the nitrogen mustard d(GpG) intrastrand crosslink and the nitrogen mustard d(ApG) interstrand crosslink.

Nitrogen mustards react, either directly or following metabolic conversion to an electrophilic species, with nitrogen atoms of purines in DNA. Monoadducts are formed primarily at the N7- position of guanine, and to a smaller degree at the N3- position of adenine. These monoadducts can react with a second nitrogen molecule, primarily in the opposite strand of DNA, to form ICLs (Figure 3) between two guanines or an adenine and a guanine, as well as other minor products [49, 50].

2.2.1. *Methodologies for measuring nitrogen mustard-induced DNA damage*

Approaches to determining DNA damage induced by nitrogen mustards include antibody-based methods, the single-cell electrophoresis (comet) assay, Southern blot, quantitative PCR, and mass spectrometry. Measurement of N,N-bis[2-(N7-guaninyl) ethyl] amine (N7-G-N7-G) crosslinks in hydrolyzed DNA by LC-MS/MS has also been reported [50]. Some of these methodologies have not as yet been employed in the context of clinical studies to evaluate susceptibility to therapeutic response, and will therefore not be discussed further.

Tilby *et al.* [51] developed a sensitive ELISA based on the use of a monoclonal antibody raised against DNA modified *in vitro* with melphalan. The assay specificity was validated, and appeared to be directed towards adducts at the N7 position of guanine, showing highest specificity for crosslinked nucleotides (N7-G-N7-G and N3-A-N7-G). A similar antiserum was employed for quantitative immunostaining of single cells [52, 53].

Measurement of ICL induced by mustards has been conducted using a modification of the single-cell electrophoresis (comet) assay, as previously described for platinum-based drugs. Additionally, Souliotis *et al.* [54] adapted an already existing method to study the formation and repair of ICL, as well as the precursor N-alkylpurine monoadducts, at specific genomic loci. In this method, DNA was initially digested with selected restriction enzymes, depending on the gene locus of interest, followed by conversion for the measurement of total N-alkylpurine adducts as single-strand breaks via heating and alkaline hydrolysis. Subsequently, size fractionation by electrophoresis, Southern blotting, and hybridization with a radioactively labeled probe related to the genomic locus of interest, allowed quantitation of the target bands by autoradiography.

2.2.2. *Clinical studies: multiple myeloma*

The formation of ICL-related DNA adducts (mainly N7-G-N7-G and N3-A-N7-G) in peripheral blood leukocytes was examined, using ELISA, in 13 patients after treatment with different doses of melphalan [55]. ICL levels, which showed a linear relationship with administered dose, were maximal by 1 hour posttreatment and remained constant during the following 24 hours. A rather low interindividual variation in the levels of adducts formed 1 hour posttreatment was observed. In contrast to the 10-fold variability reported for cisplatin-DNA adducts, for example, there was less than 2-fold variability after adjusting for the administered dose.

In a small study, Souliotis *et al.* [54] examined seven patients with multiple myeloma treated with high-dose melphalan, followed by autologous stem cell transplantation. In this study, lymphocytes were exposed to melphalan both *in vitro* and *in vivo* from patients previously treated with the same drug. Data showed that the monoadducts reached maximal levels within 2 hours, whereas the highest ICL levels were found after 8 hours of treatment. *In vivo*, about 30%–35% of the monoadducts as well as ICL were lost in the first 24 hours. Finally, of seven patients examined, three who showed the lowest levels of DNA damage did not respond to the melphalan treatment, suggesting a correlation between DNA damage and therapeutic response.

The relationship between DNA damage and therapeutic response was further evaluated in a subsequent, larger study involving 26 multiple myeloma patients treated with high-dose melphalan and autologous blood stem-cell transplantation [56]. A large interindividual variation in the peak levels of both monoadducts and ICL was observed, which paralleled closely the variation in repair capacity as reflected in the trajectory of adduct loss during 8 and 24 hours postexposure. The implication is that DNA repair constitutes the primary determinant of DNA adduct accumulation. A clear correlation between DNA damage and repair on one hand, and therapeutic response on the other, was also seen with the monoadducts (precursors of ICL), which showed the most significant differences between therapy-resistant patients and responders. Furthermore, longer progression-free survival correlated with increased peak monoadduct levels.

The possible role of ICL repair in relation to clinical responsiveness was raised by the findings of Spanswick *et al.* [57]. Plasma cells were examined from 12 patients with multiple myeloma, half of whom had no previous melphalan, and half of whom had relapsed following melphalan therapy. It was found that, the untreated group had no ICL repair by 40 hours posttreatment, but that cells from relapsed patients showed significant repair, suggesting that induction of ICL repair may have contributed to the emergence of drug resistance.

2.2.3. *Clinical studies: sarcoma*

The modified comet assay has been employed for the measurement of ICL in a number of pilot-scale studies of nitrogen mustards. Hartley *et al.* [24] demonstrated the induction of ICL in lymphocytes from patients with Ewing sarcoma within 3 hours of receiving ifosfamide, while Johnstone *et al.* [58] measured ICL in blood leukocytes and cells from tumor aspirates of 8 ifosfamide-treated breast cancer patients. In the latter study, adducts in leukocytes peaked 48 hours after ifosfamide, whereas those in tumor cells were not elevated until 3 weeks after treatment. Adduct levels in either tissue did not significantly correlate with clinical response or toxicity, although lower levels were observed in the tumor cells of those patients who subsequently relapsed relative to those that remained disease free.

Willits *et al.* [59] examined ICL formation in blood lymphocytes of 19 patients aged 2–19 years, mostly with rhabdomyosarcoma or Ewings sarcoma, up to 96 hours after infusions of ifosfamide. In all patients, adduct levels reached a peak at 72 hours from the start of infusion, and returned to pretreatment values by 3 weeks later. Patients receiving fractionated dosing had more DNA damage than patients receiving ifosfamide by continuous infusion.

2.2.4. *In vitro DNA adducts of nitrogen mustards, and their repair, as predictive markers of therapeutic response*

The utility of *in vitro*-induced DNA damage, measured in blood samples obtained from patients prior to therapy, for predicting subsequent patient responses was evaluated by Dimopoulos *et al.* [60]. In a study of 32 multiple myeloma patients, monoadduct and ICL levels at the p53 gene locus, measured at various times after *in vitro* treatment of blood samples, were strongly correlated with those found following patient treatment. Furthermore, all measures reflecting higher DNA damage and slower repair capacity were predictive of a favorable response to subsequent therapy. Highest predictive ability was obtained using the area-under-the-curve of total adducts, and selection of an optimal cut-off value led to a classification of patents with a positive predictive value for clinical response of 100%, since all patients with an equal or higher value showed a positive response. Moreover, patients with an area-under-the-curve equal to or higher than this cut-off value had significantly longer times to progression than had patients with a lower value. This result was confirmed and extended in a subsequent larger study involving a total of 85 patients [61].

Based on the above observations, a sensitive and relatively simple assay, suitable for clinical application, was developed for the measurement of locus-specific DNA adducts induced by melphalan, cisplatin and carboplatin [62]. This assay, which is a modification of previously reported methods, uses multiplex long quantitative PCR (QPCR) to amplify a 7 kb fragment within the p53 gene locus (target locus) and a small (0.5 kb) fragment within the IFB-b1 locus serving as internal standard. The selection of these particular genomic loci for examination of DNA damage did not relate to the biological function of the corresponding genes in the disease, but only provided targets for amplification with the assumption that correlating levels of adducts would occur at most other loci. In this assay, the extent of PCR amplification is inversely related to the levels of DNA adducts present in the amplified locus, as DNA adducts block polymerase-mediated DNA synthesis. Separate studies showed that the melphalan DNA adducts measured were ICL, which only partially blocked the polymerase, while the platinum drug lesions blocked the polymerase completely. Application of the QPCR assay to *in vitro* melphalan-treated peripheral blood mononuclear cells from multiple myeloma patients confirmed the correspondence between formation of higher DNA adduct levels and more favorable therapeutic responses, with a positive predictive value of 92.9% for response to melphalan therapy.

2.2.5. *Conclusions regarding the relationship between DNA damage and therapeutic response to nitrogen mustards*

- Although there are fewer studies investigating the relationship between nitrogen mustard-induced DNA damage and therapeutic response, than for the platinum-based drugs, the available evidence shows that accumulation of ICL following treatment with melphalan is an important determinant of individual therapeutic response in multiple myeloma.
- Measurement of ICL accumulation following *in vitro* treatment of cultured leukocytes with melphalan suggest that this approach may identify patients who are more likely to respond to therapy.
- As in analogous studies with platinum-based drugs, even in cases where a statistically significant difference in adduct levels between patient groups with different response to therapy was observed, extensive overlap in the corresponding adduct levels reduces the specificity of adducts as predictors of therapeutic response at the individual patient level.

2.3. ***DNA adduct formation by cancer chemotherapeutic drugs: chloroethyl nitrosoureas, methylating agents, and related drugs***

Methylating agents employed in cancer chemotherapy include drugs which, usually after metabolism, release methyldiazonium ion intermediates that react with nitrogen and oxygen in DNA, giving rise to corresponding methylated DNA adducts (Figure 4). The best known representatives of this group of drugs are the methylhydrazine derivative procarbazine, used against Hodgkin's lymphoma and certain brain tumors, the triazine dacarbazine (DTIC), used against Hodgkin's lymphoma and malignant melanoma, and temozolomide, used against some brain tumors. The group also includes the nitrosourea streptozotocin, used for metastatic cancer of the pancreatic islet cells, and N-methylnitrosourea, used in the past in cancer chemotherapy. The nitrosoureas and methylating agents both methylate DNA by a similar mechanism, however the nitrosoureas do so by direct hydrolysis and without metabolism. Finally, a number of nitrosoureas also employed in cancer chemotherapy, the most widely used of which are

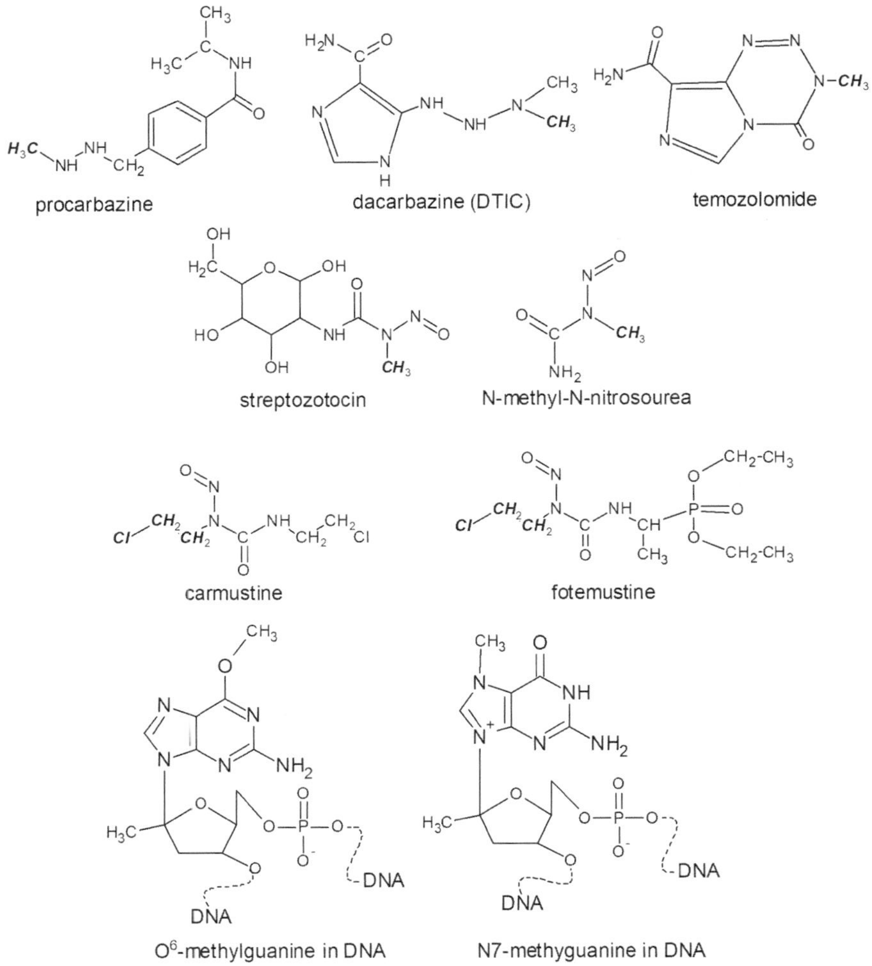

Figure 4. Structural formulas of commonly used methylating drugs (procarbazine, dacarbazine, temozolomide, streptozotocin, and N-methyl-N-nitrososurea), and related chloroethylating nitrosoureas (carmustine and fotemustine), and their main DNA adducts, O^6-me-dG and N7-me-dG.

carmustine (BCNU), lomustine (CCNU), and fotemustine, release the chloroethyldiazonium ion to bind to oxygen or nitrogen in DNA, and the chloroethyl group subsequently reacts further with a second nucleophilic atom in DNA to give rise to DNA ICLs.

Following reaction with DNA, the methylating agents mentioned above generate primarily N7-methylguanine (N7-me-dG; 70% of adducts formed), O^6-methylguanine (O^6-me-dG; 7% of adducts), and other products in smaller proportions (N3-methyladenine, O^4-methylthymine, methylphosphotriesters, and others) [63]. The quantitatively major DNA lesion N7-me-dG is not directly premutagenic, but it can undergo enzymatic or spontaneous depurination to form mutagenic apurinic sites. On the other hand, the most abundant of the oxygen-centered base methylation adducts, O^6-me-dG, is directly miscoding and plays a major role in mutagenesis and carcinogenesis. Importantly, O^6-me-dG is also a powerful cytotoxic lesion, causing cell death via a mechanism involving DNA replication followed by futile mismatch repair cycles, and for this reason it forms the main focus of attention in relation to methylating cytostatic drugs.

O^6-me-dG is repaired by a unique one-step reversal mechanism involving the direct transfer of the methyl group from DNA to a cysteine residue of the protein O^6-methylguanine-DNA-methyltransferase (MGMT or ATase), which becomes stoichiometrically and irreversibly inactivated. MGMT also repairs, by an analogous mechanism, the O^6-chloroethylguanine intermediates formed by chloroethylating nitrosoureas, and thus prevents the formation of cytotoxic ICLs. Because MGMT plays a critical role in the protection of cells against methylating and chloroethylating DNA adduct formation, its modulation (depletion) has attracted much attention as a potential means of enhancing the chemotherapeutic efficacy of these types of drugs.

Importantly, the methylated DNA adducts that are formed by the methylating cytostatic drugs are also generated by some well-known carcinogens, to which there is widespread human environmental exposure. These include N-nitrosodimethylamine (NDMA), and the tobacco-specific nitrosamine 4-(methylnitrosamino)-1-(3-pyridyl)-1-butanone (NNK), both of which are considered to contribute to the etiology of human cancer. In this context, studies with methylating drugs have provided a unique opportunity to observe the behavior of these DNA adducts in humans under highly controlled conditions.

2.3.1. *Methods for measuring alkylated DNA adducts*

Studies on methylated DNA adducts in relation to chemotherapy have made use of immunochemical analytical methods based on antisera raised against methylated nucleoside haptens (mainly O^6- and N7-me-dG) or liquid chromatography of hydrolyzed DNA with sensitive electrochemical detection (mainly for N7-me-dG) [64, 65]. A sensitive "competitive repair assay" for O^6-me-dG in DNA exploits the stoichiometric mode of action of the MGMT protein and is based on the competition between the DNA under analysis and a radioactively labeled, O^6-me-dG-containing oligonucleotide for repair by a limiting amount of MGMT [66].

2.3.2. *Clinical studies: Hodgkin's and non-Hodgkin lymphoma*

Procarbazine and dacarbazine are employed for the treatment of Hodgkin's and non-Hodgkin lymphoma, in the context of multidrug protocols. The first DNA adduct measurements in humans following treatment with methylating chemotherapeutic drugs were reported by Souliotis *et al.* [67] and involved 7 patients (subsequently expanded to include 14 patients [68]) with Hodgkin's or non-Hodgkin lymphoma, treated with a protocol that included daily doses of procarbazine taken during repeated 10-day cycles in combination with additional drugs. O^6-me-dG was measured in blood leukocyte DNA by the competitive repair assay. Daily intake of procarbazine was associated with a continuous adduct accumulation which, averaged over all the subjects examined, was linearly related to cumulative drug dose (Figure 5A). No

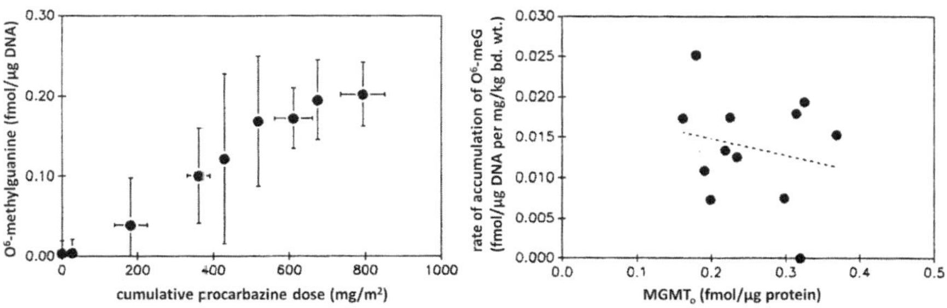

(A) (B)

Figure 5. (A) Accumulation of O^6-me-dG in blood leukocyte DNA of patients with Hodgkin's and non-Hodgkin lymphoma as a function of the cumulative dose of procarbazine. Each point represents the mean of 2–14 patient samples for up to three consecutive treatment cycles. (B) The relationship between the rate of accumulation of O^6-me-dG in different patients and the pretreatment levels of MGMT in blood lymphocytes. Reproduced from Kyrtopoulos *et al.* [68].

consistent depletion of MGMT in blood lymphocytes was observed, in line with the much lower stoichiometric amount of O^6-me-dG formed per cell (<0.5 fmol/μg DNA), relative to the amount of MGMT present (5–15 fmol/μg DNA). On the other hand, examination of 0^6-me-dG accumulation in different individuals provided evidence that interindividual variability of pretreatment MGMT levels influenced adduct accumulation (Figure 5B). The small size of the study did not allow a proper assessment of the relationship between O^6-me-dG accumulation and therapeutic response, however it was noted that, of two nonresponders, one showed the lowest adduct accumulation throughout the treatment while the second accumulated high levels of adducts up to day 7 and subsequently showed an unexplained rapid loss of adducts. Broadly similar results were obtained in a follow-up study of 23 Hodgkin's disease patients receiving similar treatment with procarbazine in which, in addition to O^6-me-dG, N7-me-dG, the major DNA adduct formed by methylating agents, was measured [69]. Similarly, here also there was no correlation between the DNA adduct levels and therapeutic response.

A similar series of studies was conducted on 12 Hodgkin's disease patients treated with a multidrug combination protocol involving a single i.v. administration of dacarbazine (in the range 100–200 mg/m^2) during 15-day cycles [70]. Rapid formation of O^6-me-dG was observed within 1 hour of treatment, followed by loss of approximately 50% of the adducts during the following 24 hours. The adduct levels were similar to those observed in the procarbazine study and, as in that study, no depletion of MGMT was observed. There was no significant excess of mean DNA adduct levels in patients showing full, partial, or no remission of clinical symptoms at the end of the same cycle or at the end of all treatment cycles (Figure 6).

2.3.3. *Clinical studies: malignant melanoma*

In contrast to the relatively low doses of dacarbazine employed in patients with lymphoma, high doses of this drug, alone or in combination with other drugs, are used in the treatment of malignant melanoma. In a study of 34 melanoma patients treated with multiple high doses of dacarbazine (1,000 mg/m^2 every 3 weeks), substantially higher O^6-me-dG

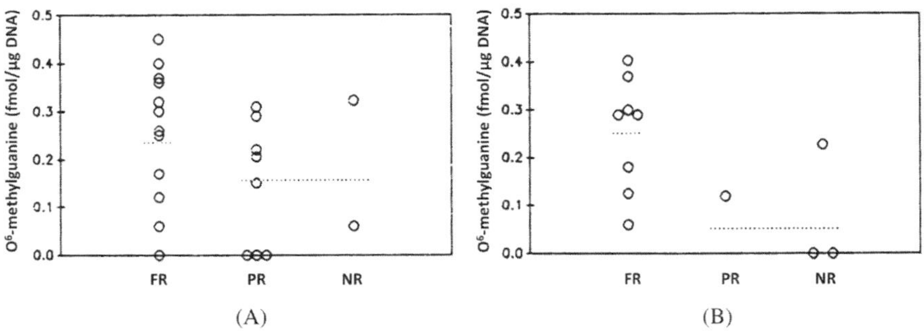

Figure 6. Scattergram of O^6-me-dG levels in non-Hodgkin's lymphoma patients 2 hours after treatment with dacarbazine, plotted as a function of clinical response at the end of a particular cycle (Figure 6A) or of the full course of treatment (Figure 6B). In 6A, the O^6-me-dG levels shown are the averages of those measured after each cycle for which data were available. The dotted lines indicate the means of the full responder (FR) and partial responder/nonresponder (PR/NR) groups. Reproduced from Kyrtopoulos *et al.* [68].

levels and extensive MGMT depletion (approximately 55%) were observed 24 hours after treatment, compared to the lymphoma patients [71]. MGMT levels were still approximately 25% below their initial levels just prior to administration of the second dose of dacarbazine, probably explaining the observation of higher levels of O^6-meG (but not N7-me-dG) after the second dose of dacarbazine. There was a negative correlation between pretreatment lymphocyte MGMT levels and O^6-meG levels at 24 hours after treatment, while co-treatment with hydroxyurea tended to be associated with lower levels of N7-meG. No correlation between the peak or 24-hour levels of O^6-me-dG or N7-me-dG and tumor responses was noted in this study. The impact of high-dose dacarbazine on MGMT levels, and the associated increased formation of O^6-me-dG on the second and subsequent cycles, were confirmed in an additional study on malignant melanoma patients, where it was also shown that co-treatment with hydroxyurea did not affect O^6-me-dG accumulation or repair [72].

O^6-me-dG and N7-me-dG were also measured in a study by van Delft *et al.* [73] involving 5 malignant melanoma patients treated with dacarbazine alone (800 mg/m^2) and 3 patients treated with 3 daily doses of 225 mg dacarbazine/m^2 in combination with other drugs. N7-me-dG appeared to be lost from DNA with a half-life of 40–96 hours while O^6-me-dG had a half-life of 25–27 hours as also previously reported. The levels of the former adduct were 20-fold higher than those of O^6-me-dG, a ratio that would be anticipated from the rates of loss of the two adducts, and not very different from that observed in studies in rodents.

The depletion of MGMT following high doses of dacarbazine (400–800 mg/m^2) has been examined in detail in a series of studies by Lee *et al.* [74–76], in an effort to utilize such treatment as a means of inhibiting MGMT-mediated adduct repair. Wide interindividual variation in the depletion and regeneration of MGMT activity was noted, with higher dacarbazine doses leading to higher MGMT depletion, as expected. Also considerable interindividual variation in the peak O^6-me-dG levels was seen, and MGMT pretreatment levels correlated with the overall adduct accumulation. However no clinical exploitation of these positive findings appears to have taken place and, as discussed in the following paragraph, efforts to modulate MGMT activity in the context of chemotherapy focused mainly on the use of small molecule inhibitors.

The initial observation that MGMT recognizes the free base O^6-benzylguanine as a substrate and transfers the benzyl group to its active site via a similar suicide mechanism to that involved in the repair of O^6-me-dG in DNA, led to extensive efforts for the development of small molecule O^6-alkylguanine analogues for use as irreversible MGMT inhibitors in chemotherapy in combination with methylating or chloroethylating drugs. Monitoring of O^6-me-dG in such studies has served to validate the potential of MGMT depletion to lead to higher adduct accumulation. For example, Watson *et al.* [77] have shown that extended dosing of malignant melanoma patients with the MGMT inhibitor lomeguatrib [O^6-(4-bromothenyl)guanine] results in complete inactivation of MGMT in peripheral blood mononuclear cells, and in persistent high levels of O^6-me-dG. The use of lomeguatrib, in combination with the methylating agents temozolomide and dacarbazine, has been tested in the context of a number of clinical trials, which have indicated increased hematological toxicity but unfortunately no improved therapeutic efficacy [78–80]. An overview of efforts to enhance the effectiveness of chemotherapy through the use of synthetic inhibitors of MGMT is beyond the scope of the current discussion and the reader is referred to relevant reviews [81, 82].

2.3.4. *Conclusions regarding the relationship between DNA damage and therapeutic response to chloroethyl nitrosoureas, methylating agents, and related drugs*

- There is no evidence for a positive correlation between the accumulation, in blood leukocytes, of methylated DNA adducts (O^6-me-dG) and therapeutic response. This applies both to low-dose procarbazine or dacarbazine in patients with Hodgkin's or non-Hodgkin lymphoma, and high-dose dacarbazine in patients with malignant melanoma. This conclusion is supported by an apparent absence of improved therapeutic efficacy in cases where the repair of O^6-me-dG was been blocked.
- Possibly the above negative result is related to the relatively small number and size of most of these studies, the use of therapeutic protocols involving multiple drugs, and/or variability in the formation, stability, and repair of O^6-me-dG in the tumor tissues.
- Despite the disappointing clinical findings, these studies have revealed a great deal regarding methylated DNA adduct processing in humans. For example: the methylated adducts are formed in humans in similar proportions as in experimental animals; O^6-me-dG is relatively short-lived as compared to N7-me-dG; there is wide interindividual variation in adduct baseline levels; and the depletion and the regeneration of MGMT impact O^6-me-dG accumulation. All of these findings are of great interest from the point of view of environmental carcinogenesis [68, 83].

3. Use of DNA Damage Response as a Biomarker of Chemotherapeutic Efficacy

With all types of DNA damage-inducing chemotherapeutic drugs, we have seen that, even though in some cases higher accumulation of DNA damage correlates with a positive patient response to chemotherapy, the distribution of adduct levels in different subjects tends to show extensive overlap between response groups. This broad interindividual variation clearly limits the utility of DNA damage measurements for identification of patients likely to benefit from the treatment of interest. A possible cause of such overlap may be variability in DNA damage responses (DDRs), i.e.

cellular events occurring downstream of DNA adduct formation. Such DDR may be related to a range of cellular functions including the repair of the damage in question, and the operation of cell cycle checkpoints, which control critical events such as DNA replication or mitosis.

Recently DDRs have begun to attract considerable attention as potential biomarkers for response to chemotherapy. We have shown that the efficiency of DNA repair, measured in blood leukocytes or circulating tumor cells treated *in vitro* with DNA-damaging drugs, can serve as a predictor of clinical response [47, 48, 56, 60]. An early cellular response to DNA damage is the phosphorylation of H2AX, a variant of histone 2A, to form a phosphoprotein known as γ-H2AX. This protein is formed rapidly upon induction of DNA double-strand breaks, which are early intermediates of the repair of many DNA lesions [84]. Thus γ-H2AX, which is detected very quickly after the induction of DNA damage and repair, is a biomarker for DNA double-strand breaks. The detection of γ-H2AX by immunostaining is a sensitive biomarker for DNA damage and the early stages of repair, which can be applied to query the response to chemotherapy [84–86]. For example, in a study of cancer patients treated with pyrrolobenzodiazepine, a DNA crosslinking agent, it was shown that γ-H2AX foci could be detected in blood lymphocytes of all patients within 1 hour after drug administration. Furthermore, the fact that significant levels of foci were still present up to 15 days later, confirmed that γ-H2AX can serve as a biomarker of pharmacodynamic response to a DNA-damaging chemotherapeutic drug in a clinical setting [87].

Subsequently, the utility of γ-H2AX as a biomarker of DNA damage predictive of response to chemotherapy was explored in clinical studies. In a small pilot study based on 11 ovarian cancer patients [88], there was a statistically significant higher accumulation of γ-H2AX foci (area-under-the-curve for 0–24 hours) in seven patients who responded to subsequent platinum chemotherapy, compared to four patients who showed resistance to therapy. Strikingly, apoptosis rates were inversely correlated with DNA repair efficiencies, being significantly higher in platinum-sensitive than in platinum-resistant patients. Very similar observations were made in a study of 26 patients with multiple myeloma treated with melphalan and supported by autologous stem cell transplantation [89]. More and larger studies will be needed to show whether γ-H2AX or other DDR biomarkers have sufficiently high specificity or sensitivity, relative to DNA adduct measurements per se, as predictors of response to chemotherapy, to allow their use in the design of personalized therapeutic protocols.

4. Concluding Remarks

The work conducted during the previous 3–4 decades has provided consistent, even if limited, support for the utility of DNA damage as a biomarker of clinical response to cancer chemotherapy. The results of studies with representatives of most categories of DNA-damaging chemotherapeutic drugs have repeatedly suggested that patients who accumulate higher DNA damage levels, measured even in a surrogate tissue such as peripheral blood leukocytes, tend to respond better to chemotherapy than patients who accumulate less DNA damage. Unfortunately, most such studies hardly exceeded the pilot study scale and, despite encouraging outcomes, were never followed by clinical studies of appropriate design and sufficient size to permit definitive conclusions. It is clear that the potential heterogeneity of patient groups in relation to their disease (tumor type, histology, stage etc.), and variability of treatment protocols (level and frequency of drug dosing, use of additional drugs), in addition to demographic factors (sex, age, etc.), impose a major burden on the size requirements of such studies, with respect to desired statistical power. In the search for biomarkers predictive of patient response to chemotherapy, current attention appears to focus more on mechanism-based approaches, shifting from DNA damage per se to downstream DDRs. Proper validation of the utility of corresponding markers will require appropriate clinical studies which can realistically be conducted only in the context of collaborative networks.

Abbreviations

AAS	atomic absorption spectroscopy
AMS	accelerator mass spectrometry
BCNU	carmustine
carbo-Pt	carboplatin, diamminecyclobutanedicarboxylato-platinum (II)

CCNU	lomustine
Cis-Pt	cisplatin, cis-diamminedichloroplatinum (II)
DDR	DNA damage response
ELISA	enzyme-linked immunosorbent assay
GC-MS/MS	gas chromatography with tandem mass spectrometry (MS/MS)
γ-H2AX	phosphorylated histone 2AX
ICL	interstrand crosslink
ICP-MS	inductively coupled plasma mass spectrometry
IHC	immunohistochemistry
MGMT	O^6-methylguanine-DNA-methyltransferase
N7-G-N7-G crosslink	N,N-bis[2-(N7-guaninyl)ethyl] amine crosslink
N7-me-dG	N7-methyl-deoxyguanine
NSLC	non-small cell lung cancer
NDMA	N-nitrosodimethylamine
NNK	4-(methylnitrosamino)-1-(3-pyridyl)-1-butanone
O^6-me-dG	O^6-methylguanine
oxaliplatin	[(1*R*,2*R*)-cyclohexane-1,2-diamine](ethanedioato-O,O′)platinum(II)
Pt-d(ApG)	cis-diammineplatinum adduct formed on the N7 positions of adjacent deoxyadenine and deoxyguanine
Pt-d(GpG)	cis-diammineplatinum adduct formed on the N7 positions of two adjacent deoxyguanines
QPCR	multiplex long quantitative PCR

References

1. Pearl LH, Schierz AC, Ward SE, Al-Lazikani B, Pearl FMG. Therapeutic opportunities within the DNA damage response. *Nat Rev Cancer* 2015; 15(3): 166–180.
2. Goldstein M, Kastan MB. The DNA damage response: implications for tumor responses to radiation and chemotherapy. *Annu Rev Med* 2015; 66: 129–143.
3. Georgiadis P, Hebeis DG, Valavanis I, Liampa I, Bergdahl IA, Johansson A *et al.* Omics for prediction of environmental health effects: blood

leukocyte-based cross-omic profiling reliably predicts diseases associated with tobacco smoking. *Sci Rep* 2016; 6: 20544.

4. Franklin JG, Paus MD, Pluetschow A, Specht L. Chemotherapy, radiotherapy and combined modality for Hodgkin's disease, with emphasis on second cancer risk. *Cochrane Database Syst Rev* 2005; (4): CD003187.
5. Curreri SA, Fung C, Beard CJ. Secondary malignant neoplasms in testicular cancer survivors. *Urol Oncol* 2015; 33(9): 392–398.
6. Ahmad S. Platinum-DNA interactions and subsequent cellular processes controlling sensitivity to anticancer platinum complexes. *Chem Biodivers* 2010; 7(3): 543–566.
7. Sar DG, Montes-Bayón M, Blanco-González E, Sanz-Medel A. Quantitative methods for studying DNA interactions with chemotherapeutic cisplatin. *TrAC Trends Anal Chem* 2010; 29(11): 1390–1398.
8. Cimino GD, Pan C, Henderson PT. Personalized medicine for targeted and platinum-based chemotherapy of lung and bladder cancer. *Bioanalysis* 2013; 5(3): 369–391.
9. Poirier MC, Lippard SJ, Zwelling LA, Ushay HM, Kerrigan D, Thill CC *et al.* Antibodies elicited against cis-diamminedichloroplatinum(II)-modified DNA are specific for cis-diamminedichloroplatinum(II)-DNA adducts formed *in vivo* and *in vitro*. *Proc Natl Acad Sci USA* 1982; 79(21): 6443–6447.
10. Reed E, Gupta-Burt S, Litterst CL, Poirier MC. Characterization of the DNA damage recognized by an antiserum elicited against cis-diamminedichloroplatinum (II)-modified DNA. *Carcinogenesis* 1990; 11(12): 2117–2121.
11. Shamkhani H, Anderson LM, Henderson CE, Moskal TJ, Runowicz CD, Dove LF *et al.* DNA adducts in human and patas monkey maternal and fetal tissues induced by platinum drug chemotherapy. *Reprod Toxicol Elmsford N* 1994; 8(3): 207–216.
12. Fichtinger-Schepman AM, Baan RA, Luiten-Schuite A, van Dijk M, Lohman PH. Immunochemical quantitation of adducts induced in DNA by cis-diamminedichloroplatinum (II) and analysis of adduct-related DNA-unwinding. *Chem Biol Interact* 1985; 55(3): 275–288.
13. Terheggen PM, Floot BG, Lempers EL, van Tellingen O, Begg AC, den Engelse L. Antibodies against cisplatin-modified DNA and cisplatin-modified (di)nucleotides. *Cancer Chemother Pharmacol* 1991; 28(3): 185–191.
14. Parker RJ, Gill I, Tarone R, Vionnet JA, Grunberg S, Muggia FM *et al.* Platinum-DNA damage in leukocyte DNA of patients receiving carboplatin

and cisplatin chemotherapy, measured by atomic absorption spectrometry. *Carcinogenesis* 1991; 12(7): 1253–1258.

15. Gupta-Burt S, Shamkhani H, Reed E, Tarone RE, Allegra CJ, Pai LH *et al.* Relationship between patient response in ovarian and breast cancer and platinum drug-DNA adduct formation. *Cancer Epidemiol Biomark Prev Publ Am Assoc Cancer Res Cosponsored Am Soc Prev Oncol* 1993; 2(3): 229–234.
16. Poirier MC, Reed E, Shamkhani H, Tarone RE, Gupta-Burt S. Platinum drug-DNA interactions in human tissues measured by cisplatin-DNA enzyme-linked immunosorbent assay and atomic absorbance spectroscopy. *Environ Health Perspect* 1993; 99: 149–154.
17. Reed E, Parker RJ, Gill I, Bicher A, Dabholkar M, Vionnet JA *et al.* Platinum-DNA adduct in leukocyte DNA of a cohort of 49 patients with 24 different types of malignancies. *Cancer Res* 1993; 53(16): 3694–3699.
18. Motzer RJ, Reed E, Perera F, Tang D, Shamkhani H, Poirier MC *et al.* Platinum-DNA adducts assayed in leukocytes of patients with germ cell tumors measured by atomic absorbance spectrometry and enzyme-linked immunosorbent assay. *Cancer* 1994; 73(11): 2843–2852.
19. Bonetti A, Apostoli P, Zaninelli M, Pavanel F, Colombatti M, Cetto GL *et al.* Inductively coupled plasma mass spectroscopy quantitation of platinum-DNA adducts in peripheral blood leukocytes of patients receiving cisplatin- or carboplatin-based chemotherapy. *Clin Cancer Res Off J Am Assoc Cancer Res* 1996; 2(11): 1829–1835.
20. Harrington CF, Le Pla RC, Jones GDD, Thomas AL, and Farmer PB. Determination of cisplatin 1,2-intrastrand guanine-guanine DNA adducts in human leukocytes by high-performance liquid chromatography coupled to inductively coupled plasma mass spectrometry. *Chem Res Toxicol* 2010; 23(8): 1313–1321.
21. Henderson PT, Li T, He M, Zhang H, Malfatti M, Gandara D *et al.* A microdosing approach for characterizing formation and repair of carboplatin-DNA monoadducts and chemoresistance. *Int J Cancer* 2011; 129(6): 1425–1434.
22. Welters MJ, Maliepaard M, Jacobs-Bergmans AJ, Baan RA, Schellens JH, Ma J *et al.* Improved 32P-postlabelling assay for the quantification of the major platinum-DNA adducts. *Carcinogenesis* 1997; 18(9): 1767–1774.
23. Zwelling LA, Anderson T, Kohn KW. DNA-protein and DNA interstrand cross-linking by cis- and trans-platinum(II) diamminedichloride in L1210 mouse leukemia cells and relation to cytotoxicity. *Cancer Res* 1979; 39(2 Pt 1): 365–369.

24. Hartley JM, Spanswick VJ, Gander M, Giacomini G, Whelan J, Souhami RL *et al.* Measurement of DNA cross-linking in patients on ifosfamide therapy using the single cell gel electrophoresis (comet) assay. *Clin Cancer Res Off J Am Assoc Cancer Res* 1999; 5(3): 507–512.
25. Reed E, Yuspa SH, Zwelling LA, Ozols RF, Poirier MC. Quantitation of cis-diamminedichloroplatinum II (cisplatin)-DNA-intrastrand adducts in testicular and ovarian cancer patients receiving cisplatin chemotherapy. *J Clin Invest* 1986; 77(2): 545–550.
26. Reed E, Ozols RF, Tarone R, Yuspa SH, Poirier MC. Platinum-DNA adducts in leukocyte DNA correlate with disease response in ovarian cancer patients receiving platinum-based chemotherapy. *Proc Natl Acad Sci USA* 1987; 84(14): 5024–5028.
27. Reed E, Ostchega Y, Steinberg SM, Yuspa SH, Young RC, Ozols RF *et al.* Evaluation of platinum-DNA adduct levels relative to known prognostic variables in a cohort of ovarian cancer patients. *Cancer Res* 1990; 50(8): 2256–2260.
28. Reed E, Ozols RF, Fasy T, Yuspa SH, and Poirier MC. Biomonitoring of cisplatin-DNA adducts in cancer patients receiving cisplatin chemotherapy. *Prog Clin Biol Res* 1986; 209B: 247–252.
29. Poirier MC, Reed E, Ozols RF, Fasy T, Yuspa SH. DNA adducts of cisplatin in nucleated peripheral blood cells and tissues of cancer patients. *Prog Exp Tumor Res* 1987; 31: 104–113.
30. Reed E, Ozols RF, Tarone R, Yuspa SH, Poirier MC. The measurement of cisplatin-DNA adduct levels in testicular cancer patients. *Carcinogenesis* 1988; 9(10): 1909–1911.
31. Fichtinger-Schepman AM, van der Velde-Visser SD, van Dijk-Knijnenburg HC, van Oosterom AT, Baan RA, Berends F. Kinetics of the formation and removal of cisplatin-DNA adducts in blood cells and tumor tissue of cancer patients receiving chemotherapy: comparison with *in vitro* adduct formation. *Cancer Res* 1990; 50(24): 7887–7894.
32. Boffetta P, Fichtinger-Schepman AM, Weiderpass E, van Dijk-Knijnenburg HC, Stoter G, van Oosterom AT *et al.* Cisplatin-DNA adducts and protein-bound platinum in blood of testicular cancer patients. *Anticancer Drugs* 1998; 9(2): 125–129.
33. Ma J, Verweij J, Planting AS, de Boer-Dennert M, van Ingen HE, van der Burg ME *et al.* Current sample handling methods for measurement of platinum-DNA adducts in leucocytes in man lead to discrepant results in DNA adduct levels and DNA repair. *Br J Cancer* 1995; 71(3): 512–517.
34. Schellens JH, Ma J, Planting AS, van der Burg ME, van Meerten E, de Boer-Dennert M *et al.* Relationship between the exposure to cisplatin,

DNA-adduct formation in leucocytes and tumour response in patients with solid tumours. *Br J Cancer* 1996; 73(12): 1569–1575.

35. Blommaert FA, Michael C, Terheggen PM, Muggia FM, Kortes V, Schornagel JH *et al.* Drug-induced DNA modification in buccal cells of cancer patients receiving carboplatin and cisplatin combination chemotherapy, as determined by an immunocytochemical method: interindividual variation and correlation with disease response. *Cancer Res* 1993; 53(23): 5669–5675.
36. van de Vaart PJ, Belderbos J, de Jong D, Sneeuw KC, Majoor D, Bartelink H, Begg AC. DNA-adduct levels as a predictor of outcome for NSCLC patients receiving daily cisplatin and radiotherapy. *Int J Cancer* 2000; 89(2): 160–166.
37. Fichtinger-Schepman AM, van Oosterom AT, Lohman PH, Berends F. cis-Diamminedichloroplatinum(II)-induced DNA adducts in peripheral leukocytes from seven cancer patients: quantitative immunochemical detection of the adduct induction and removal after a single dose of cis-diamminedichloroplatinum(II). *Cancer Res* 1987; 47(11): 3000–3004.
38. Schellens JHM, Planting AS, van Zandwijk N, Ma J, Maliepaard M, van der Burg ME *et al.* Adaptive intrapatient dose escalation of cisplatin in combination with low-dose vp16 in patients with nonsmall cell lung cancer. *Br J Cancer* 2003; 88(6): 814–821.
39. Hoebers FJP, Pluim D, Verheij M, Balm AJ, Bartelink H, Schellens JH, Begg AC. Prediction of treatment outcome by cisplatin-DNA adduct formation in patients with stage III/IV head and neck squamous cell carcinoma, treated by concurrent cisplatin-radiation (RADPLAT). *Int. J. Cancer* 2006; 119(4): 750–756.
40. Hoebers FJP, Pluim D, Hart AA, Verheij M, Balm AJ, Fons G *et al.* Cisplatin-DNA adduct formation in patients treated with cisplatin-based chemoradiation: lack of correlation between normal tissues and primary tumor. *Cancer Chemother Pharmacol* 2008; 61(6): 1075–1081.
41. Ledermann JA, Gabra H, Jayson GC, Spanswick VJ, Rustin GJ, Jitlal M *et al.* Inhibition of carboplatin-induced DNA interstrand cross-link repair by gemcitabine in patients receiving these drugs for platinum-resistant ovarian cancer. *Clin Cancer Res Off J Am Assoc Cancer Res* 2010; 16(19): 4899–4905.
42. Parker RJ, Eastman A, Bostick-Bruton F, Reed E. Acquired cisplatin resistance in human ovarian cancer cells is associated with enhanced repair of cisplatin-DNA lesions and reduced drug accumulation. *J Clin Invest* 1991; 87(3): 772–777.
43. Fichtinger-Schepman AM, van Oosterom AT, Lohman PH, Berends F. Interindividual human variation in cisplatinum sensitivity, predictable in an *in vitro* assay? *Mutat Res* 1987; 190(1): 59–62.

44. Veal GJ, Dias C, Price L, Parry A, Errington J, Hale J, Pearson AD *et al.* Influence of cellular factors and pharmacokinetics on the formation of platinum-DNA adducts in leukocytes of children receiving cisplatin therapy. *Clin Cancer Res Off J Am Assoc Cancer Res* 2001; 7(8): 2205–2212.
45. Welters MJ, Braakhuis BJ, Jacobs-Bergmans AJ, Kegel A, Baan RA, van der Vijgh WJ *et al.* The potential of platinum-DNA adduct determination in ex vivo treated tumor fragments for the prediction of sensitivity to cisplatin chemotherapy. *Ann Oncol Off J Eur Soc Med Oncol ESMO* 1999; 10(1): 97–103.
46. Oshita F, Yamamoto N, Fukuda M, Ohe Y, Tamura T, Eguchi K *et al.* Correlation of therapeutic outcome in non-small cell lung cancer and DNA damage assayed by polymerase chain reaction in leukocytes damaged *in vitro*. *Cancer Res* 1995; 55(11): 2334–2337.
47. Nel I, Gauler TC, Eberhardt WE, Nickel AC, Schuler M, Thomale J *et al.* Formation and repair kinetics of Pt-(GpG) DNA adducts in extracted circulating tumour cells and response to platinum treatment. *Br J Cancer* 2013; 109(5): 1223–1229.
48. Wang L-E, Yin M, Dong Q, Stewart DJ, Merriman KW, Amos CI *et al.* DNA repair capacity in peripheral lymphocytes predicts survival of patients with non-small-cell lung cancer treated with first-line platinum-based chemotherapy. *J Clin Oncol Off J Am Soc Clin Oncol* 2011; 29(31): 4121–4128.
49. Lawley PD, Phillips DH. DNA adducts from chemotherapeutic agents. *Mutat Res* 1996; 355(1–2): 13–40.
50. Malayappan B, Johnson L, Nie B, Panchal D, Matter B, Jacobson P *et al.* Quantitative high-performance liquid chromatography-electrospray ionization tandem mass spectrometry analysis of bis-N7-guanine DNA-DNA cross-links in white blood cells of cancer patients receiving cyclophosphamide therapy. *Anal Chem* 2010; 82(9): 3650–3658.
51. Tilby MJ, Styles JM, Dean CJ. Immunological detection of DNA damage caused by melphalan using monoclonal antibodies. *Cancer Res* 1987; 47(6): 1542–1546.
52. Rothbarth J, Koevoets C, Tollenaar RA, Tilby MJ, van de Velde CJ, Mulder GJ *et al.* Immunohistochemical detection of melphalan-DNA adducts in colon cancer cells *in vitro* and human colorectal liver tumours *in vivo*. *Biochem Pharmacol* 2004; 67(9): 1771–1778.
53. Frank AJ, Proctor SJ, Tilby MJ. Detection and quantification of melphalan-DNA adducts at the single cell level in hematopoietic tumor cells. *Blood* 1996; 88(3): 977–984.

54. Souliotis VL, Dimopoulos MA, Sfikakis PP. Gene-specific formation and repair of DNA monoadducts and interstrand cross-links after therapeutic exposure to nitrogen mustards. *Clin Cancer Res Off J Am Assoc Cancer Res* 2003; 9(12): 4465–4474.
55. Tilby MJ, Newell DR, Viner C, Selby PJ, Dean CJ. Application of a sensitive immunoassay to the study of DNA adducts formed in peripheral blood mononuclear cells of patients undergoing high-dose melphalan therapy. *Eur J Cancer Oxf Engl 1990* 1993; 29A(5): 681–686.
56. Dimopoulos MA, Souliotis VL, Anagnostopoulos A, Papadimitriou C, Sfikakis PP. Extent of damage and repair in the p53 tumor-suppressor gene after treatment of myeloma patients with high-dose melphalan and autologous blood stem-cell transplantation is individualized and may predict clinical outcome. *J Clin Oncol Off J Am Soc Clin Oncol* 2005; 23(19): 4381–4389.
57. Spanswick VJ, Craddock C, Sekhar M, Mahendra P, Shankaranarayana P, Hughes RG *et al.* Repair of DNA interstrand crosslinks as a mechanism of clinical resistance to melphalan in multiple myeloma. *Blood* 2002; 100(1): 224–229.
58. Johnstone EC, Lind MJ, Griffin MJ, Boddy AV. Ifosfamide metabolism and DNA damage in tumour and peripheral blood lymphocytes of breast cancer patients. *Cancer Chemother Pharmacol* 2000; 46(6): 433–441.
59. Willits I, Price L, Parry A, Tilby MJ, Ford D, Cholerton S *et al.* Pharmacokinetics and metabolism of ifosfamide in relation to DNA damage assessed by the COMET assay in children with cancer. *Br J Cancer* 2005; 92(9): 1626–1635.
60. Dimopoulos MA, Souliotis VL, Anagnostopoulos A, Bamia C, Pouli A, Baltadakis I *et al.* Melphalan-induced DNA damage *in vitro* as a predictor for clinical outcome in multiple myeloma. *Haematologica* 2007; 92(11): 1505–1512.
61. Gkotzamanidou M, Sfikakis PP, Kyrtopoulos SA, Bamia C, Dimopoulos MA, Souliotis VL. Chromatin structure, transcriptional activity and DNA repair efficiency affect the outcome of chemotherapy in multiple myeloma. *Br J Cancer* 2014; 111(7): 1293–1304.
62. Stefanou DT, Episkopou H, Kyrtopoulos SA, Bamias A, Gkotzamanidou M, Bamia C *et al.* Development and validation of a PCR-based assay for the selection of patients more likely to benefit from therapeutic treatment with alkylating drugs. *Br J Clin Pharmacol* 2012; 74(5): 842–853.
63. Fu D, Calvo JA, Samson LD. Balancing repair and tolerance of DNA damage caused by alkylating agents. *Nat Rev Cancer* 2012: 12(2): 104–120.

64. Wild CP, Lu SH, Montesano R. Radioimmunoassay used to detect DNA alkylation adducts in tissues from populations at high risk for oesophageal and stomach cancer. *IARC Sci Publ* 1987; (84): 534–537.
65. van Delft JH, Steenwinkel MJ, de Groot AJ, van Zeeland AA, Eberle-Adamkiewicz G, Rajewsky MF *et al.* Determination of N7- and O^6-methylguanine in rat liver DNA after oral exposure to hydrazine by use of immunochemical and electrochemical detection methods. *Fundam Appl Toxicol Off J Soc Toxicol* 1997; 35(1): 131–137.
66. Souliotis VL Kyrtopoulos SA. A novel, sensitive assay for O^6-methyl- and O^6-ethylguanine in DNA, based on repair by the enzyme O^6-alkylguanine-DNA-alkyltransferase in competition with an oligonucleotide containing O^6-methylguanine. *Cancer Res* 1989; 49(24 Pt 1): 6997–7001.
67. Souliotis VL, Kaila S, Boussiotis VA, Pangalis GA, Kyrtopoulos SA. Accumulation of O^6-methylguanine in human blood leukocyte DNA during exposure to procarbazine and its relationships with dose and repair. *Cancer Res* 1990; 50(9): 2759–2764.
68. Kyrtopoulos SA, Souliotis VL, Valavanis C, Boussiotis VA, Pangalis GA. Accumulation of O^6-methylguanine in human DNA after therapeutic exposure to methylating agents and its relationship with biological effects. *Environ Health Perspect* 1993; 99: 143–147.
69. Bianchini F, Weiderpass E, Kyrtopoulos S, Souliotis VL, Henry-Amar M, Wild CP *et al.* Detection of DNA methylation adducts in Hodgkin's disease patients treated with procarbazine. *Biomark Biochem Indic Expo Response Susceptibility Chem* 1996; 1(4): 226–231.
70. Souliotis VL, Boussiotis VA, Pangalis GA, Kyrtopoulos SA. *In vivo* formation and repair of O^6-methylguanine in human leukocyte DNA after intravenous exposure to dacarbazine. *Carcinogenesis* 1991; 12(2): 285–288.
71. Philip PA, Souliotis VL, Harris AL, Salisbury A, Tates AD, Mitchell K *et al.* Methyl DNA adducts, DNA repair, and hypoxanthine-guanine phosphoribosyl transferase mutations in peripheral white blood cells from patients with malignant melanoma treated with dacarbazine and hydroxyurea. *Clin Cancer Res Off J Am Assoc Cancer Res* 1996; 2(2): 303–310.
72. Saunders MP, Salisbury AJ, O'Byrne KJ, Souliotis VL, Varcoe SM, Talbot DC *et al.* A phase II study evaluating the effect of tamoxifen on DNA repair in melanoma patients treated with dacarbazine. *Anticancer Res* 1997; 17(6D): 4677–4680.
73. van Delft JH, Luiten-Schuite A, Souliotis VL, Kyrtopoulos SA, Ouwerkerk J, Keizer HJ *et al.* N7-Methylguanine and O(6)-methylguanine levels in DNA of

white blood cells from cancer patients treated with dacarbazine. *Biomark Biochem Indic Expo Response Susceptibility Chem* 1996; 1(2): 94–98.

74. Lee SM, Thatcher N, Margison GP. O^6-alkylguanine-DNA alkyltransferase depletion and regeneration in human peripheral lymphocytes following dacarbazine and fotemustine. *Cancer Res* 1991: 51(2): 619–623.
75. Lee SM, Thatcher N, Dougal M, Margison GP. Dosage and cycle effects of dacarbazine (DTIC) and fotemustine on O^6-alkylguanine-DNA alkyltransferase in human peripheral blood mononuclear cells. *Br J Cancer* 1993; 67(2): 216–221.
76. Lee SM, Margison GP, Thatcher N, O'Connor PJ, Cooper DP. Formation and loss of O^6-methyldeoxyguanosine in human leucocyte DNA following sequential DTIC and fotemustine chemotherapy. *Br J Cancer* 1994; 69(5): 853–857.
77. Watson AJ, Luiten-Schuite A, Souliotis VL, Kyrtopoulos SA, Ouwerkerk J, Keizer HJ *et al.* O(6)-methylguanine-DNA methyltransferase depletion and DNA damage in patients with melanoma treated with temozolomide alone or with lomeguatrib. *Br J Cancer* 2009; 100(8), 1250–1256.
78. Ranson M, Hersey P, Thompson D, Beith J, McArthur GA, Haydon A *et al.* Randomized trial of the combination of lomeguatrib and temozolomide compared with temozolomide alone in chemotherapy naive patients with metastatic cutaneous melanoma. *J Clin Oncol Off J Am Soc Clin Oncol* 2007; 25(18): 2540–2545.
79. Kefford RF, Thomas NP, Corrie PG, Palmer C, Abdi E, Kotasek D *et al.* A phase I study of extended dosing with lomeguatrib with temozolomide in patients with advanced melanoma. *Br J Cancer* 2009; 100(8): 1245–1249.
80. Tawbi HA, Villaruz L, Tarhini A, Moschos S, Sulecki M, Viverette F *et al.* Inhibition of DNA repair with MGMT pseudosubstrates: phase I study of lomeguatrib in combination with dacarbazine in patients with advanced melanoma and other solid tumours. *Br J Cancer* 2011; 105(6): 773–777.
81. Kaina B, Margison GP, Christmann M. Targeting O^6-methylguanine-DNA methyltransferase with specific inhibitors as a strategy in cancer therapy. *Cell Mol Life Sci CMLS* 2010; 67, (21): 3663–3681.
82. Jiang G, Li L-T, Xin Y, Zhang L, Liu Y-Q, Zheng J-N. Strategies to improve the killing of tumors using temozolomide: targeting the DNA repair protein MGMT. *Curr Med Chem* 2012; 19(23): 3886–3892.
83. Souliotis VL, Valavanis C, Boussiotis VA, Pangalis GA, Kyrtopoulos SA. Comparative study of the formation and repair of O^6-methylguanine in humans and rodents treated with dacarbazine. *Carcinogenesis* 1996; 17(4): 725–732.

84. Dickey JA, Redon CA, Nakamura AJ, Baird BJ, Sedelnikova OA, Bonner WM. H2AX: functional roles and potential applications. *Chromosoma* 2009; 118(6): 683–692.
85. Ivashkevich A, Redon CE, Nakamura AJ, Martin RF, Martin OA. Use of the γ-H2AX assay to monitor DNA damage and repair in translational cancer research. *Cancer Lett* 2012; 327(1–2): 123–133.
86. Willers H, Gheorghiu L, Liu Q, Efstathiou JA, Wirth LJ, Krause M *et al.* DNA damage response assessments in human tumor samples provide functional biomarkers of radiosensitivity. *Semin Radiat Oncol* 2015; 25(4): 237–250.
87. Wu J, Clingen PH, Spanswick VJ, Mellinas-Gomez M, Meyer T, Puzanov I *et al.* γ-H2AX foci formation as a pharmacodynamic marker of DNA damage produced by DNA cross-linking agents: results from 2 phase I clinical trials of SJG-136 (SG2000). *Clin Cancer Res Off J Am Assoc Cancer Res* 2013; 19(3): 721–730.
88. Stefanou DT, Bamias A, Episkopou H, Kyrtopoulos SA, Likka M, Kalampokas T *et al.* Aberrant DNA damage response pathways may predict the outcome of platinum chemotherapy in ovarian cancer. *PloS One* 2015; 10(2): e0117654.
89. Gkotzamanidou M, Terpos E, Bamia C, Munshi NC, Dimopoulos MA, Souliotis VL. DNA repair of myeloma plasma cells correlates with clinical outcome; the effect of the non-homologous end-joining inhibitor SCR7. *Blood* 2016; 128(9): 1214–1225.

Chapter 9

Antiretroviral Drugs, Genotoxicity, and Carcinogenesis

Ofelia A. Olivero

National Cancer Institute, Center for Cancer Research,
National Institutes of Health,
Bethesda, MD, 20892, USA

Carcinogen-DNA Interactions Section, Laboratory of Cancer
Biology and Genetics, 37 Convent Dr. MSC 4255,
Bldg 37 Rm 4032B, Bethesda, MD 20892, USA
oliveroo@exchange.nih.gov

Abstract

Nucleoside analogs were first approved by the U.S. Food and Drug Administration for use against the human immunodeficiency virus-1 (HIV-1) in 1987. Since then, these agents, now commonly referred to as nucleoside reverse transcriptase inhibitors (NRTIs), have become essential components of the highly active antiretroviral therapy (HAART) drug combinations used for treatment of HIV-1 infections. Incorporation of the drug into host nuclear (nDNA) and mitochondrial DNA (mtDNA) may be largely responsible for dose-limiting toxicities. Zidovudine (AZT, 3′-azido-3′-deoxythymidine), the first NRTI approved for the therapy of HIV-1, is incorporated into DNA, causing mutations, and inducing micronuclei (MN), chromosomal aberrations, sister chromatid exchanges, shortened telomeres, and aneuploidy in cultured cells. Genomic instability would be predicted as a consequence of these events. Animal models, such as mice and monkeys, as well as *ex vivo* models, such as human placental perfusion, have provided insights into our understanding of AZT-induced genotoxicity, and are discussed here. Detailed information on the mechanisms underlying NRTI-associated antiretroviral efficacy, toxicity, and metabolic resistance were not available when AZT was first approved for use as an antiretroviral agent. Our increase in knowledge, acquired through these models and observations in the clinic, may lead to intervention strategies to attenuate toxicity without altering drug efficacy.

1. Introduction

In this chapter, we review the genotoxicity of nucleoside analogs, a class of antiretroviral drugs used widely in the therapy of acquired immune deficiency syndrome (AIDS). The mechanism of action of the antiretroviral nucleoside reverse transcriptase inhibitors (NRTIs) is based on both an arrest of DNA replication and an inhibition of the viral reverse transcriptase (RT) [1,2]. In cultured cells, animal models, and humans, these drugs become incorporated into DNA, and the subsequent arrest of DNA replication results in both simple mutagenesis and aberrations in chromosomal functioning that can produce aneuploidy (reviewed in [3]).

Patients infected with the HIV-1 undergo extensive therapy with antiretroviral drug cocktails that typically contain 1-3 NRTIs, as well as non-nucleoside reverse-transcriptase inhibitors (NNRTIs), and protease inhibitors (PIs) [4]. The NRTI, Zidovudine® (3′azido-3′-deoxythymidine, AZT), was the first drug shown to be effective against HIV-1 [5] and is still a frequent component of highly active antiretroviral therapy (HAART). Currently, an effective HAART cocktail includes three or four different types of drugs, with the exact regimen tailored to the individual. HIV-1 is an RNA virus (retrovirus), and because it is unable to integrate into DNA, the conversion of RNA to DNA, which is not performed by mammalian cells, is accomplished by a viral protein called reverse transcriptase (RT). Although highly specific for HIV-1 RT, NRTIs exhibit some affinity for cellular polymerases [1,6,7]. Among the multiple therapeutic approaches to counteract viral activity, many have targeted the viral RT and proven to be very effective [5].

2. Classes of Antiretroviral Drugs Used for HIV-1 Therapy

Most NRTIs are nucleoside analogs that lack a ribose 3′-OH (Figure 1, circles). Examples of this class of drugs include zidovudine (AZT), lamivudine (3TC), didanosine (ddI), stavudine (d4T), and abacavir (ABC). They act as DNA replication chain terminators because the 3′-OH of the deoxyribose sugar is lacking, so it is impossible to extend the nascent DNA chain by forming a 5′ to 3′ phosphodiester bond with the incoming nucleotide. In addition, NRTIs are able to disrupt the activity of RT, thereby preventing entrance of the viral genome information into the host DNA. For the host genome, the end result can be the introduction of mutations occurring because the polymerase is disrupted [8], with subsequent induction of error-prone repair, and/or large-scale chromosome breaks, with loss of a partial chromosome (micronucleus [MN] formation), or loss of a whole chromosome (aneuploidy) [9].

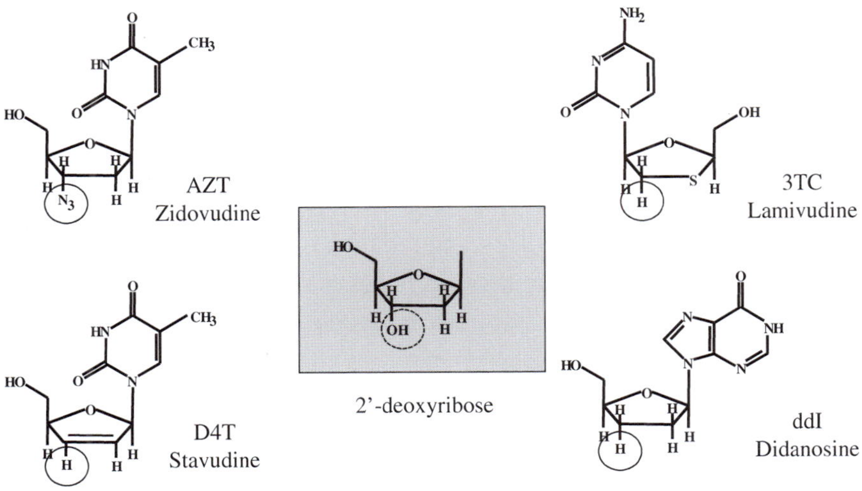

Figure 1. Structures of the NRTIs most commonly used for therapy of HIV-1. The OH group at the 3′ position of deoxyribose (dotted circle) is absent in NRTIs (solid circles), which results in arrest of DNA replication. In place of the 3′-OH group, AZT has an azido group, 3TC and ddI have hydrogen, and d4T has an unsaturated 2′,3′-dideoxyribose.

The non-nucleoisde reverse transcriptase inhibitors (NNRTIs) inhibit RT by binding to an allosteric site on the enzyme and inhibiting RT noncompetitively. NNRTIs affect the RT handling of substrate (nucleotides) by binding near the active site. The NNRTI Nevirapine (NVP) is widely used for the treatment of HIV-1 infection even though chronic use has been associated with severe liver and skin toxicity [10–12]. NVP is the most widely used antiretroviral drug in developing countries, both in combination therapy and to prevent mother-to-child transmission of HIV. Antunes *et al.* [13] investigated the formation of NVP-dT adducts under biomimetic conditions. The authors suggest that the adduct N3-thymidine of NVP (N3-NVP-dT) may be formed *in vivo* and can play a role in the hepatotoxicity and/or putative hepatocarcinogenicity of NVP [13].

3. NRTIs become Incorporated into Nuclear, Mitochondrial, and Telomeric DNA

Incorporation of AZT into eukaryotic nDNA by host polymerases was first suggested 30 years ago [1, 14], and since that time numerous laboratories, employing diverse techniques, have provided unequivocal documentation that AZT becomes phosphorylated and incorporated into host nuclear DNA (nDNA and mitochondrial (mtDNA in place of thymidine (Table 1). Radiolabeled AZT, with subsequent analysis by high-performance liquid chromatography (HPLC), was used to detect AZT-DNA in bone marrow [15] and human T-lymphoblastoid cells [16]. In 1994, Olivero *et al.* [17] used a radioimmunoassay (RIA) to show AZT-DNA incorporation, and the data were subsequently validated using radiolabeled compound. The RIA was used to demonstrate similar levels of AZT-DNA incorporation in mouse, hamster, and human cell lines exposed to the same AZT dose [17], and to demonstrate that AZT incorporates into the DNA of multiple organs of AZT-exposed mice and monkeys, and DNA of human leukocytes [18–22] (Table 1). Furthermore, similar levels of AZT-DNA incorporation were detected in all species examined [19,23]. In assays using ^{32}P-postlabeling combined with HPLC [24], AZT-DNA incorporation was observed in nDNA of lung and liver from neonatal mice. In assays using radiolabeled AZT with subsequent analysis by HPLC [15], AZT-DNA was found in mouse bone marrow nDNA [15]. Furthermore, AZT-DNA incorporation

Table 1. Values for AZT incorporation into nuclear and mitochondrial DNA of various species

Species exposed to AZT	Target	Molecules of AZT/10^6 nucleotides (range)	Reference
Human	Maternal PBMC and fetal cord blood WBC: nuclear DNA (nDNA)	183–344	[23]
Human	Cultured TK6 cells, nDNA	21–27	[25]
CD1 mice — transplacental exposure	nDNA — fetal various organs	23–226	[19]
CD1 mice — transplacental exposure	Mitochondrial DNA (mtDNA) — fetal various organs	28–302	[19]
Nonhuman primate — *E. patas* transplacental exposure	nDNA — fetal various organs	60–739	[19]
Nonhuman primate — *E. patas* transplacental exposure	mtDNA — fetal various organs	22–324	[19]
Nonhuman primate — *M. mulatta* transplacental exposure	nDNA — fetal various organs	29–1944	[26]
Human	AZT-perfused fresh human placentas, nDNA	12–42	[27]

was observed in CEM T-lymphoblastoid cells and in the chronic myelogenous leukemia cell line K562 [16].

Incorporation of AZT into mtDNA has also been reported in multiple organs (liver, lung, reproductive organs, and skin) of CD1 mouse pups exposed transplacentally to AZT [19]. In addition, AZT was found in mtDNA from human myeloid HL60 cells after exposure to radiolabeled AZT [28]. Furthermore, inhibition of the mtDNA polymerase by AZT-triphosphate was suggested as the cause of AZT-induced myopathy [29,30].

Preferential incorporation of AZT into telomeric DNA was first shown in cultured cells, and resulted in reversible telomeric shortening, as well as the potential for premature senescence [31–34]. Chinese hamster ovary (CHO) cells, exposed to AZT for 24 hours were processed to obtain metaphase spreads, which were incubated with anti-AZT antibodies exhibiting a preferential localization of AZT within the telomeric region or chromosomal ends [35]. Furthermore, telomeric DNA attrition was observed in mice and monkeys transplacentally exposed to AZT [33], and in human endometrial cancer HeLa cells exposed long-term to AZT [34]. Uncapped telomeres confer a lack of chromosomal stability and promote incorrect exchanges of chromosomal material at interphase. Once mitosis takes place, and the daughter cells separate during telophase, abnormal DNA bridges may result from the adhesiveness of the disrupted telomeres.

3.1. *Genotoxic consequences of AZT exposure: nDNA*

Repair of AZT incorporated into nDNA occurs through the activity of a 3′ to 5′ exonuclease, which is able to remove a variety of replication terminating dideoxynucleosides. This has been demonstrated in leukemic H9 cells [36]. Furthermore, evidence of removal of AZT from DNA by an exonucleolytic repair enzyme [31] was demonstrated *in vitro*. Human hepatoma HepG2 cells exposed to AZT for 2 weeks showed significantly increased gene expression for genes associated with nucleotide excision repair mechanisms, for example XPC was one of the upregulated elements [32]. The same authors demonstrated that in XPC knocked-down cells, 100 μM AZT exposure significantly increased AZT incorporation into DNA, decreased the total number of viable cells, increased the release of lactate dehydrogenase, and induced apoptosis. Furthermore, a more extensive G2/M cell cycle arrest was observed, in contrast with nontransfected HepG2 cells, showing that XPC plays an essential role in the repair of AZT-induced DNA damage [32].

NRTIs are chemical mutagens [8, 22, 25, 33–35, 37, 38]. Meng *et al.* investigated the potential mechanisms of AZT mutagenicity and carcinogenicity by quantifying the incorporation of AZT into cellular DNA, measuring AZT-induced thymidine kinase (TK) mutant frequencies (Mfs), and determining the percentage of loss of heterozygosity (LOH) in spontaneous or AZT-induced TK mutants in the human lymphoblastoid cell line TK6. The authors reported a positive correlation between

AZT-DNA incorporation and AZT-induced TK Mfs, suggesting that AZT incorporation into cellular DNA has a direct role on the generation of mutations [25]. Combination of AZT with other NRTIs generally yielded enhanced rates of HPRT and TK Mfs, suggesting that NRTI-induced mutagenicity may be driven by cumulative dose, and raising the question of whether AZT/3TC has a greater mutagenic effect than AZT alone in perinatally exposed children [37].

Additional types of AZT-DNA damage include cell cycle arrest, MN, sister chromatid exchanges, chromosomal aberrations, telomere shortening [34], centrosomal amplification (CA), aneuploidy [39, 40], and abnormal ciliogenesis [41], most of which are discussed below.

Cell cycle arrest occurs as a consequence of AZT exposure. Cells containing DNA damage are typically arrested in mitosis to allow repair processes to remove the abnormal nucleotides before the cell cycle continues. Impairment of cell cycle arrest, resulting in the inability of the cell to accomplish DNA repair, typically results in increased mutation and cancer rates. Cell cycle delay induced by AZT, was reported by Wu *et al.* to occur in human chronic myeloid leukemia cells [42] and in HL60 cells [43]. Other studies [44, 45] have shown that AZT caused a decrease in the proliferative response to phytohemagglutinins, and an S-phase delay, in peripheral blood mononuclear cells (PBMCs) and human leukemic CEM cells. A cytostatic effect in human colon carcinoma WiDr cells was induced at 12 hours, peaked at 24 hours, and reversed to baseline at 72 hours [46]. The authors proposed that AZT-induced cytostasis is a widespread transient and reversible effect, observed in many cell lines, and in eight of nine tumor cell lines examined in their laboratory.

A correlation between changes in the gene expression of cell cycle-related genes and S-phase arrest was demonstrated in HeLa cells [47]. In experiments carried out using different microarray platforms, the authors examined the correlation of up-/downregulation of some key genes in cell cycle G1/S progression, and S-phase accumulation in cells exposed to the combination AZT/3TC for 24 hours. Some of these studies suggested that the cyclin D–Rb pathway could be mediating the cell cycle arrest [47].

Chromatin bridges and MN with or without chromosome loss typically occur as a consequence of telomeric attrition. The material involved in the formation of chromatin bridges may break away, and remain lost in the cytoplasm of any or both of the daughter cells, as an MN, a small piece of DNA. It has been observed that MN differ in size. A large MN may include intact chromosomes, as demonstrated by staining with kinetochore-specific antibodies, which detect centromeric regions of the chromosome. Kinetochore-specific antibodies were isolated from patients with the CREST (calcinosis, Raynaud phenomenon, esophageal dysmotility, sclerodactyly, and telangiectasia) syndrome, [37]. When normal human mammary epithelial cells (NHMECs) exposed to 200-μM AZT for 24 hours were incubated with the CREST antibody, CREST positive MN, indicating the presence of whole chromosomes, and CREST negative MN, indicating the presence of a partial chromosome, were both observed [20] (Figure 2). Missegregation of chromosomes in mitosis has been documented by scoring kinetochore-positive (CREST positive) MN [39].

Centrosomal amplification (CA) occurs when the normal cross-talk between the cell cycle and the centrosomal cycle become dysfunctional, disrupting normal cell division and homeostasis. The centrosome, an

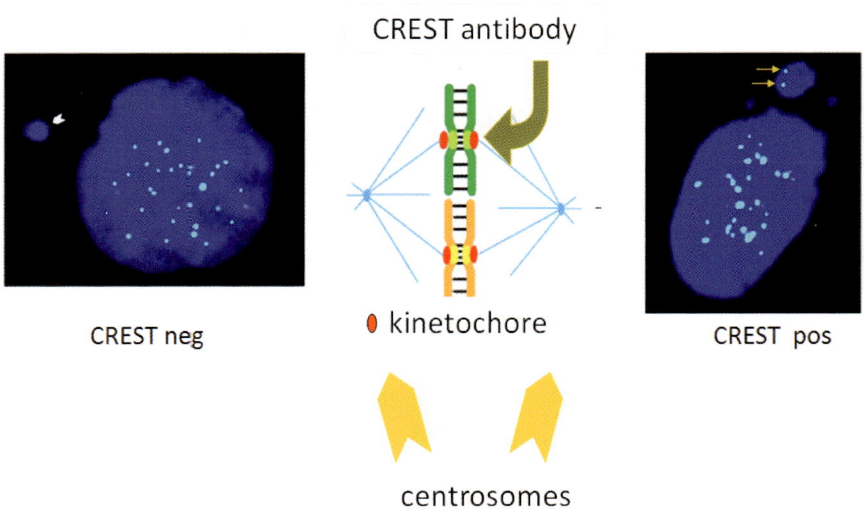

Figure 2. Micronuclei (MN), with and without whole chromosomes, in CHO cells exposed to 800 μM AZT for 24 hours. Arrowhead (left) indicates the presence of a MN that does not contain kinetochore material (CREST negative), and arrows (right) indicate an MN with two CREST-positive kinetochores. In the center, a schematic of chromosomes and mitotic spindles depicts the chromosomal region the CREST antibody is specific for.

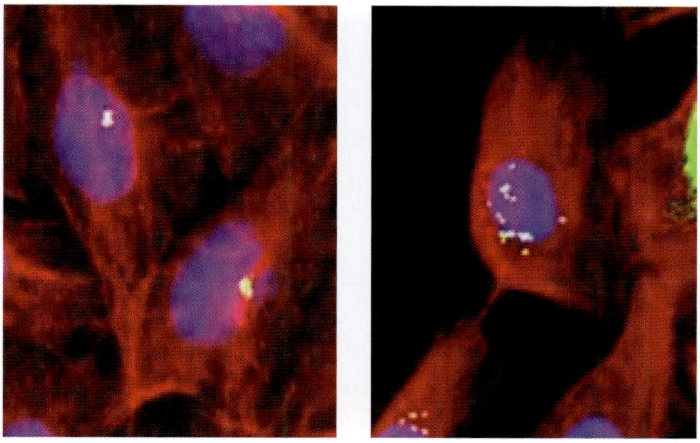

(A) (B)

Figure 3. Localization of centrosomes in NHMEC cells stained with anti-pericentrin antibody (white). Cells are also stained with anti-β tubulin antibody (red) and DAPI for visualizing DNA (blue). Untreated NHMECs (A), show a normal centrosome in an undividing cell. NHMECs exposed to 200 μM AZT for 24 hours (B), show centrosomal amplification and fragmentation.

organelle that regulates migration of chromosomes to the daughter cells, is also a microtubule organizer. During the cell cycle, asynchronous centrosomal division may produce CA. Furthermore, because the centrosome-associated protein matrix contains multiple proteins, centrosomal activity has been implicated in cell cycle transitions, which include G1 to S-phase, G2 to mitosis, and metaphase to anaphase [48, 49]. Resting cells have one centrosome (Figure 3A), which duplicates in synchrony with the S-phase of the cell cycle, to generate two new centrosomes, each of which is comprised of a pair of orthogonally placed centrioles. Most human carcinomas have an abnormal centrosomal number, which contributes to the genomic instability of transformed cells [50–53]. Additionally, it has been reported that malfunction of the centrosome induces delay in the G1–S boundary of the cell cycle [54, 55].

Some genotoxic agents, generating an array of both structural and numerical chromosomal aberrations, can cause CA (Figure 3B). Employing immunohistochemistry, Borojerdi *et al.* [39] explored the ability of AZT to act as a centrosome disruptor. Using CHO cells and human NHMEC strains exposed to AZT for 24 hours, these authors showed CA by anti-pericentrin staining (Figure 3B). Additionally, they demonstrated aberrations in tubulin polymerization in AZT-exposed cells bearing multiple centrosomes. In NHMECs [39, 56], pericentrin-positive signals were scored for the total number of cells with visible centrosomes, and the results demonstrated a dose-increase in pericentrin-positive signals in treated cells. Typical consequences of CA include chromosomal instability and aneuploidy.

AZT exposure may produce abnormal ciliogenesis. A primary cilium arises from the centrosome in a quiescent or postmitotic cell, serving as a sensory organelle that communicates mechanical and chemical stimuli from the environment to the interior of the cell. Cilium formation may, therefore, become a useful end-point signaling exposure to genotoxins or aneugens. Since cilia are derived from centrioles and aneugens induce CA, the production of multiple cilia arising from centrioles may be indicative of exposure to an aneugen. Divi *et al.* [41] demonstrated a five-fold increase in production of multiple cilia in hTERT immortalized retinal epithelial cells exposed to 20 μM AZT, compared to unexposed controls (Figure 4). They also observed CA in cells with multiple cilia.

3.2. *Genotoxic consequences of AZT exposure: mtDNA*

Mitochondrial dysfunction induced by NRTIs is considered to result from direct inhibition of the mitochondrial polymerase γ (Pol γ), as well as DNA incorporation of the drug and arrest of DNA replication [57, 58]. Clinical manifestations of NRTI use in human pregnancy [59], especially in HIV-1-uninfected children born to HIV-1-infected mothers, include neurological and cardiac effects evident in a subset of children. However, morphological and molecular manifestations of mitochondrial compromise, which have been frequently observed in children exposed to NRTIs *in utero* [60–62], include aberrant levels of mtDNA, mitochondrial morphological damage, and altered oxidative phosphorylation.

In a comprehensive review, Lee *et al.* report that some toxic side effects observed with NRTI therapy are correlated with the kinetics of incorporation into mtDNA and inhibition of the mitochondrial Pol γ [63]. Others tested a series of FDA-approved NRTIs and provided a method to screen for potential mitochondrial toxicity [64]. Furthermore, they defined

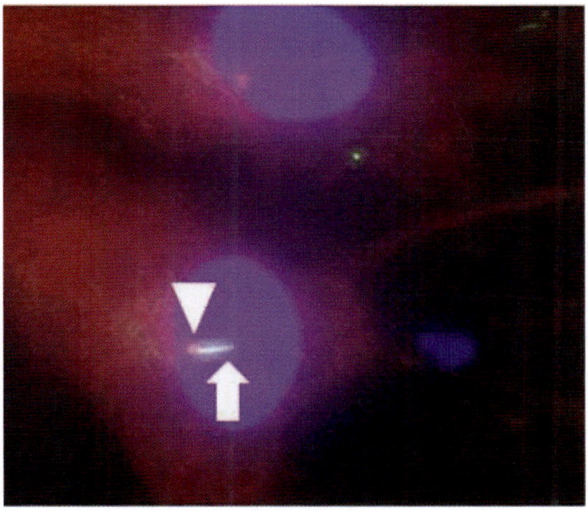

(A)

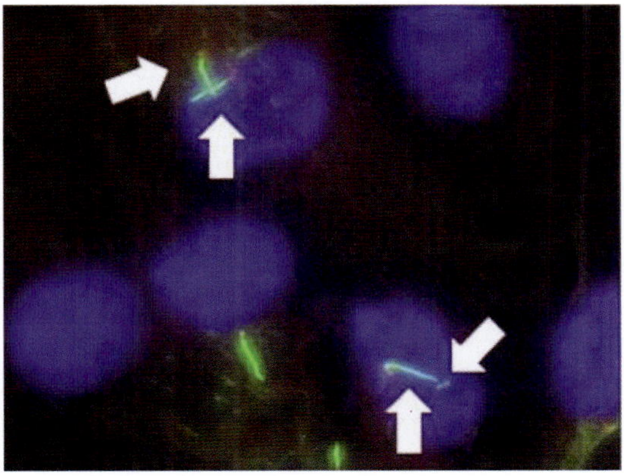

(B)

Figure 4. Formation of cilia in RPE cells treated with AZT. (A) Normal RPE cell showing a primary cilium stained with anti-acetylated tubulin (arrow), anti-pericentrin to label centrosome (arrowhead), and DAPI to stain nuclei (B) RPE cells treated with 20 μM AZT for 72 hours, stained as in (A), and exhibiting more than one cilium per cell (arrows).

a toxicity index for chain terminators to account for relative rates of incorporation versus removal [64]. Whereas incorporation of NRTIs into mtDNA has been widely documented, and it is assumed to be the primary cause of many of the dose-limiting NRTI toxicities, it is not considered to be directly linked with cancer induction.

3.3. *Tumorigenic consequences of AZT exposure in mice*

Transplacental carcinogenesis studies in mice have shown that AZT is a moderately-strong transplacental carcinogen. In the earliest study, pregnant CD-1 mice were exposed to 25 mg AZT/day for the last week of gestation [19], offspring were born naturally, and allowed to grow to adulthood without further exposure. Enhanced tumorigenicity was observed in multiple organs of AZT-exposed mice [19, 65, 66]. One-year-old offsprings of the AZT-exposed CD-1 dams showed increased tumor incidences (up to 60% for liver and lung tumors), and shortened tumor latency, compared to the unexposed mouse offspring [19]. Diwan *et al.* [66] reporting on 2-year-old mouse offspring from the same study, described an uncharacteristic increase in reproductive organ (ovary, seminal vesicle, testicular) and mammary gland tumors in the AZT-exposed mice, compared to the unexposed mice [66]. Parallel to the tumor study, incorporation of AZT into nDNA and mtDNA was examined in multiple tissues from the mouse offspring at birth [67], using an AZT-specific RIA. Values for AZT-DNA incorporation are shown in Table 1.

Studies in a different mouse strain showed similar carcinogenic effects. Female B6C3F1 mice were gavaged daily with 0, 80, 240, or 480 mg AZT/kg body weight during the last 7 days of gestation. At 2 years postpartum, mouse offspring showed significant dose-related trends and increases in hemangiosarcoma of the liver, and hepatic carcinoma in the highest dose AZT group [68].

In a recent report from the National Toxicology Program, Leakey and Olson [69] tested the effects of AZT on the offspring of female mice genetically modified to be sensitive to cancer induction. Mice null for *p53* were given AZT by gavage during the last week of pregnancy, and the pups were given AZT orally for 30–45 weeks following birth. Exposure to AZT *in utero* and after birth caused increases in the rates of liver cancer in the male pups given AZT for 45 weeks, and in malignant lymphomas in both male and female pups exposed to AZT for 30 weeks after birth. The authors concluded that AZT caused liver cancers in male pups exposed to AZT before and following birth, and added that malignant lymphomas in male and female pups may have been related to AZT exposure before and following birth [69].

3.4. *Genotoxic and tumorigenic consequences of AZT exposure in primates*

A transplacental nonhuman primate model was used to model the fetal consequences of NRTI use in human pregnancy. These studies employed a breeding colony of *Erythrocebus patas* (patas) monkeys, a species of small Old World monkeys with placentation similar to the human, and NRTI pharmacokinetics in pregnancy essentially identical to those found in humans [70]. The pregnant patas dams were exposed to clinically-relevant antiretroviral drug combinations in human-equivalent drug protocols, and their offspring were given the same drugs for the first 6 weeks of life. Patas mesenchymal cells, grown out of the patas' bone marrow taken from offspring at birth, 1 and 3 years of age, were examined for genotoxic end points, including CA, MN, and MN containing whole chromosomes (MN+C) (Table 2). Cells from unexposed, age-matched infants were compared to cells from infants exposed to AZT/3TC, AZT/3TC/ABC or AZT/3TC/NVP. This *ex vivo* approach, revealed persistent genotoxicity with typically 5- to 10-fold higher values for the genotoxic biomarkers in the NRTI-exposed patas offspring up to 3 years of age, compared to the unexposed patas (Table 2). These studies were designed to model human clinical protocols as accurately as possible because the infrequency of bone marrow draws in children makes it unlikely that these results will be repeated in children born to HIV-1-infected mothers.

Aneuploidy has been found in children born to HIV-infected mothers. Andre-Schmutz *et al.* [71] demonstrated an increase in aneuploidy in cord blood from infants of HIV-1 infected mothers receiving AZT along with other antiretroviral drugs. The authors karyotyped CD34(+) hematopoietic stem/progenitor cells and found a higher proportion of aneuploid cells in the group exposed *in utero* to AZT, compared with the children exposed to other antiretroviral drugs. Gene expression studies in these same children revealed significant alterations in cell cycle control genes, mitotic checkpoints, and DNA repair genes, compared to the controls [71].

Cancer risk in children born to HIV-1 infected mothers. Taken together, the patas and human studies provide a powerful statement for fetal hematopoietic genotoxicity, which is induced by *in utero* and perinatal exposure to antiretroviral drugs. However, the most important question

Table 2. Comparison of values for %CA, %MN and %MN+C (mean ± SE), among patas treatment groups

Age	Exposure (n)	%CA	%Total MN	%MN + C
Birth	Control (3)	0.57 ± 0.35	1.14 ± 0.16	0.28 ± 0.09
	AZT/3TC (3)	4.50 ± 0.27	2.34 ± 0.36	0.83 ± 0.07
		***P* = 0.0009**	***P* = 0.04**	***P* = 0.009**
1 year	Control (3)	0.74 ± 0.05	0.58 ± 0.07	0.11 ± 0.06
	AZT/3TC/ABC (2)	4.40 ± 0.18[a]	6.23 ± 0.13[a]	3.02 ± 0.22[a]
		***P* = 0.0001**	***P* < 0.0001**	***P* = 0.0005**
3 years	Control (3)	0.49 ± 0.11	0.70 ± 0.21	0.18 ± 0.15
	AZT/3TC/ABC (3)	1.90 ± 0.09[a]	1.47 ± 0.36[a,b]	0.72 ± 0.19[a]
		***P* = 0.0006**	***P* = 0.13**	***P* = 0.05**
3 years	Control (3)	0.49 ± 0.11	0.70 ± 0.21	0.18 ± 0.15
	AZT/3TC/NVP (3)[c]	3.75 ± 1.55[c]	3.11 ± 0.18[b]	1.09 ± 0.14
		***P* = 0.07**	***P* = 0.0009**	***P* = 0.003**

[a]Differences for AZT/3TC/ABC at 1 year vs. 3 years are significant to *P* = *0.0008, 0.002,* and *0.0004,* for %CA, %MN and %MN+C, respectively, demonstrating reduction in AZT/3TC/ABC genotoxicity between 1 and 3 years of age.

[b]At 3 years of age, values for %MN in cultured bone marrow cells from monkeys exposed to AZT/3TC/ABC were significantly lower than those exposed to ABC/3TC/NVP (*P* = 0.01).

[c]For %CA there were 2 monkeys for this group.

remaining relates to potential cancer risk in the children exposed to antiretroviral drugs *in utero,* and born HIV-1-uninfected. Up to now, most of the antiretroviral drug combinations used in human pregnancy do not appear to confer an elevated cancer risk. Two studies in the United States have reported that the relative risk (RR) of cancer for HIV-1-uninfected children exposed to NRTIs *in utero* was generally not significant [72, 73]. In the French Perinatal Cohort Study Group, children exposed *in utero* to AZT/ddI were shown to be at increased risk of leukemia, but because the number of subjects was small the result was not considered robust [74]. A subsequent analysis of cancers in children exposed to ddI revealed an increased risk (hazard ratio 5.5 [2.1–14.4]) when exposure was in the first trimester of pregnancy [75]. However, since ddI is no longer used in pregnancy, the overall results are reassuring.

3.5. *AZT incorporates into DNA of human placenta and infant blood DNA*

In an attempt to determine if the genotoxicity of AZT, and other nucleoside analogs, would compromise human health, placental perfusion experiments were performed. In this model we perfused of human placentas *ex vivo* with three AZT doses. Drug was administered on the maternal site of the organ, and perfusates were recovered on the fetal side at 15 minutes intervals for a period of 2 hours. At 120 minutes the maternal and fetal AZT concentrations reached equivalent levels [27]. Incorporation of AZT into placental DNA showed a dose-response in placentas perfused for 2 hours with AZT [27]. These observations proved that humans can metabolize AZT leading to DNA incorporation. We therefore examined mother–infant pairs exposed to different regimens of HAART, to address directly the human genotoxicity of NRTI use in pregnancy. Peripheral blood from the mother and umbilical cord blood cells from the fetus were isolated at birth, DNA was extracted, and AZT-RIA was performed on the samples (Figure 5). The figure

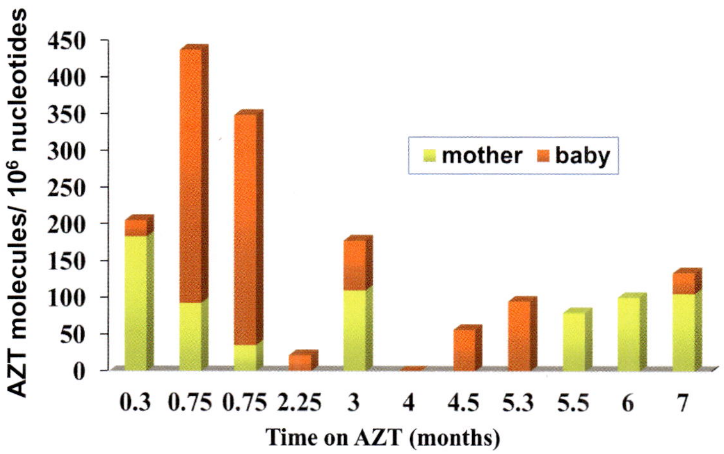

Figure 5. Bar graph of AZT-DNA incorporation in mother (yellow) peripheral blood, and infant (orange) cord blood. AZT-DNA incorporation is expressed as AZT molecules/10^6 nucleotides. Each bar represents exposure of the mother/infant pair for different periods of time. The time indicated on the abscissa is the period just before birth. 0.3 months represents the last 10 days before birth, and 7 months represents the last 7 months of pregnancy.

shows a lack of correlation between the length of exposure and the extent of AZT-DNA incorporation. Furthermore, there was no correlation between the mother and her infant, with respect to extent of AZT-DNA incorporation. Finally, the highest levels of AZT-DNA incorporation were found in two infants whose mothers received AZT only during the last 3 weeks of gestation, whereas infants born after ≥3 months of AZT exposure have lower levels of AZT-DNA incorporation. It is possible to speculate that infants exposed for longer times may develop a resistance mechanism, as this occurs in cultured cells. *In vitro* studies in human lymphoblastoid cells indicated that the expression of the enzyme TK1, responsible for the initial phosphorylation of AZT, is decreased at early passages of cells exposed chronically, and completely abolished at later passages [76]. A similar decrease in TK1 enzyme activity [77, 78], with lack of expression of active TK1 enzyme, was seen in NHMECs that do not incorporate AZT into DNA [79].

4. Conclusions

The purpose of this book is to explore the relationship between DNA damage and human cancer risk. This chapter describes extensive human exposure, occurring *in utero* and resulting in multiple genotoxic events, that has, some 25 years after the initiation of widespread use in human pregnancy, not resulted in an increase in human cancer risk. It is, however, likely that insufficient time has passed to be sure that increased cancer risk in the offspring will not result from NRTI exposures during human pregnancy. The multiple positive genotoxic end points observed in human infants and nonhuman primates, after *in utero* exposure to NRTIs, would suggest that there are biological consequences to these exposures that may result in persistent unwanted sequelae.

Understanding the genotoxicity of the drugs used to treat HIV-1 is paramount in the future of AIDS therapy. Even though the risk–benefit ratio requires that the drugs be used in pregnancy, the underlying mechanisms and their consequences should be studied in depth in order to improve therapy and drug design, and reveal as much as possible about the potential toxic consequences.

Abbreviations:

3TC - (-)	2′,3′-dideoxy 3′-thiacytidine
ABC	Abacavir
AIDS	Acquired Immune Deficiency Syndrome
AZT	zidovudine, 3′-azido-3′-deoxythymidine
AZT-DNA	azidothymidine-DNA
CA	centrosomal amplification
CHO	Chinese hamster ovary
CREST	calcinosis, Raynaud phenomenon, esophageal dysmotility, sclerodactyly, and Telangiectasia
D4T	stavudine
ddI	didanosine
HAART	highly active antiretroviral therapy
HIV-1	human immunodeficiency virus-1
HPLC	high-performance liquid chromatography
LOH	loss of heterozygosity
mtDNA	mitochondrial DNA
MN	micronucleus/micronuclei
MN+C	micronucleus + centrosome
Mf	mutant frequency
nDNA	nuclear DNA
NHMEC	normal human mammary epithelial cell
NRTI	nucleoside reverse transcriptase inhibitor
NNRTI	nonnucleoside reverse transcriptase inhibitor
NVP	Nevirapine
N3-NVP-dT	N3-thymidine adduct of NVP
PBMC	peripheral blood mononuclear cells
PI	protease inhibitor
RIA	radioimmunoassay
RT	reverse transcriptase
TK	thymidine kinase
TK6	human lymphoblast–thymidine kinase heterozygote cell line
WBC	white blood cells

References

1. Furman PA, Fyfe JA, St Clair MH, Weinhold K, Rideout JL, Freeman GA, *et al.* Phosphorylation of 3′-azido-3′-deoxythymidine and selective interaction of the 5′-triphosphate with human immunodeficiency virus reverse transcriptase. *Proc Natl Acad Sci USA* 1986; 83: 8333–8337.
2. Cheng YC, Dutschman GE, Bastow KF, Sarngadharan MG, Ting RYC. Human immunodeficiency virus reverse transcriptase. General properties and its interactions with nucleoside triphosphate analogs. *J Biol Chem* 1987; 262: 2187–2189.
3. Olivero OA. Mechanisms of genotoxicity of nucleoside reverse transcriptase inhibitors. *Environ.Mol.Mutagen.* 2006; 48: 215–223.
4. Palella FJ, Jr., Delaney KM, Moorman AC, Loveless MO, Fuhrer J, Satten GA, *et al.* Declining morbidity and mortality among patients with advanced human immunodeficiency virus infection. HIV Outpatient Study Investigators. *N Engl J Med* 1998; 338: 853–860.
5. Mitsuya H, Weinhold KJ, Furman PA, St Clair MH, Lehrman SN, Gallo RC, *et al.* 3′-Azido-3′-deoxythymidine (BW A509U): an antiviral agent that inhibits the infectivity and cytopathic effect of human T-lymphotropic virus type III/lymphadenopathy-associated virus in vitro. *Proc Natl Acad Sci USA* 1985; 82: 7096–7100.
6. Lim SE, Copeland WC. Differential incorporation and removal of antiviral deoxynucleotides by human DNA polymerase gamma. *J Biol Chem* 2001; 276: 23616–23623.
7. Parker WB, White EL, Shaddix SC, Ross LJ, Buckheit RW, Jr., Germany JM, *et al.* Mechanism of inhibition of human immunodeficiency virus type 1 reverse transcriptase and human DNA polymerases alpha, beta, and gamma by the 5′-triphosphates of carbovir, 3′-azido-3′-deoxythymidine, 2′,3′-dideoxyguanosine and 3′-deoxythymidine. A novel RNA template for the evaluation of antiretroviral drugs. *J Biol Chem* 1991; 266: 1754–1762.
8. Meng Q, Grosovsky AJ, Shi X, Walker VE. Mutagenicity and loss of heterozygosity at the APRT locus in human lymphoblastoid cells exposed to 3′-azido-3′-deoxythymidine. *Mutagenesis* 2000; 15: 405–410.
9. Olivero OA, Borojerdi JP, Semino-Mora C, Ward Y, Poirier MC. Genomic instability induced by AZT in cultured normal human mammary epithelial cells (NHMECs) generates aneuploidy. *Retrovirology* 2006; 3 Suppl 1: 48.
10. Wit FW, Kesselring AM, Gras L, Richter C, van der Ende ME, Brinkman K, *et al.* Discontinuation of nevirapine because of hypersensitivity reactions in

patients with prior treatment experience, compared with treatment-naive patients: the ATHENA cohort study. *Clin Infect Dis* 2008; 46: 933–940.

11. Chu KM, Boulle AM, Ford N, Goemaere E, Asselman V, Van Cutsem G. Nevirapine-associated early hepatotoxicity: incidence, risk factors, and associated mortality in a primary care ART programme in South Africa. *PLOS ONE* 2010; 5: e9183.
12. Bottaro EG, Huberman MJ, Iannella Mdel C, Vesperoni FA, Scapellato PG, Errea S, *et al.* Nevirapine-associated toxicity in clinical practice in Buenos Aires, Argentina. *J Int Assoc Physicians AIDS Care (Chic)* 2010; 9: 306–312.
13. Antunes AM, Wolf B, Oliveira MC, Beland FA, Marques MM. 2′-Deoxythymidine adducts from the anti-HIV drug nevirapine. *Molecules* 2013; 18: 4955–4971.
14. Copeland WC, Chen MS, Wang TS. Human DNA polymerases alpha and beta are able to incorporate anti-HIV deoxynucleotides into DNA. *J Biol Chem* 1992; 267: 21459–21464.
15. Sommadossi JP, Carlisle R, Zhou Z. Cellular pharmacology of 3′-azido-3′-deoxythymidine with evidence of incorporation into DNA of human bone marrow cells. *Mol Pharmacol* 1989; 36: 9–14.
16. Avramis VI, Markson W, Jackson RL, Gomperts E. Biochemical pharmacology of zidovudine in human T-lymphoblastoid cells (CEM). *AIDS* 1989; 3: 417–422.
17. Olivero OA, Beland FA, Poirier MC. Immunofluorescent localization and quantitation of 3′-azido-2′, 3′-dideoxythymidine (AZT) incorporated into chromosomal DNA of human, hamster and mouse cell lines. *Int J Oncol* 1994; 4: 49–54.
18. Agarwal RP, Olivero OA. Chromosomal aberrations in H 9 cells chronically exposed to 3′azido-2′-3′-dideoxythymidine (AZT) *in vitro. Proc Am Assoc Cancer Res* 1994:—.
19. Olivero OA, Anderson LM, Diwan BA, Haines DC, Harbaugh SW, Moskal TJ, *et al.* Transplacental effects of 3′-azido-2′,3′-dideoxythymidine (AZT): tumorigenicity in mice and genotoxicity in mice and monkeys. *J Natl Cancer Inst* 1997; 89: 1602–1608.
20. Olivero OA, Shearer GM, Chougnet CA, Kovacs AA, Landay AL, Baker R, *et al.* Incorporation of zidovudine into leukocyte DNA from HIV-1-positive adults and pregnant women, and cord blood from infants exposed in utero *AIDS* 1999; 13: 919–925.
21. Olivero OA, Reddy MK, Pietras SM, Poirier MC. Plasma drug levels compared with DNA incorporation of 3′-azido-3′-deoxythymidine (AZT) in

adult cynomolgus (*Macaca fascicularis*) monkeys. *Exp Biol Med* 2001; 226: 446–449.

22. Poirier MC, Olivero OA, Walker DM, Walker VE. Perinatal genotoxicity and carcinogenicity of anti-retroviral nucleoside analog drugs. *Toxicol Appl Pharmacol* 2004; 199: 151–161.
23. Olivero OA, Shearer GM, Chougnet CA, Kovacs AA, Baker R, Stek AM, *et al.* Incorporation of zidovudine into cord blood DNA of infants and peripheral blood DNA of their HIV-1-positive mothers. *Ann N Y Acad Sci* 2000; 918: 262–268.
24. Fang JL, Beland FA. Development of a novel (32)P-postlabeling method for the analysis of 3′-azido-3′-deoxythymidine. *Cancer Lett* 2000; 153: 25–33.
25. Meng Q, Su T, Olivero OA, Poirier MC, Shi X, Ding X, *et al.* Relationships between DNA incorporation, mutant frequency, and loss of heterozygosity at the TK locus in human lymphoblastoid cells exposed to 3′-azido-3′-deoxythymidine. *Toxicol Sci* 2000; 54: 322–329.
26. Poirier MC, Patterson TA, Slikker W, Jr., Olivero OA. Incorporation of 3′-azido-3′-deoxythymidine (AZT) into fetal DNA and fetal tissue distribution of drug after infusion of pregnant late-term rhesus macaques with a human-equivalent AZT dose. *J Acquir Immune Defic Syndr* 1999; 22: 477–483.
27. Olivero OA, Parikka R, Poirier MC, Vahakangas K. 3′-azido-3′-deoxythymidine (AZT) transplacental perfusion kinetics and DNA incorporation in normal human placentas perfused with AZT. *Mutat Res-Fund Mol M* 1999; 428: 41–47.
28. Nusbaum NJ, Joseph PE. AZT incorporation into mitochondria: study in a human myeloid cell line. *DNA Cell Biol* 1996; 15: 363–366.
29. Arnaudo E, Dalakas M, Shanske S, Moraes CT, DiMauro S, Schon EA. Depletion of muscle mitochondrial DNA in AIDS patients with zidovudine-induced myopathy. *Lancet* 1991; 337: 508–510.
30. Lewis W, Simpson JF, Meyer RR. Cardiac mitochondrial DNA polymerase-gamma is inhibited competitively and noncompetitively by phosphorylated zidovudine. *Circ Res* 1994; 74: 344–348.
31. Vazquez-Padua MA, Starnes MC, Cheng YC. Incorporation of 3′-azido-3′-deoxythymidine into cellular DNA and its removal in a human leukemic cell line. *Cancer Commun* 1990; 2: 55–62.
32. Wu Q, Beland FA, Chang CW, Fang JL. Role of DNA repair pathways in response to Zidovudine-induced DNA damage in immortalized human liver THLE2 cells. *Int J Biomed Sci* 2013; 9: 18–25.

33. Walker DM, Poirier MC, Campen MJ, Cook DL, Jr., Divi RL, Nagashima K, *et al.* Persistence of mitochondrial toxicity in hearts of female B6C3F1 mice exposed in utero to 3′-azido-3′-deoxythymidine. *Cardiovasc Toxicol* 2004; 4: 133–153.
34. IARC. *Some Antiviral and Antineoplastic Drugs, and Other Pharmaceutical Agents*. World Health Organization, International Agency for Research On Cancer; 2000.
35. Walker DM, Rucher R, Funk K, Hardisty JH, Walker VE. Transplacental carcinogenicity of 3′-azido-3′-deoxythymine in B6C3F1 mice and F344 rats. *The Toxicologist* 2004; 73.
36. Skalski V, Liu SH, Cheng YC. Removal of anti-human immunodeficiency virus 2′,3′-dideoxynucleoside monophosphates from DNA by a novel human cytosolic 3′-->5′ exonuclease. *Biochem Pharmacol* 1995; 50: 815–821.
37. Torres SM, Walker DM, Carter MM, Cook DL, Jr., McCash CL, Cordova EM, *et al.* Mutagenicity of zidovudine, lamivudine, and abacavir following in vitro exposure of human lymphoblastoid cells or in utero exposure of CD-1 mice to single agents or drug combinations. *Environ Mol Mutagen* 2007; 48: 224–238.
38. Meng Q, Su T, O'Neill JP, Walker VE. Molecular analysis of mutations at the HPRT and TK loci of human lymphoblastoid cells after combined treatments with 3′-azido-3′-deoxythymidine and 2′,3′-dideoxyinosinedagger. *Environ Mol Mutagen* 2002; 39: 282–295.
39. Borojerdi JP, Ming J, Cooch C, Ward Y, Semino-Mora C, Yu M, *et al.* Centrosomal amplification and aneuploidy induced by the antiretroviral drug AZT in hamster and human cells. *Mutat Res* 2009; 665: 67–74.
40. Yu M, Ward Y, Poirier MC, Olivero OA. Centrosome amplification induced by the antiretroviral nucleoside reverse transcriptase inhibitors lamivudine, stavudine, and didanosine. *Environ Mol Mutagen* 2009; 50: 718–724.
41. Divi KV, Ward Y, Poirier MC, Olivero OA. Use of ciliogenesis to detect aneugens: the role of primary Cilia. *Curr Protoc Toxicol* 2015; 66: 3.13.1–3.13.18.
42. Wu YW, Xiao Q, Jiang YY, Fu H, Ju Y, Zhao YF. Synthesis, in vitro anticancer evaluation, and interference with cell cycle progression of N-phosphoamino acid esters of zidovudine and stavudine. *Nucleosides Nucleotides Nucleic Acids* 2004; 23: 1797–1811.
43. Roskrow M, Wickramasinghe SN. Acute effects of 3′-azido-3′-deoxythymidine on the cell cycle of HL60 cells. *Clin Lab Haematol* 1990; 12: 177–184.

44. Heagy W, Crumpacker C, Lopez PA, Finberg RW. Inhibition of immune functions by antiviral drugs. *J Clin Invest* 1991; 87: 1916–1924.
45. Viora M, Di Genova G, Rivabene R, Malorni W, Fattorossi A. Interference with cell cycle progression and induction of apoptosis by dideoxynucleoside analogs. *Int J Immunopharmacol* 1997; 19: 311–321.
46. Chandrasekaran B, Kute TE, Duch DS. Synchronization of cells in the S phase of the cell cycle by 3′-azido- 3′-deoxythymidine: implications for cell cytotoxicity. *Cancer Chemother Pharmacol* 1995; 35: 489–495.
47. Olivero OA, Tejera AM, Fernandez JJ, Taylor BJ, Das S, Divi RL, *et al.* Zidovudine induces S-phase arrest and cell cycle gene expression changes in human cells. *Mutagenesis* 2005; 20: 139–146.
48. D'Assoro AB, Lingle WL, Salisbury JL. Centrosome amplification and the development of cancer. *Oncogene* 2002; 21: 6146–6153.
49. Doxsey S, Zimmerman W, Mikule K. Centrosome control of the cell cycle. *Trends Cell Biol* 2005; 15: 303–311.
50. Carroll PE, Okuda M, Horn HF, Biddinger P, Stambrook PJ, Gleich LL, *et al.* Centrosome hyperamplification in human cancer: chromosome instability induced by p53 mutation and/or Mdm2 overexpression. *Oncogene* 1999; 18: 1935–1944.
51. Lingle WL, Lutz WH, Ingle JN, Maihle NJ, Salisbury JL. Centrosome hypertrophy in human breast tumors: implications for genomic stability and cell polarity. *Proc Natl Acad Sci USA* 1998; 95: 2950–2955.
52. Pihan GA, Purohit A, Wallace J, Knecht H, Woda B, Quesenberry P, *et al.* Centrosome defects and genetic instability in malignant tumors. *Cancer Res* 1998; 58: 3974–3985.
53. Satoh MS, Lindahl T. Enzymatic repair of oxidative DNA damage. *Cancer Res* 1994; 54: 1899s–1901s.
54. Hinchcliffe EH, Miller FJ, Cham M, Khodjakov A, Sluder G. Requirement of a centrosomal activity for cell cycle progression through G1 into S phase. *Science* 2001; 291: 1547–1550.
55. Mikule K, Delaval B, Kaldis P, Jurcyzk A, Hergert P, Doxsey S. Loss of centrosome integrity induces p38-p53-p21-dependent G1-S arrest. *Nat Cell Biol* 2007; 9: 160–170.
56. Yu M, Poirier MC, Olivero OA. Genomic Instability and centrosomal amplification induced by the antiretroviral AZT. *Environ Mol Mutagen* 2008; 49: 548.
57. Lewis W, Dalakas MC. Mitochondrial toxicity of antiviral drugs. *Nat Med* 1995; 1: 417–422.
58. Copeland WC. Defects in mitochondrial DNA replication and human disease. *Crit Rev Biochem Mol Biol* 2012; 47: 64–74.

59. Foster C, Lyall H. HIV and mitochondrial toxicity in children. *J Antimicrob Chemother* 2008; 61: 8–12.
60. Poirier MC, Divi RL, Al-Harthi L, Olivero OA, Nguyen V, Walker B, *et al.* Long-term mitochondrial toxicity in HIV-uninfected infants born to HIV-infected mothers. *J Acquir Immune Defic Syndr* 2003; 33: 175–183.
61. Divi RL, Walker VE, Wade NA, Nagashima K, Seilkop SK, Adams ME, *et al.* Mitochondrial damage and DNA depletion in cord blood and umbilical cord from infants exposed in utero to Combivir. *AIDS* 2004; 18: 1013–1021.
62. Brogly SB, Ylitalo N, Mofenson LM, Oleske J, Van Dyke R, Crain MJ, *et al.* In utero nucleoside reverse transcriptase inhibitor exposure and signs of possible mitochondrial dysfunction in HIV-uninfected children. *AIDS* 2007; 21: 929–938.
63. Lee H, Hanes J, Johnson KA. Toxicity of nucleoside analogues used to treat AIDS and the selectivity of the mitochondrial DNA polymerase. *Biochemistry* 2003; 42: 14711–14719.
64. Johnson AA, Ray AS, Hanes J, Suo Z, Colacino JM, Anderson KS, *et al.* Toxicity of antiviral nucleoside analogs and the human mitochondrial DNA polymerase. *J Biol Chem* 2001; 276: 40847–40857.
65. Zhang Z, Diwan BA, Anderson LM, Logsdon D, Olivero OA, Haines DC, *et al.* Skin tumorigenesis and Ki-ras and Ha-ras mutations in tumors from adult mice exposed in utero to 3′-azido-2′,3′-dideoxythymidine. *Mol Carcinog* 1998; 23: 45–51.
66. Diwan BA, Riggs CW, Logsdon D, Haines DC, Olivero OA, Rice JM, *et al.* Multiorgan transplacental and neonatal carcinogenicity of 3′-azido-3′-deoxythymidine in mice. *Toxicol Appl Pharmacol* 1999; 15: 82–99.
67. Olivero OA, Poirier MC. Preferential incorporation of 3′-azido-2′, 3′-dideoxythymidine (AZT) into telomeric DNA and Z-DNA-containing regions of chinese hamster ovary cells. *Mol Carcinog* 1993; 8: 81–88.
68. Walker DM, Malarkey DE, Seilkop SK, Ruecker FA, Funk KA, Wolfe MJ, *et al.* Transplacental carcinogenicity of 3′-azido-3′-deoxythymidine in B6C3F1 mice and F344 rats. *Environ Mol Mutagen* 2007; 48: 283–298.
69. Olson GL, Julian. Toxicology and carcinogenicity studies of 3′-Azido-3′-deoxythymidine (CAS NO. 30516-87-1) in genetically modified C3B6. 129F1-Trp53tm1Brd N12 haploinsufficient mice. *Natl Toxicol Program Genet Modif Model Rep* 2013: 1–236.
70. Divi RL, Doerge DR, Twaddle NC, Shockley ME, St.Claire MC, Harbaugh JW, *et al.* Metabolism and pharmacokinetics of the combination Zidovudine plus Lamivudine in the adult Erythrocebus patas monkey determined by

liquid chromatography-tandem mass spectrometric analysis. *Toxicol Appl Pharmacol* 2007; 226: 206–211, doi:10.1016/j.taap.2007.09.007

71. Andre-Schmutz A, Dal-Cortivo L, Six E, Kaltenbach S, Cocchiarella F, Lechenadec J, *et al.* Genotoxic signature in cord blood cells of newborns exposed in utero to a zidovudine-based antiretroviral combination. *J Infect Dis* 2013; 208: 235–243, (Submitted 2012 as a companion to this paper).
72. Brogly S, Williams P, Seage GR, 3rd, Van Dyke R. In utero nucleoside reverse transcriptase inhibitor exposure and cancer in HIV-uninfected children: an update from the pediatric AIDS clinical trials group 219 and 219C cohorts. *J Acquir Immune Defic Syndr* 2006; 41: 535–536.
73. Ivy W, 3rd, Nesheim SR, Paul SM, Ibrahim AR, Chan M, Niu X, *et al.* Cancer among children with perinatal exposure to HIV and antiretroviral medications — New Jersey, 1995–2010. *J Acquir Immune Defic Syndr* 2015; 70: 62–66.
74. Benhammou V, Warszawski J, Bellec S, Doz F, Andre N, Lacour B, *et al.* Incidence of cancer in children perinatally exposed to nucleoside reverse transcriptase inhibitors. *AIDS* 2008; 22: 2165–2177.
75. Hleyhel M, Goujon S, Delteil C, Vasiljevic A, Luzi S, Stephan JL, *et al.* Risk of cancer in children exposed to didanosine in utero. *AIDS* 2016; 30: 1245–1256.
76. Vazquez IL, Olivero OA, Poirier MC. Altered AZT metabolism may induce cellular drug resistance in human cells. *Environ Mol Mutagen* 2004; 44: 234. Abstract # 186.
77. Antonelli G, Turriziani O, Verri A, Narciso P, Ferri F, D'Offizi G, *et al.* Long-term exposure to zidovudine affects *in vitro* and *in vivo* the efficiency of phosphorylation of thymidine kinase. *AIDS Res Hum Retroviruses* 1996; 12: 223–228.
78. Jacobson MA, Gundacker H, Hughes M, Fischl M, Volberding P. Zidovudine side effects as reported by black, Hispanic, and white/non-Hispanic patients with early HIV disease: combined analysis of two multicenter placebo-controlled trials. *J Acquir Immune Defic Syndr Hum Retrovirol* 1996; 11: 45–52.
79. Olivero OA, Ming JM, Das S, Vazquez IL, Richardson DL, Weston A, *et al.* Human inter-individual variability in metabolism and genotoxic response to zidovudine. *Toxicol Appl Pharmacol* 2008; 228: 158–164.

Chapter 10

Cancer Chemoprevention *via* Reduction of DNA Damage

Gary D. Stoner

The Ohio State University
and Medical College of Wisconsin, 7475 Amsterdam Road,
Manhattan, Montana 59741, USA

Abstract

Most environmental carcinogens produce DNA damage that leads to mutational events in important cellular genes. Ultimately, the accumulation of mutations in multiple genes leads to loss of growth control and cancer development. Investigations over the past several decades have resulted in the identification of numerous naturally-occurring and synthetic compounds that exhibit the ability to inhibit the development of cancer by a variety of mechanisms. The ability to reduce DNA damage and ensuing mutational events associated with this damage is an important role for these compounds, generally referred to as chemopreventive agents. This chapter discusses mechanisms by which chemopreventive agents reduce DNA damage and adduct formation using specific compounds and vitamins as examples. The principal criteria for selecting these compounds for discussion is the fact that data were available on the mechanisms by which they reduce not only DNA damage but also tumorigenesis in specific organs. Protocols for the identification of agents that reduce DNA damage in animal model systems are given and some results from human clinical trials are presented. It is important to recognize that there may be many naturally-occurring chemopreventive agents that have not as yet been identified and many of them undoubtedly have the ability to reduce DNA damage. Probably the most effective agents are those that have a dual role of reducing DNA damage and mutational events as well as cellular and molecular events associated with tumor promotion/progression.

1. Introduction

Cancer chemoprevention can be defined as the inhibition or reversal of carcinogenesis by the administration of one or more chemical entities, either as naturally occurring constituents of the diet or as individual drugs [1]. During the past several decades, numerous compounds have been shown to exhibit chemopreventive activity in animals but, unfortunately, relatively few are effective in humans. In 1985, Lee Wattenberg, a pioneer in and major contributor to the field of cancer chemoprevention, described a classification scheme for chemopreventive agents based essentially on the time period during which agents appear to exhibit activity in animal models of carcinogenesis [2]. In this scheme, chemopreventive agents are classified as blocking agents and suppressing agents. Blocking agents are defined as inhibitors of tumor initiation and suppressing agents as inhibitors of tumor promotion/progression. Through various mechanisms, blocking agents reduce DNA damage caused by chemical carcinogens and oxidative radicals, thereby preventing the mutational changes required for tumor initiation. In contrast, suppressing agents act through multiple mechanisms to influence cellular events associated with tumor progression such as proliferation, differentiation, apoptosis, inflammation, angiogenesis, tissue invasion, immune response, and others. Although chemopreventive agents have been classified as either blocking agents or suppressing agents, it should be recognized at the outset that the most promising agents are those that exhibit both blocking and suppressing activities, preferably in several animal tumor models.

2. Design of Chemoprevention Studies: DNA Adduct Endpoint

This chapter is concerned with the identification of blocking agents, or agents that inhibit tumor development by reducing DNA damage. Ideally, putative blocking agents should be identified in animals by initially determining whether or not they reduce DNA damage in the target tissue in which the carcinogen under study induces cancer. A typical scheme for assessing whether or not an agent inhibits DNA adduct formation in a specific organ is shown in Figure 1. Essentially, a short-term study is undertaken in which the putative blocking agent is administered to a suitable number of animals (usually 6–10 per group) at two or three dose levels in either the diet, drinking water (if stable in drinking water), or by gavage, for 1 or 2 weeks before administration of the carcinogen. The dose levels of the carcinogen and the characteristics of the diet are described in the bioassay section. Typically, the carcinogen is administered intraperitoneally (i.p.), or by another suitable route, and at one or more time points after carcinogen administration (usually up to 24 hours) the target tissue is removed, DNA isolated and purified, and the formation of carcinogen-DNA adducts quantified using high-performance liquid chromatography (HPLC) and confirmed by mass spectrometry [3], or by the ^{32}P-postlabeling assay [4] using appropriate adduct standards. DNA adducts in tissues can also be measured by immunoassays and immunohistochemical methods using antisera specific for DNA adducts or carcinogen-modified DNA, and by accelerator

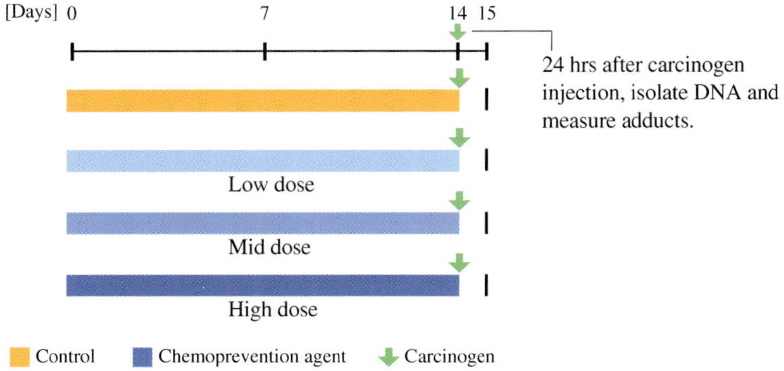

Figure 1. Protocol for DNA adduct study.

mass spectrometry, which detects adducts *in vivo* using low-level radiolabeled (^{3}H or ^{14}C) compounds [3].

Because the liver is a major organ for carcinogen metabolism, the effect of the chemotherapeutic agent is often measured in the liver and in the target organ, when the liver itself is not the target organ. Statistical analysis is performed to determine if the agent under study causes a significant reduction in the binding of the carcinogen to DNA, or preferably and more specifically, if there is a reduction in the formation of carcinogen-DNA adducts. Interestingly, in many investigations, the effect of putative chemopreventive agents on levels of carcinogen-DNA damage has been measured only *after* the agent has been evaluated for its ability to inhibit tumor development in animals. Perhaps one reason for this is, as will be described below, that there is not always a close correlation between the ability of blocking agents to reduce DNA damage and their ability to inhibit tumor development in target organs.

3. Design of Chemoprevention Studies: Tumor Endpoint

A typical protocol for evaluating the ability of a putative blocking agent to inhibit tumor development in a target organ(s) is shown in Figure 2. After approximately 2 weeks in quarantine at the test site, relatively young animals (usually 5–8-week-old rodents), preferably both male and female, are randomly divided into an untreated control group, a group treated with

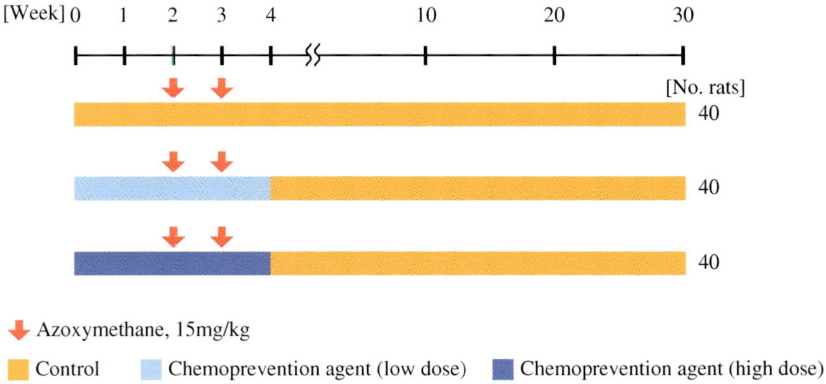

Figure 2. Protocol for blocking colon cancer in rats.

agent only, a group treated with carcinogen only, and groups treated with carcinogen plus agent. The number of animals per group is determined by statistical analysis of prior data with the specific animal model system to determine how many animals would be required to demonstrate a significant reduction in the tumor response in the carcinogen/agent group(s) versus the group treated with carcinogen only. Ideally, the agent is tested at three dose levels to achieve a dose–response effect; typically at 0.8, 0.5, and 0.2 of a predetermined maximum tolerated dose (MTD) [5].

As for the DNA damage studies described earlier, putative blocking agents are typically evaluated for antitumor potential by administration of the agent 1 or 2 weeks prior to, and during, treatment of animals with carcinogen. The rationale for administering the agent prior to and during, treatment with carcinogen is to permit the agent to bind to and inhibit P450 enzymes involved in the metabolic activation of the carcinogen to DNA damaging metabolites, and/or to stimulate phase II enzymes involved in metabolic detoxification of the carcinogen. Importantly, the agent is administered in a synthetic diet, such as the American Institute of Nutrition 76A (AIN-76A) or the AIN-93 diet, which presumably contain no known cancer-inhibitory agents. Alternative methods of treatment include administration of the agent by gavage, or in the drinking water if the agent is stable in water. Intravenous or i.p. administration of agents is not practical for these studies because chemopreventive agents are likely to be given to humans over long periods of time. Because the agent is being evaluated for blocking activity only, agent treatment stops either at the time carcinogen treatment ceases, or shortly thereafter. During the length of the bioassay, animal weight data are collected on a weekly basis and the animals are examined frequently for clinical signs of toxicity, which might include things such as tremor, hair loss, lethargy, and so forth. At the end of the bioassay, the tumor response, defined as the percentage of animals with tumors (tumor incidence), number of tumors per animal (tumor multiplicity), and size of tumors in animals treated with agent plus carcinogen, is compared to that in animals treated with carcinogen only, to determine if the agent causes a significant and dose-dependent reduction in the tumor response. Ideally, the tumor response in untreated control animals is zero or no more than might be expected to occur spontaneously, and the response in animals treated with agent only is similar to untreated controls. This confirms lack of carcinogenic potential of the agent itself.

Overall, this approach has been very useful for identifying blocking agents that reduce DNA damage and inhibit tumorigenesis. However, it should be recognized that DNA damage occurs during all stages of the carcinogenesis process, not just the initiation stage. Thus, agents that reduce DNA damage are likely to be more effective when administered during the entire period of a carcinogenesis bioassay or, when translating to humans, during their entire lifetime. Literally hundreds of chemopreventive agents exhibit blocking activity by reducing DNA damage and it is not possible to document all of them in this chapter. Therefore, using specific agents as examples, I will describe mechanisms by which blocking agents exhibit the ability to reduce DNA damage and adduct formation.

4. Mechanisms for Inhibition of DNA Damage

It is well known that most environmental procarcinogens must first be metabolically activated to electrophilic forms that damage DNA. The positively charged electrophilic species binds to DNA, forming adducts that result in base mispairing and mutation in key genes such as the oncogenes and tumor suppressor genes. Agents that block these events can be assigned to one or more of four major categories: *(1) inhibitors of Phase I cytochrome P450 enzymes; (2) inducers of Phase I cytochrome P450 enzymes; (3) inducers of Phase II detoxifying enzymes such as uridine diphosphate (UDP)-glucuronyltransferase, glutathione S-transferase (GST) and glutathione peroxidase; and (4) scavengers of electrophiles and oxidative free radicals* [1, 2, 5, 6]. Each of these types of mechanism is discussed below, but it is well to recognize that many blocking agents exhibit activity in more than one of these categories and the net effect is a significant reduction in DNA damage at one or more organ sites.

4.1. *Inhibitors of phase I cytochrome P450 enzymes*

Inhibitors of cytochrome P450 enzymes (CYP-450s) prevent the metabolic activation of carcinogens to their ultimate carcinogenic forms. Some of the most potent inhibitors of CYP P450s are the isothiocyanates. Isothiocyanates are present as glucosinolates in cruciferous vegetables such as broccoli, watercress, Brussel sprouts, cauliflower, and radishes, and they are derived from hydrolysis of the glucosinolates by the enzyme myrosinase [7].

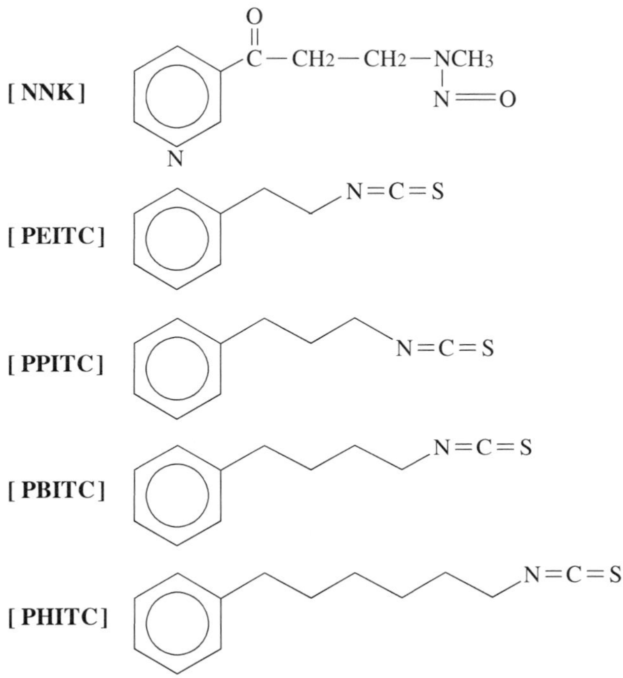

Figure 3. NNK and isothiocyanates.

Structures of some of the most potent isothiocyanates with inhibitory activity toward CYP 450 enzymes are shown in Figure 3. Phenethyl isothiocyanate (PEITC) is a naturally occurring compound, whereas phenylpropyl isothiocyanate (PPITC), phenylbutyl isothiocyanate (PBITC), and phenylhexyl isothiocyanate (PHITC) are synthetic compounds. Because of their structural relationship to N-nitrosamine carcinogens, both naturally occurring and synthetic isothiocyanates are among the most potent chemopreventive agents for N-nitrosoamine-induced cancers in rodents. Most studies have focused on the ability of the isothiocyanates to inhibit lung carcinogenesis by the nicotine-derived, tobacco-specific carcinogen, 4-(methylnitrosamino)-1-(3-pyridyl)-1-butanone (NNK, Figure 3). NNK is an abundant carcinogen in tobacco smoke and it induces lung and other cancers in multiple animal species [8]. In an initial study, PEITC, derived from the naturally occurring glucosinolate, gluconasturtin, at a concentration of 3 mmol/kg diet, inhibited NNK-induced lung tumors in F344 rats by about 50% [9]. This reduction in lung tumorigenesis correlated closely

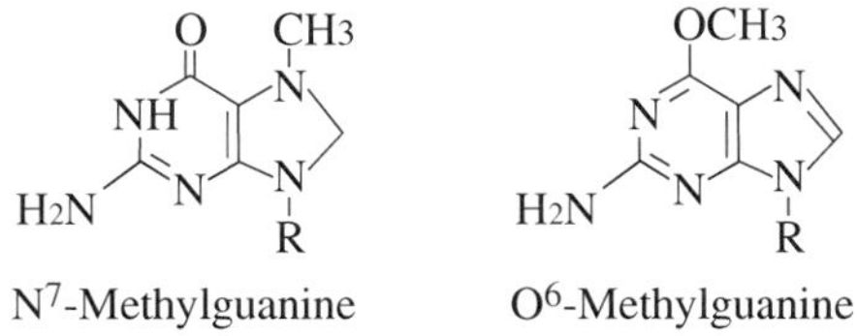

Figure 4. NNK methylated guanine adducts.

with a 50% reduction in the formation of NNK-induced methylated adducts (Figure 4) in lung DNA.

Studies in the laboratory of C. S. Yang and colleagues have provided much of the available knowledge of the effects of PEITC on CYP P450 enzyme activities likely related to the metabolism of NNK and other nitrosamines [10–13]. They observed that the administration of a single dose of PEITC to F344 rats resulted in decreased liver N-nitrosodimethylamine demethylase activity (mainly due to P450 2E1) and increased liver pentoxyresorufin O-dealkylase (PROD) and P450 2B1 proteins [11). The liver microsomal ethoxyresorufin O-dealkylase (EROD) activity (mainly due to P450 1A) and erythromycin N-demethylase (mainly due to P450 3A) were also decreased and then recovered. PEITC reduced rates of oxidative metabolism of NNK in liver microsomes by 40%–60%, and in lung microsomes by 60%–70%. PEITC also had effects on phase II enzyme activities in that the single dose treatment resulted in a fivefold increase in NAD(P) H:quinone oxidoreductase activity and a 1.5-fold increase in GST activity. In contrast, hepatic UDP-glucuronosyltransferase and sulfotransferase activities were decreased. The study concluded that PEITC selectively effects xenobiotic metabolism enzymes in liver and lung and is especially effective in inhibiting the P450-dependent oxidation of NNK in lung. In a later report by the same laboratory, human P450 1A2 was also found to be involved in the metabolic activation of NNK and its activity was decreased *in vitro* by PEITC [13]. The effects of other isothiocyanates on the metabolism of NNK in rodent lung and liver have been investigated and the reader is referred to the publication by Guo *et al.* [14] for more extensive information.

In addition to altering metabolism, Phase I CYP inhibitors have been shown to reduce tumorigenesis. For example, the effect of isothiocyanate

Table 1. Effects of isothiocyanates on DNA methylation and tumarigenesis by NNK in A/J mouse lung.

Treatment	O^6-Methylguanine (µmol/mol guanine)	Lung tumors/ mouse
NNK	41.2 ± 3.8	7.9 ± 0.4
PEITC	17.2 ± 0.2	4.1 ± 0.8
PPTTC	13.8 ± 0.4	0.2 ±0.1
PBITC	13.4 ± 0.5	0.2 ± 0.1
PPEITC	11.1 ± 0.2	0.3 ± 0.1
PHITC	9.9 ± 0.1	0.1 ± 0.1

* Data taken from Morse *et al. Cancer Res.* 51: 1846–1850.

chain length on NNK-induced lung tumorigenesis in strain A/J mice was investigated using dietary PEITC, PPITC, PBITC, and PHITC [15]. The relative potency of the isothiocyanates in inhibiting lung tumorigenesis was in the order of PHITC > PBITC > PPITC > PEITC, and the tumor data correlated closely with the relative ability of these compounds to inhibit NNK-induced methylated adducts in lung DNA [16, 17] (Table 1). PHITC was remarkably potent, inhibiting NNK-induced lung tumorigenicity by >80% when administered at a dose 50-fold lower than NNK itself. The relative ability of the isothiocyanates to inhibit lung tumorigenesis appeared to be due their relative binding affinity to CYP P450 enzymes responsible for the metabolic activation of NNK [14].

At the same dietary concentration used in the rat lung bioassay (3 mmol/kg diet), PEITC inhibited *N*-nitrosomethylbenzylamine (NMBA)-induced esophageal squamous cell papillomas in F344 rats by nearly 100% [18]. This may have human relevance in that the carcinogen NMBA is found in high-risk regions for esophageal squamous cell carcinoma in China [19]. In explant cultures of rat esophagus, PEITC (10–100 µM) produced a dose-dependent (53%–97%) inhibition of the binding of NMBA metabolites to esophageal DNA and in the levels of DNA methylation at the N^7 (20%–89%) and O^6 (55%–93%) positions of guanine [18]. It is well known that the O^6-methylguanine adduct is the most important for tumorigenesis because it is poorly repaired. One of the P450s responsible for the metabolism of NMBA in rat esophagus *in vivo* appears to be P450 2A3 [20]; however, it is not known whether PEITC inhibits 2A3 activity in rat esophagus.

Somewhat different results from mouse lung were obtained in rat esophagus when isothiocyanates of increasing chain length were evaluated for their ability to inhibit NMBA-tumorigenesis [21]. As in mouse lung, PPITC was more active than PEITC, however, PBITC was inactive; these data correlated closely with levels of methylated adducts in DNA. In another study, PHITC was found to promote NMBA-induced tumorigenesis in rat esophagus and this was related to a significant *increase* in the number of dysplastic foci in esophagi treated with PHITC + NMBA versus NMBA alone [22]. PHITC appeared to promote esophageal tumorigenesis through the induction of cytochrome P450 2A3 [23] and by stimulation of cell proliferation, presumably through some underlying toxicity, but this conclusion was tentative [24]. Unfortunately, the effect of PHITC on DNA methylation in NMBA-treated rat esophagus was not determined. Interestingly, PHITC was also found to promote azoxymethane (AOM)-induced tumorigenesis in rat colon [25]. These data illustrate the importance of utilizing more than one animal model system in evaluating the efficacy of chemopreventive agents.

Recently, PEITC was evaluated for its potential to inhibit the metabolic activation of NNK in 82 active tobacco smokers [26]. Smokers were administered PEITC (10 mg in 1 mL of olive oil, 4×/day for 1 week) and urine was collected to determine the NNK metabolic activation ratio ([pyridine-D4] hydroxyl acid/total [pyridine-D4]NNAL). PEITC reduced the metabolic activation ratio of NNK by 7.7% suggesting that the isothiocyanate might be chemopreventive against NNK-induced lung cancer in active smokers.

Chemoprevention has also been examined for the carcinogenic polycyclic aromatic hydrocarbons (PAHs), the metabolism of which is described in this volume (see chapters by S. S. Hecht, R. M. Santella, and A. Weston). Major human exposures occur by inhalation tobacco, as well as indoor and outdoor ambient environments, and by ingestion of charcoal-broiled meats and other dietary sources [27]. DNA adducts of carcinogenic PAHs have been detected in multiple human tissues including leukocytes, lung, oral cavity, and in the esophagus of rural Chinese who have high rates of esophageal cancer [28–31]. Benzo(a)pyrene (BP) is the prototypical PAH, and its metabolism and carcinogenic potential have

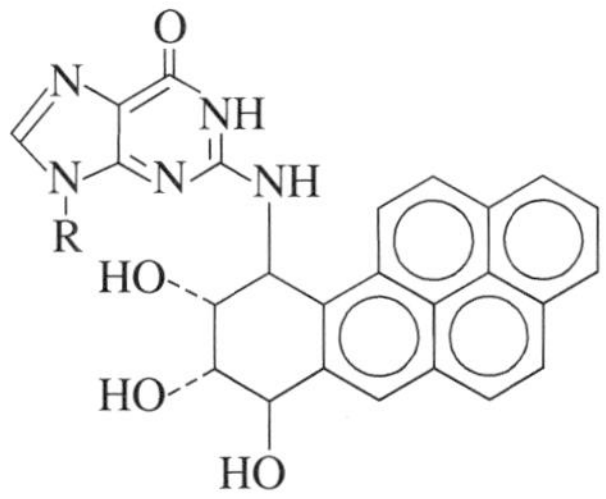

Figure 5. Benzo(a)pyrene — guanine adduct.

been investigated extensively. BP has a complicated pattern of metabolism, only a fraction of which leads to the formation of the DNA adduct associated with cancer induction, the r7, t8, t9-trihydroxy-c-10-(N^2deoxyguanosyl)-7, 8, 9, 10-tetrahydro-benzo[*a*]pyrene (BPdG, Figure 5) [27, 32, 33]. Numerous chemoprevention studies have been conducted to identify agents that inhibit BP carcinogenesis and some of the agents found to be active inhibitors include benzyl isothiocyanate [34], β-naphthoflavone [35], butylated hydroxyanisole [36], curcumin [37], ellagic acid (EA) [38], quercetin [39], resveratrol [40], and sulforaphane [41]. Numerous agents have been shown to inhibit either the binding of BP to DNA or the formation of BP-DNA adducts *in vitro* and *in vivo*, far more than can be documented in this chapter. As an example, treatment of cultured human bronchial epithelial BEAS-2B cells with [^{3}H]BP resulted in increased binding of the carcinogen to DNA for 72 hours, and this correlated with increased CYP1A1/1B1 and microsomal epoxide hydrolase (mEH) enzyme activities [42]. Simultaneous treatment of the cells with BP and 5,7-dimethoxyflavone (DMF), 3′,4′-DMF or resveratrol resulted in >75% inhibition of BP-DNA binding, and this correlated with reductions in the catalytic activity of CYP1A1/1B1, with 3′,4′-DMF being the most potent inhibitor. Only 3′,4′-DMF decreased mEH activity, 5,7-DMF had no effect, and resveratrol increased mEH activity [42].

With respect to tobacco carcinogenesis, the concept of combined chemoprevention with agents that inhibit DNA adduct formation from both N-nitrosamine carcinogens such as NNK and NNAL and PAHs such as BP has been proposed by Hecht [43]. This strategy would appear to be a very logical approach to reduce tobacco-induced DNA damage in the lung and other target organs for tobacco-induced cancer.

4.2. *Inducers of cytochrome P450 enzymes*

Wattenberg classified compounds that characteristically induce pronounced *increases* in CYP P450 activity as Type B inhibitors (2). Type B inhibitors also enhance the activity of major conjugating systems such as GST and UDP-glucuronosyltransferase. Because these agents can both activate and detoxify chemical carcinogens, their net effect is the more rapid elimination of carcinogen metabolites from the body. Type B inhibitors have been shown to inhibit tumorigenesis resulting from treatment of animals with BP, 7,12-dimethylbenz[a]anthracene (DMBA), *N*-2-fluorenylacetamide (2-AAF), urethane, and aflatoxin (AFB1) but, unfortunately, in none of these studies was the effect of the inhibitor on the extent of DNA adduct formation by the carcinogen measured. The Type B inhibitor, indole-3-carbinol (I-3-C), is a potent inducer of P450 enzymes and is chemopreventive at multiple organ sites [44–46]. However, compounds that induce P450 enzymes may be protective in one situation and promoting in another. For example, when dietary I-3-C was given to trout before and during administration of AFB1, the number of liver tumors was reduced [47]. In contrast, when I-3-C was administered after carcinogen exposure, there was a significant dose-dependent increase in the number of tumors. Presumably, the inhibitory effect when I-3-C was given before and during carcinogen administration was due to reduced DNA adduct formation, whereas the promoting effect may have been due to effects on postinitiation events such as rates of cell proliferation or apoptosis. In a multiorgan rat model in which DMBA, AFB1, and AOM were used to induce mammary, liver, and colon carcinogenesis, respectively, I-3-C treatment led to different effects in different organs [48]. The rats were treated initially with the three carcinogens; dietary I-3-C was begun 1 week after the last carcinogen treatment and continued for 25 weeks. I-3-C treatment resulted in a delay in latency of mammary tumor formation, but did not affect mammary tumor incidence or multiplicity. In the colon, the protocol produced a 40% decrease in aberrant crypt foci formation; however, in the liver, it strongly induced GST-positive foci in both carcinogen-treated (fourfold increase in volume percent foci) and in vehicle control rats (a 69-fold increase). These results suggest that, in chemoprevention studies, it is important to evaluate the putative preventative agent in both anti-initiation and antipromotion/progression protocols, and to examine for tumor-inhibitory/enhancing effects at multiple organ sites. This may be especially important for Type B inhibitors but undoubtedly pertains to all chemoprevention agents under investigation. The development of DNA adduct data in such studies could be most useful in interpreting the results.

4.3. *Inducers of phase II enzymes*

Inducers of phase II detoxifying enzymes such as GST, UDP-glucuronyltransferase, glutathione peroxidase, NAD(P)H:quinone reductase (QR), epoxide hydrolase (EH), and heme oxygenase are preferred to cytochrome P450 inducers because they are less likely to be carcinogenic themselves. Perhaps the most elegant studies to demonstrate a relationship between phase II enzyme induction, reduced DNA adduct formation, and inhibition of tumorigenesis have been done with oltipraz, a dithiolethione investigated for its effects on aflatoxin carcinogenesis in the liver. In initial publications by Kensler *et al.* [49, 50], in rats fed aflatoxins, the effect of dietary administration of oltipraz (0.01%–0.1% of diet) on aflatoxin metabolism, DNA adduct formation (Figure 6), and hepatic carcinogenesis, as measured by the volume of liver occupied by γ-glutamyltranspeptidase-positive foci, was determined. The effects of oltipraz on all parameters measured were impressive. Treatment with oltipraz at all doses reduced by

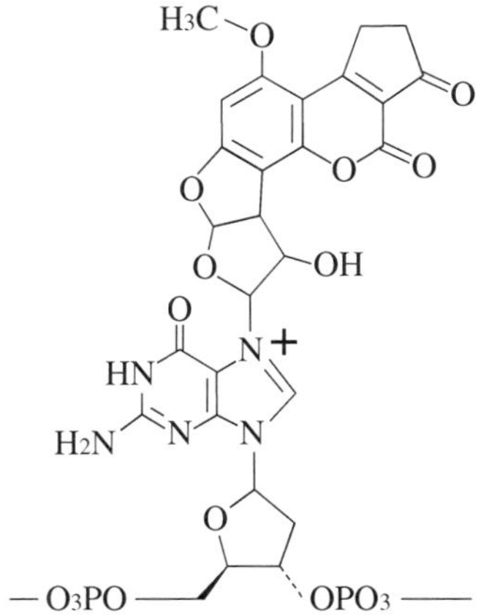

AFB1-N7-Gua

Figure 6. Aflatoxin guanine adduct.

Table 2. Effects of oltipraz on aflatoxin-induced γ-glutamyl transpeptidase-positive foci in rat liver.

Dose of oltipraz (% in diet)	Area (cm^2) of liver examined/rat	No. of foci/cm^2	Foci vol (% of total liver)
Experiment 1			
Control	4.90 ± 0.30[a]	1.60 ± 0.28	0.109 ± 0.025
0.1	5.18 ± 0.27	0.11 ± 0.09[b]	0.003 ± 0.003
Experiment 2			
Control	7.68 ± 0.25	1.41 ± 0.29	0.435 ±0.192
0.01	7.26 ± 0.25	0.44 ± 0.11[b]	0.031 ± 0.008[b]
0.02	16.08 ± 0.47	0.21 ± 0.10[b]	0.008 ± 0.005[b]
0.04	15.49 ± 0.25	0.17 ± 0.06[b]	0.008 ± 0.003[b]
0.075	18.35 ± 0.52	0.03 ± 0.01[b]	0.002 ± 0.001[b]

[a]Mean ± SE.
[b]Differs from control, at P< 0.01.
*Data taken from Kensler *et al. Cancer Res.* 47: 4271–4277.

>90% the volume of liver occupied by γ-glutamyltranspeptidase-positive foci (Table 2). The levels of aflatoxin-DNA adducts in the liver were reduced between 40% and 80% in rats fed increasing doses of oltipraz. These results suggested that there may not be a 1:1 correlation between oltipraz-caused reduction in hepatic tumorigenesis and its ability to reduce the formation of AFB1-DNA adducts in the liver; that is, the levels of reduction in DNA adduct formation may underestimate the reduction in hepatic tumor formation. A major observation in this study was the marked ability of the higher doses of oltipraz to increase the specific activity of GSTs, presumably leading to more efficient detoxification of aflatoxin. At the lowest dietary concentration (0.01%), the only inductive effects observed were on the activities of selected CYP-450 monooxygenases. The authors concluded that the protection afforded by oltipraz may be due to its ability to stimulate detoxification pathways for AFB1 as well as modify its oxidative metabolism. A later study by the same group confirmed the ability of dietary oltipraz to reduce the volume of γ-glutamyl-transpeptidase-positive foci in the liver of AFB1-treated rats and, importantly, the progressive development of hepatocellular carcinomas

[51]. Oltipraz afforded complete protection against the development of both AFB1-induced hepatocellular adenomas and carcinomas. Subsequent studies by Kensler, Groopman, and others confirmed that oltipraz and related dithiolethiones act principally by inducing the activities of phase II enzymes, especially the glutathione-*S*-transferases, in the liver and other tissues in both rodents and humans [50, 52, 53]. Mechanistically, oltipraz increased phase II conjugation of the ultimate carcinogenic species, aflatoxin-8,9-oxide, in AFB1-treated rats yielding higher rates of excretion of aflatoxin–mercapturic acid in the urine [52]. In a subsequent study by this same group, rats were fed either a control diet or a diet supplemented with 0.03% oltipraz and then treated with AFB1 to assess the use of serum and urinary biomarkers as indices of chemopreventive efficacy [54]. In rats fed oltipraz, the overall reductions in the levels of hepatic DNA adducts, urinary aflatoxin-N7-guanine adducts, and serum aflatoxin-albumin adducts, over 2 weeks of AFB1 exposure, were 76%, 62%, and 66%, respectively.

These similarities in reductions of the levels of urinary and serum biomarkers relative to target organ DNA adduct burden indicated that these biomarkers could be useful in assessing the efficacy of oltipraz in humans exposed to aflatoxin. Indeed, a clinical trial of oltipraz was conducted in residents of Qidong, China, who are at increased risk for liver cancer due, in part, to consumption of AFB1-contaminated foods [55]. Treatment with 125-mg oltipraz daily resulted in increased phase II conjugation of aflatoxin-8,9-oxide, yielding higher rates of excretion of aflatoxin–mercapturic acid in urine [56]. The increased formation of the urinary aflatoxin–mercapturic acid was inversely associated with levels of aflatoxin-DNA adducts formed in liver and excreted in urine. Serum aflatoxin–albumin levels were reduced only in subjects treated with high-dose oltipraz (500 mg weekly) but these reductions were not consistent during treatment [57]. Because the metabolism of aflatoxin leads to the formation of oxidized DNA as well as carcinogen-DNA adducts, the urinary excretion of the oxidized nucleoside, 8-oxo-2′-deoxyguanosine (8-oxo-dG) was measured in the China trial and found not to be affected by treatment of the residents with oltipraz [58].

Another compound investigated for its ability to inhibit AFB1-induced rat hepatic tumorigenesis is the synthetic triterpenoid

1-[2-cyano-3-,12-dioxooleana-1,9(11)-dien-28-oyl]imidazole (CDDO-Im) [59]. At a dietary concentration of 1 µmol/kg body weight, CDDO-Im produced an 85% reduction in the formation of preneoplastic liver lesions (GST-P-positive foci) in AFB1-treated rats and a >99% reduction at 100 µmol/kg body weight. This correlated reasonably well with reduction in the levels of aflatoxin-DNA adducts which were ~40%–90% over the CDDO-Im concentrations of 1–100 µmol/kg body weight. Because the triterpenoids are known to be strong inducers of phase II enzymes, changes in mRNA levels of genes involved in aflatoxin metabolism were measured in rat liver following a single dose of 1 µmol/kg body weight of CDDO-Im. GSTA2, GSTA5, EPHX1, and AFAR transcripts were found to be elevated 6 hours after CDDO-Im treatment and this induction of phase II genes was nuclear factor E2-related factor 2 (Nrf2) dependent. The authors concluded that at low micromolar doses, CDDO-Im induced cytoprotective genes, inhibited DNA adduct formation, and dramatically blocked hepatic tumorigenesis in rats. In fact, its activity in the rat hepatocarcinogenesis model was found to exceed that of oltipraz about 100-fold [59].

The monoterpene compounds perillyl alcohol and d-limonene are found in numerous dietary sources and were shown by Elegbede and Gould [60] to reduce the formation of aflatoxin-DNA adducts in rat liver. The precise mechanism(s) by which these compounds reduced adduct formation were not investigated in this study; however, limonene has been shown to induce phase II hepatic drug metabolizing enzymes [61]. Chlorophyllin, a mixture of semisynthetic water-soluble derivatives of chlorophyll, was evaluated for its ability to influence aflatoxin metabolism in residents of Qidong, China [62]. Ingestion of 100 mg of chlorophyllin a day for 3 months resulted in an overall 55% reduction in urinary levels of aflatoxin-N^7-guanine adducts. The mechanism(s) of action of chlorophyllin does not appear to be due to effects on phase II enzymes involved in aflatoxin metabolism. Rather, it may act as an "interceptor molecule" that binds to directly to AFB1, diminishing its absorption into blood and increasing its elimination in feces [63]. Intervention with chlorophyllin, or supplementation of diets with foods rich in chlorophyll, may be a practical means to reduce risk for the development of liver and other cancers caused by aflatoxin or other environmental agents. Chlorophyllin also appears to be an ideal chemopreventive agent because of its widespread availability at low cost and its lack of known toxicities.

Most, if not all, of the compounds that induce phase II enzymes appear to act through the Keap1-Nrf2 signaling pathway [59]. This pathway has been described in detail in the chapter by Kensler and Groopman of this volume and will be mentioned only briefly here. Compounds that induce phase II genes interact with cysteines in Keap 1, permitting the transcription factor Nrf2 to escape proteosomal degradation and accumulate in the nucleus [64]. Nrf2 then binds with small Maf proteins to antioxidant response elements (AREs) in the promoter regions of phase II genes and induces their expression. Thus, expression levels of the phase II genes are tightly regulated by NRF2. Other categories of genes similarly regulated by Nrf2 include antioxidative/anti-inflammatory genes, stress response genes, and carcinogen-metabolizing enzymes [59, 65]. Because multiple genes in these categories are associated with cancer development, the Keap1-Nrf2 signaling pathway is an important target for chemoprevention.

4.4. *Scavengers of electrophiles and oxidative free radicals*

The final category of naturally occurring and synthetic agents that intervene in the carcinogenesis process by reducing DNA damage are the scavengers of electrophiles and of oxidative-free radicals. Scavenging agents are compounds that physically react with the metabolically activated (electrophilic) forms of carcinogens and oxygen-free radicals. EA, *N*-acetylcysteine (NAC), and chlorophyllin (mentioned earlier) are three examples of compounds that can bind to carcinogens and/or their metabolites and reduce their ability to form adducts with DNA. EA reacts directly with r7,t8-dihydroxy-t-9,10-epoxy-7,8,9,10-tetrahydrobenzo[a]pyrene (BPDE) to form both *cis*- and *trans*-adducts [66]. Because BPDE is the ultimate carcinogenic metabolite of BP, its removal by conjugation with ellagic acid may well account for inhibition of BPDE-induced mutagenicity and carcinogenicity [67, 68]. As stated previously, chlorophyllin appears to serve as an "interceptor molecule" that binds directly to aflatoxin and, potentially, other carcinogens reducing its absorption into blood and resulting in its rapid elimination through the feces [63]. It seems likely that many other naturally occurring and synthetic compounds are capable of binding to carcinogens and/or their metabolites and reducing their ability to elicit DNA damage, but this mechanism of their action has not been identified because agents are not typically evaluated for binding activity.

Aerobic metabolism of environmental toxins and multiple substrates in the diet, and endogenous defense reactions mediated by inflammatory leukocytes, can result in the formation of reactive oxygen and nitrogen species such as superoxide anion radical, hydrogen peroxide, hydroxyl radical, singlet oxygen, nitric oxide, and peroxy-nitrite [69]. These reactive species can bind to and damage important cellular macromolecules such as DNA, lipids, proteins, and carbohydrates, contributing to multiple diseases including cardiovascular disease, arthritis, diabetes, and cancer, as well as to the aging process. Depending upon the nature of the reactive oxygen species (ROS), several types of DNA damage, including modification and loss of nucleotide bases, strand breaks, or DNA–protein cross-linking, are possible. In humans, it has been estimated that oxidative stress — a sustained attack of ROS on DNA — causes 10^4–10^5 hits per cell per day [70]. Oxidative damage affects primarily the guanine base in DNA, inducing the formation of mutagenic 8-oxo-7,8-dehydro-2′-deoxyguanosine (8-oxo-dG or 8-OH-dG) (Figure 7). Occurring in both nuclear and mitochondrial DNA, 8-oxo-dG is the most commonly measured biomarker of oxidative damage [71]; however, other DNA adducts are also formed [72]. Swenberg [73] measured steady-state endogenous DNA damage in mammalian cells and found that 8-oxo-dG occurs at an average frequency of 2400 adducts per cell. 8-oxo-dG can be measured in urine, lymphocytes, and in internal tissues such as liver, lungs, and kidneys. A detailed description of methods

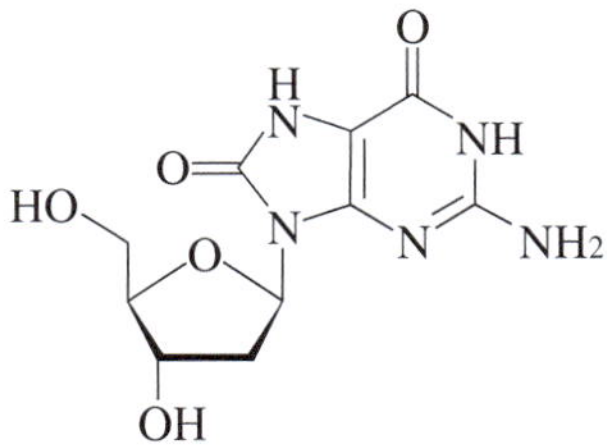

Figure 7. 8-Oxo-2′-deoxyguanosine.

to measure 8-oxo-dG in biological samples can be found in the review by Knasmüller *et al.* [74]. Although 8-oxo-dG is efficiently removed by excision DNA repair, it can be promutagenic leading to G to T transversion mutations [69]. Such mutations have been detected in genes involved in the multistage process of cancer development including key genes in multiple signal transduction pathways. In addition, Nishida *et al.* [75] examined 8-oxo-dG levels and promoter methylation of 11 tumor suppressor genes in liver biopsy samples from patients with chronic hepatitis C and found that increased levels of 8-oxo-dG was highly correlated with promoter methylation of the suppressor genes potentially leading to their reduced levels of expression. These results suggest that oxidative adducts in DNA may be involved in epigenetic events in carcinogenesis.

A very large number of dietary constituents are capable of scavenging ROS and potentially reducing the formation of oxidative adducts in DNA. Perhaps, the initial studies on antioxidants as inhibitors of carcinogenesis were done with the carotenoids (precursors of vitamin A), vitamin C, and vitamin E. Carotenoids scavenge ROS, activate signaling pathways leading to antioxidant response activation, and upregulate the expression of detoxifying enzymes [69]. The carotenoid, β-carotene, efficiently scavenges singlet molecular oxygen ($^1O_2$) and peroxyl radicals, especially at low oxygen tension; however, under certain conditions such as high oxygen tension it may act as a prooxidant [69]. This may be the reason for the increased occurrence of lung cancer in active smokers administered β-carotene [76, 77]. The effects of β-carotene on the levels 8-oxo-dG in cultured lung epithelial cells following incubation with hydrogen peroxide and in ferret lung after chronic exposure to BP have been examined [78]. β-carotene increased *in vitro* hydroxyl radical formation and enhanced 8-OHdG adduct formation in cultured lung epithelial cells but had no effect on 8-oxo-dG levels in ferret lung. This result may have been due to increased excision DNA repair capacity in ferret lung. In contrast, intake of β-carotene suppressed the formation of 8-oxo-dG adducts in the lungs of mice exposed to diesel exhaust particles suggesting that its effects on 8-oxo-dG adduct formation may be dependent upon the agent used to produce the adducts [79].

Lycopene, a carotenoid found primarily in tomatoes, watermelon, and other plant-based foods, is the most potent carotenoid antioxidant through its ability to quench singlet oxygen [80]. Lycopene is thought to be protective against prostate cancer and one study showed that consumption of tomato sauce-based pasta by prostate cancer patients for 3 weeks resulted in significantly reduced oxidative damage (ratio of 8-oxo-dG to dG) in both leukocyte and prostate tissue DNA [81]. This result suggests a possible role for tomato sauce or lycopene in the prevention/treatment of prostate cancer.

An extensive literature exists regarding investigations of the chemopreventive activity of vitamin A (retinol) and its oxidation products, retinal, and retinoic acid, which occur in various isomeric forms. Collectively, these oxidation products are termed retinoids and most of their antiproliferative activity is mediated by the compounds, all-*trans*-retinoic acid and various *cis*-retinoic acid isomers. A large number of synthetic retinoids have also been evaluated for chemopreventive activity. These agents are thought to function principally through binding to two groups of nuclear receptors, the retinoic acid receptor (RAR), and the retinoid X receptor (RXR) families. These receptors act either directly on retinoid response elements or indirectly by modifying the responses of other transcription factors, ultimately regulating the expression of several hundred genes [69]. Thus, the retinoids appear to influence multiple signaling events associated with tumor progression. However, retinol and its derivatives have also been shown in several studies to influence the formation of DNA adducts. For example, vitamin A and several of its derivatives were shown to inhibit the formation of the BP metabolites, BP-7,8-diol and BPDE (the ultimate carcinogenic metabolite), and of BPdG adducts in an *in vitro* reaction catalyzed by rat liver microsomes [82]. The inhibitory effect was through modulation of the activities of microsomal enzymes. Similarly, retinol, retinal, retinoic acid, and retinyl acetate were shown to inhibit the formation of AFB1-DNA adducts in an *in vitro* liver microsomal enzyme assay [83]. Finally, pretreatment of mice with retinol decreased the formation of DNA adducts in the kidney caused by two mycotoxins, ochratoxin A and zearalenone, that have been implicated in numerous mycotoxicosis in farm animals [84]. These studies clearly indicate that vitamin A and its derivatives have the ability to reduce DNA damage caused by various carcinogens and toxins.

Another vitamin with the ability to scavenge ROS is vitamin C [85]. It acts as a potent antioxidant in nonenzymatic reactions through its ability to scavenge free radicals (hydroxyl, peroxyl, superoxide anion, and nitrogen dioxide) and nonradical species (hypochlorous acid, singlet oxygen, nitrooxide, and peroxynitrite) [70]. Vitamin C is readily bioavailable and the effects of vitamin C supplementation on biomarkers of oxidative DNA damage has been measured in numerous studies [86]. Most studies assessed levels of 8-oxo-Gua, 8-oxo-dG, or strand breaks in DNA of leukocytes isolated from subjects exposed to oxidative stress. Most human studies have demonstrated an inverse correlation between plasma levels of vitamin C and levels of oxidized DNA base lesions in leukocyte cells. Several studies have found that vitamin C-mediated decreases in the level of oxidized DNA adducts is linked to an increase of these adducts in plasma and urine suggesting that vitamin C does not block oxidative DNA damage per se, but rather stimulates DNA repair [86].

Vitamin D has not been thoroughly evaluated for antioxidant activity. It is not known whether it scavenges oxidative radicals, but it has been shown to positively influence the levels of the antioxidant enzymes, catalase, and superoxide dismutase [87]. It has not been shown to reduce the levels of oxidative DNA adducts in tissues; however, in combination with vanadium, 1-alpha,25-dihydroxyvitamin D3 reduced 1,2-dimethylhydrazine-induced O^6-methylguanine adducts in the DNA of rat colon and colon tumorigenesis [88]. In this study, it was not clear whether the chemopreventive effect was due to vanadium, vitamin D3, or the combination. In addition, treatment of rats with 1-alpha,25-dihydroxyvitamin D3 before and during administration of the carcinogen 2-acetylaminofluorene (2-AAF) reduced the occurrence of chromosomal aberrations, and the formation of N-(deoxyguanosin-8-yl)-2-aminofluorene DNA adducts and DNA strand breaks [89]. The major effects of vitamin D on carcinogenesis appear to be associated with its antiproliferative, proapoptotic, and differentiation properties [85].

The term vitamin E is used for a group of naturally occurring tocopherols (α, β, γ, or δ) and their corresponding tocotrienols (α, β, γ, or δ). Various fat-soluble components of vitamin E such as α- and γ-tocopherol play an essential role in cellular defense against various oxidants [85]. They are capable of scavenging the peroxyl radical and terminating the chain oxidation of polyunsaturated fatty acids (PUFAs) [90, 91]. Alpha-tocopherol tends to be the most active antioxidant of the four tocopherols; however, γ-tocopherol is a more effective antioxidant against reactive nitrogen species due to its ability to form 5-nitro-γ-tocopherol, a reaction not feasible with α-tocopherol [92]. Results from preclinical studies with α-tocopherol as a chemopreventive agent against breast, prostate, and colon cancer have not been consistent. However, γ-tocopherol, the most common form of vitamin E in the US diet, has demonstrated anti-inflammatory and antitumor activity in animal models of breast, colon, and prostate cancers [93, 94]. Like vitamins A and D, the anticancer activities of vitamin E include its ability to induce apoptosis, inhibit cell proliferation and angiogenesis, and reduce inflammation. Human studies have been conducted to determine if vitamin E, alone and in combination with other vitamins, is associated with levels of DNA damage. One study found that low plasma levels of vitamin E were associated with increased levels of PAH-DNA adducts in circulating mononuclear cells of male cigarette smokers who were null for glutathione S-transferase M_1 ($GSTM_1$) [95]. This was not the case for smokers with wild-type $GSTM_1$. In another study, PAH-DNA adducts and 8-oxo-dG adducts were measured in oral epithelial cells and in lymphocytes of heavy cigarette smokers given the combination of vitamins C, E, and β-carotene for 6 months. Results from this study indicated that the combined vitamin treatment did not reduce the levels of either PAH-DNA or 8-oxo-dG adducts in oral cells or in lymphocytes of the smokers [96]. Thus, it appears that additional studies are necessary to further evaluate the ability of antioxidant vitamins to intervene in the formation of carcinogen-derived- and oxidative-DNA adducts in humans.

Another series of compounds that have antioxidant potential and have been shown to reduce both carcinogen-induced and oxidative DNA adducts in tissues are the polyphenolic phytochemicals. There is an enormous literature on the chemopreventive activity of the polyphenols including their ability to reduce DNA damage, and the summary by Kunnumakkara *et al.* [97] is a credible initial discussion. Perhaps, the most extensively investigated polyphenols, with respect to inhibition of DNA damage and adduct formation, are resveratrol, EA, silymarin, the catechins in tea, turmeric, and its active constituent, curcumin. Collectively, these polyphenols have been shown to prevent oral, esophagus, gastric, colon, breast, liver, leukemia, neuroblastoma, glioma, lung, skin, bladder, and prostate cancers in preclinical animal models and several either have been examined or are currently under evaluation in humans. All of these compounds inhibit hydrogen peroxide-induced DNA damage in multiple *in vitro* model systems and many also inhibit DNA adduct formation from carcinogens such as AFB1, BP, DMBA, methylnitrosourea (MNU), ultraviolet light, and others [97]. Resveratrol alone inhibits chemically induced DNA damage in at least 39 *in vitro* and *in vivo* preclinical model systems [97]. A major question is that of how effective are the polyphenols in protecting against DNA damage in humans, particularly in view of the fact that many of them have fairly limited bioavailability. A limited number of human trials suggest that silymarin, silibinin, the green tea polyphenols, resveratrol, and curcumin exhibit chemopreventive effects in humans, but more definitive information will likely be obtained both from ongoing and future studies.

5. Summary

Epidemiologic data provide evidence that the consumption of vegetables and fruit reduces risk for the development of cancer in humans. Vegetables and fruit are known to contain a large number of individual compounds that have been shown in preclinical models to exhibit chemopreventive potential in one or more organ sites, and most of them function through a variety of mechanisms to inhibit both the initiation and promotion/progression of cancer. The ability to reduce DNA damage and ensuing mutational events associated with this damage is clearly an important role of these chemopreventive agents. This chapter discusses mechanisms by which these agents reduce DNA damage and adduct formation using specific compounds and vitamins as examples. One of the principal criteria for selecting the specific compounds for discussion is the fact that data were available on the mechanisms by which they reduce not only DNA damage but also tumorigenesis in specific organs. This chapter presents only a fraction of the available data on agents that are known to have the ability to reduce DNA damage, and undoubtedly numerous additional chemopreventive agents in our diet have not, as yet, been identified. The available data are derived from vegetables and fruit that have been studied, but there are many more fruits and vegetables, and herbs that may contain even more effective chemopreventive agents than those discovered to date. Ultimately, for effective chemoprevention, selected agents must be cost effective, easily isolated or synthesized, minimally toxic or nontoxic at effective concentrations, and stable under physiological conditions, with favorable pharmacokinetic profiles, known mechanism(s) of action, and the potential to reduce DNA damage as well as tumor risk, preferably at multiple organ sites. Although today's investigations are largely focused on the ability of putative chemopreventive agents to affect tumor progression, clearly there is merit in investigating reduction in DNA damage and mutagenesis, because agents that can do both are likely to be the most effective.

Abbreviations

2-AAF	N-2-fluorenylacetamide
AFB1	aflatoxin B_1
AIN	American Institute of Nutrition
AOM	azoxymethane
ARE	antioxidant response element
BP	benzo(a)pyrene
BPdG	r7, t8, t9-trihydroxy-c-10-(N^2deoxyguanosyl)-7, 8, 9, 10-tetrahydro-benzo[*a*]pyrene
BPDE	r7,t8-dihydroxy-t-9,10-epoxy-7,8,9, 10-tetrahydrobenzo[*a*]pyrene
CDDO-IM	1-[2-cyano-3-,12-dioxooleana-1,9(11)-dien-28-oyl]imidazole
DMBA	7,12-dimethylbenz[a]anthracene
EA	ellagic acid
EH	epoxide hydrolase
EROD	ethoxyresorufin O-dealkylase
GST	glutathione-S-transferase
HPLC	high-performance liquid chromatography
I-3-C	indole-3-carbinol
i.p.	intraperitoneal
mEH	microsomal epoxide hydrolase
MTD	maximum tolerated dose
NAC	N-acetylcysteine

NMBA	N-nitrosomethylbenzylamine
NNK	4-(methylnitrosamino)-1-(3-pyridyl)-1-butanone
NNN	N-nitrosonornicotine
Nrf2	nuclear factor E2-related factor 2
PEITC	2-phenylethyl isothiocyanate
PPITC	3-phenylpropyl isothiocyanate
PBITC	4-phenylbutyl isothiocyanate
PHITC	6-phenylhexyl isothiocyanate
PROD	pentoxyresorufin O-dealkylase
QR	quinone reductase
RAR	retinoic acid receptor
RXR	retinoid X receptor
UDP	uridine diphosphate
N7-me-dG	N7-methyldeoxyguanosine
O^6-me-dG	O6-methyldeoxyguanosine
8-oxo-dG or 8-OH-dG	8-oxo-7,8-dehydro-2′-deoxyguanosine
N7-dG-AFB1	N7-deoxyguanosine-aflatoxin B1
5, 7 DMF	5, 7-dimethoxyflavone
3′, 4′ DMF	3′, 4′-dimethoxyflavone

References

1. Stoner GD, Morse MA, Kelloff GJ. Perspectives in cancer chemoprevention. *Env Health Persp* 1997; 105: 945–954.
2. Wattenberg LW. Chemoprevention of cancer. *Cancer Res* 1985; 45: 1–8.
3. Poirier MC, Santella RM, Weston A. Carcinogen macromolecular adducts and their measurement. *Carcinogenesis* 2000; 21: 353–359.
4. Phillips DH. Detection of DNA modifications by the 32P-postlabeling assay. *Mutat Res* 1997; 378: 1–12.
5. Morse MA, Stoner GD. Cancer chemoprevention: principles and prospects. *Carcinogenesis* 1993; 14: 1737–1746.
6. Kelloff GJ, Boone CW, Steele VE, Crowell JA, Lubet R, Sigman CC. Chemopreventive drug development: perspectives and progress. *Cancer Epidemiol Biomarkers Prev* 1994; 3: 85–98.
7. Thornalley PJ. Isothiocyanates: mechanism of cancer chemopreventive action. *Anticancer Drugs* 2002; 13: 331–338.

8. Hecht SS. Metabolic activation and detoxification of tobacco-specific nitrosamines — a model for cancer prevention strategies. *Drug Metab Rev* 1994; 26: 373–390.
9. Morse MA, Wang C-X, Stoner GD, Mandal S, Conran PB, Amin SG, *et al.* Inhibition of 4-(methylnitrosamino)-1-(3-pyridyl)-1-butanone-induced DNA adduct formation and tumorigenicity in the lung of F344 rats by dietary phenethyl isothiocyanate. *Cancer Res* 1989; 49: 549–553.
10. Smith TJ, Guo Z-Y, Thomas PE, Chung F-L, Morse MA, Eklind K, *et al.* Metabolism of 4-(methylnitrosamino)-1-(3-pyridyl)-1-butanone in mouse lung microsomes and its inhibition by isothiocyanates. *Cancer Res* 1990; 50: 6817–6822.
11. Guo Z, Smith TJ, Wang E, Sadrieh N, Ma Q, Thomas PE, *et al.* Effects of phenethyl isothiocyanate, a carcinogenesis inhibitor, on xenobiotic-metabolizing enzymes and nitrosamine metabolism in rats. *Carcinogenesis* 1992; 13: 2205–2210.
12. Smith TJ, Guo Z, Li C, Ning SM, Thomas PE, Yang CS. Mechanisms of inhibition of 4-(methylnitrosamino)-1-(3-pyridyl)-1-butanone (NNK) bioactivation in mouse by dietary phenethyl isothiocyanate. *Cancer Res* 1993; 53: 3276–3282.
13. Smith TJ, Guo Z, Guengerich FP, Yang CS. Metabolism of 4-(methylnitrosamino)-1-(3-pyridyl)-1-butanone (NNK) by human cytochrome P450 1A2 and its inhibition by phenethyl isothiocyanate. *Carcinogenesis* 1996; 17: 809–813.
14. Guo Z, Smith TJ, Wang E, Eklind KI, Chung F-L, Yang CS. Structure-activity relationships of arylalkyl isothiocyanates for the inhibition of 4-(methylnitrosamino)-1-(3-pyridyl)-1-butanone metabolism and the modulation of xenobiotic-metabolizing enzymes in rats and mice. *Carcinogenesis* 1992; 14: 1167–1173.
15. Morse MA, Eklind KI, Amin SG, Hecht SS, Chung F-L. Effects of alkyl chain length on the inhibition of NNK-induced lung neoplasia in A/J mice by arylalkyl isothiocyanates. *Carcinogenesis* 1989; 10: 1757–1759.
16. Morse MA, Amin SG, Hecht SS, Chung F-L. Effects of aromatic isothiocyanates on tumorigenicity, O^{6}-methylguanine formation, and metabolism of the tobacco-specific nitrosamine 4-(methylnitrosamino)-1-(3-pyridyl)-1-butanone in A/J mouse lung. *Cancer Res* 1989; 49: 2894–2897.
17. Morse MA, Eklind KI, Hecht SS, Jordan KG, Choi C-I, Desai DH, *et al.* Structure-activity relationships for inhibition of 4-(methylnitrosamino_1-3-pyridyl)-1-butanone lung tumorigenesis by arylalkyl isothiocyanates in A/J mice. *Cancer Res* 1991; 51: 1846–1850.

18. Stoner GD, Morrissey DT, Heur Y-H, Daniel EM, Galati AJ, Wagner SA. Inhibitory effects of phenethyl isothiocyanate on *N*-Nitrosobenzylmethylamine carcinogenesis in rat esophagus. *Cancer Res* 1991; 51: 2063–2068.
19. Yang CS. Research on esophageal cancer in China: a review. *Cancer Res* 1980; 40: 2633–2644.
20. Gopalakrishnan R, Morse MA, Lu J, Weghorst, CM, Sabourin CL, Stoner GD, *et al.* Expression of cytochrome P450 2A3 in rat esophagus: relevance to N-nitrosobenzylmethylamine. *Carcinogenesis* 1999; 20: 885–891.
21. Wilkinson JT, Morse MA, Kresty LA, Stoner GD. Effect of alkyl chain length on inhibition of N-nitrosomethylbenzylamine-induced esophageal tumorigenesis and DNA methylation by isothiocyanates. *Carcinogenesis* 1995; 16: 1011–1015.
22. Stoner GD, Siglin JC, Morse MA, Desai DH, Amin SG, Kresty LA, *et al.* Enhancement of esophageal carcinogenesis in male F344 rats by dietary phenylhexyl isothiocyanate. *Carcinogenesis* 1995; 16: 2473–2476.
23. Morse MA, Lu J, Gopalakrishnan R, Peterson LA, Wani G, Stoner GD. Mechanism of enhancement of esophageal tumorigenesis by phenylhexyl isothiocyanate. *Cancer Lett* 1997; 112: 119–125.
24. Hudson TS, Carlton PS, Gupta A, Stoner GD, Morse MA. Investigation of the enhancement of NMBA-induced esophageal tumorigenesis by 6-phenylhexyl isothiocyanate. *Cancer Lett* 2001; 162: 19–26.
25. Rao CV, Rivenson A, Simi B, Zang E, Hamid R, Kelloff GJ, *et al.* Enhancement of experimental colon carcinogenesis by dietary phenylhexyl isothiocyanate. *Cancer Res* 1995; 55: 4311–4318.
26. Yuan JM, Stepanov I, Murphy SE, Wang R, Allen S, Jensen J, *et al.* Clinical trial of 2-phenethyl isothiocyanate as an inhibitor of metabolic activation of a tobacco-specific lung carcinogen in cigarette smokers. *Cancer Prev Res* 2016; 9: 396–405.
27. Gelboin HV. Benzo(a)pyrene metabolism, activation, and carcinogenesis: role and regulation of mixed-function oxidases and related enzymes. *Physiol Rev* 1980; 60: 1107–1166.
28. Hecht SS. Oral cell DNA adducts as potential biomarkers for lung cancer susceptibility in cigarette smokers. *Chem Res Toxicol* 2017; 30: 367–375.
29. Beland FA, Churchwell MI, Von Tungeln LS, Chen S, Fu PP, Culp SJ, *et al.* High-performance liquid chromatography electrospray ionization tandem mass spectrometry for the detection and quantitation of benzo(a)pyrene-DNA adducts. *Chem Res Toxicol* 2005; 18: 1306–1315.
30. Alexandrov K, Cascorbi I, Rojas M, Bouvier G, Kriek E, Bartsch H. CYP1A1 and GSTM1 genotypes affect benzo(a)pyrene DNA adducts in

smokers' lung: comparison with aromatic/hydrophobic adduct formation. *Carcinogenesis* 2002; 23: 1969–1977.

31. Pratt MM, John K, MacLean AB, Afework S, Phillips DH, Poirier MC. Polycyclic aromatic hydrocarbon (PAH) exposure and DNA adduct *semi*-quantitation in archived human tissues. *Int J Environ Res Public Health* 2011; 8: 2675–2691.
32. Shimada T, Martin MV, Pruess-Schwartz D, Marnett LJ, Guengerich FP. Roles of individual cytochrome-P450 enzymes in the bioactivation of benzo(a)pyrene, 7,8-dihydroxy-7,8-dihydrobenzo(a)pyrene, and other dihydrodiol derivatives of polycyclic aromatic hydrocarbons. *Cancer Res* 1989; 49: 6304–6312.
33. Shimada T, Gillam EM, Oda Y, Tsumura F, Sutter TR, Guengerich FP, *et al.* Characterization of microsomal cytochrome P450 enzymes involved in the oxidation of xenobiotic chemicals in human fetal liver and adult lungs. *Drug Metab Dispos* 1996; 24: 515–522.
34. Wattenberg LW. Inhibitory effects of benzyl isothiocyanate administered shortly before diethylnitrosamine or benzo(a)pyrene on pulmonary and forestomach neoplasia in A/J mice. *Carcinogenesis* 1987; 8: 1971–1973.
35. Wattenberg LW, Leong JL. Inhibition of the carcinogenic action of benzo(a) pyrene by flavones. *Cancer Res* 1970; 30: 1922–1925.
36. Wattenberg LW. Inhibition of chemical carcinogen-induced pulmonary neoplasia by butylated hydroxyanisole. *J Natl Cancer Inst* 1973; 50: 1541–1544.
37. Singh SV, Hu X, Srivastava SK, Singh M, Xia H, Orchard JL, *et al.* Mechanism of inhibition of benzo(a)pyrene-induced forestomach cancer in mice by dietary curcumin. *Carcinogenesis* 1998; 19: 1357–1360.
38. Lesca P. Protective effects of ellagic acid and other plant phenols on benzo(a)pyrene-induced neoplasia in mice. *Carcinogenesis* 1983; 4: 1651–1653.
39. Jin NZ, Zhu YP, Zhou JW, Mao L, Zhao RC, Fang TH, *et al.* Preventive effects of quercetin against benzo[a]pyrene-induced DNA damages and pulmonary precancerous pathologic changes in mice. *Basic Clin Pharmacol Toxicol* 2006; 98: 593–598.
40. Huderson AC, Myers JN, Niaz MS, Washington MK, Ramesh A. Chemoprevention of benzo(a)pyrene-induced colon polyps in Apc^{Min} mice by resveratrol. *J Nutr Biochem* 2013; 24: 713–724.
41. Conaway CC, Wang C-X, Pittman B, Yang Y-M, Schwartz JE, Tian D, *et al.* Phenethyl isothiocyanate and sulforaphane and their *N*-acetylcysteine

conjugates inhibit malignant progression of lung adenomas induced by tobacco carcinogens in A/J mice. *Cancer Res* 2005; 65: 8548–8557.

42. Tsuji PA, Walle T. Inhibition of benzo[a]pyrene-activating enzymes and DNA binding in human bronchial epithelial BEAS-2B cells by methoxylated flavonoids. *Carcinogenesis* 2006; 27: 1579–1585.
43. Hecht SS. Approaches to the chemoprevention of lung cancer based on carcinogens in tobacco smoke. *Environ Health Perspect* 1997; 105: 955–963.
44. Wattenberg LW, Loub WD. Inhibition of polycyclic aromatic hydrocarbon-induced neoplasia by naturally occurring indoles. *Cancer Res* 1978; 38: 1410–1413.
45. Dashwood RH, Arbogast DN, Fond AT, Hendricks JD, Bailey GS. Mechanisms of anti-carcinogenesis by indole-3-carbinol: detailed *in vivo* DNA binding dose-response studies after dietary administration with aflatoxin B_1. *Carcinogenesis* 1989; 9: 427–432.
46. Morse MA, LaGreca SD, Amin SG, Chung F-L. Effects of indole-3-carbinol on lung tumorigenesis and DNA methylation induced by 4-(methylnitrosamino)-1-(3-pyridyl)-1-butanone (NNK), and on the metabolism and disposition of NNK in A/J mice. *Cancer Res* 1990; 50: 2613–2617.
47. Bailey GS, Hendricks JD, Shelton DW, Nixon JE, Pawlowski NE. Enhancement of carcinogenesis by the natural anticarcinogen indole-3-carbinol. *J Natl Cancer Inst* 1987; 78: 931–934.
48. Stoner G, Casto B, Ralston S, Roebuck B, Pereira C, Bailey G. Development of a multi-organ rat model for evaluating chemopreventive agents: efficacy of indole-3-carbinol. *Carcinogenesis* 2002; 23: 265–272.
49. Kensler TW, Egner P, Davidson NE, Pikul A, Groopman JD, MacMillan DL, *et al.* Protection against aflatoxin tumorigenesis in rats fed a dithiolthione. *Proc Am Assn Cancer Res* 1986; 27: 122.
50. Kensler TW, Egner PA, Dolan PM, Groopman JD, Roebuck BD. Mechanism of protection against aflatoxin tumorigenicity in rats fed 5-(2-pyrazinyl)-4-methyl-1,2-dithiol-3-thione (oltipraz) and related 1,2-dithiol-3-thiones and 1,2-dithiol-3-ones. *Cancer Res* 1987; 47: 4271–4277.
51. Roebuck BD, Liu YL, Rogers AE, Groopman JD, Kensler TW. Protection against aflatoxin B1-induced hepatocarcinogenesis in F344 rats by 5-(2-pyrazinyl)-4-methyl-1,2-dithiole-3-thione (oltipraz): predictive role for short-term molecular dosimetry. *Cancer Res* 1991; 51: 5501–5506.
52. Kwak MK, Egner PA, Dolan PM, Ramos-Gomez M, Groopman JD, Itoh K, *et al.* Role of phase 2 enzyme induction in chemoprotection by dithiolethiones. *Mutat Res* 2001; 480–481: 305–315.

53. Gupta E, Olopade OI, Ratain MJ, Mick R, Baker TM, Berezin FK, *et al.* Pharmacokinetics and pharmacodynamics of oltipraz as a chemopreventive agent. *Clin Cancer Res* 1995; 10: 1133–1138.
54. Groopman JD, DeMatos P, Egner PA, Love-Hunt A, Kensler TW. Molecular dosimetry of urinary aflatoxin-N7-guanine and serum aflatoxin-albumin adducts predicts chemoprotection by 1,2-dithiole-3-thione in rats. *Carcinogenesis* 1992; 13: 101–106.
55. Jacobson LP, Zhang BC, Zhu YR, Wang JB, Wu Y, Zhang QN, *et al.* Oltipraz chemoprevention trial in Qidong, People's Republic of China: study design and clinical outcomes. *Cancer Epidemiol Biomarkers Prev* 1997; 6: 257–265.
56. Wang JS, Shen X, He X, Zhu YR, Zhang BC, Wang JB, *et al.* Protective alterations in phase 1 and 2 metabolism of aflatoxin B1 by oltipraz in residents of Qidong, People's Republic of China. *J Natl Cancer Inst* 1999; 91: 347–354.
57. Kensler TW, He X, Otieno M, Egner PA, Jacobson LP, Chen B, *et al.* Oltipraz chemoprevention trial in Qidong, People's Republic of China: modulation of serum aflatoxin adduct biomarkers. *Cancer Epidemiol Biomarkers Prev* 1998; 7: 127–134.
58. Glintborg B, Weimann A, Kensler TW, Poulson HE. Oltipraz chemoprevention trial in Qidong, People's Republic of China: unaltered oxidative biomarkers. *Free Radical Biol Med* 2006; 41: 1010–1014.
59. Yates MS, Kwak M-K, Egner PA, Groopman JD, Bodreddigari S, Sutter TR, *et al.* Potent protection against aflatoxin-induced tumorigenesis through induction of Nrf2-regulated pathways by the triterpenoid 1-[2-cyano-3-12-dioxooleana-1,9(11)-dien-28-oyl]imidazole. *Cancer Res* 2006; 66: 2488–2494.
60. Elegbede JA, Gould MN. Monoterpenes reduced adducts formation in rats exposed to aflatoxin B_1. *African J Biotech* 2002; 1: 46–49.
61. Elegbede JA, Maltzman TH, Elson CE, Gould MN. Effects of anticarcinogenic monoterpenes on phase II hepatic metabolizing enzymes. *Carcinogenesis* 1993; 14: 1221–1223.
62. Egner PA, Wang J-B, Zhu Y-R, Zhang B-C, Wu Y, Zhang Q-N, *et al.* Chlorophyllin intervention reduces aflatoxin-DNA adducts in individuals at high risk for liver cancer. *Proc Natl Acad Sci* 2001; 98: 14601–14606.
63. Breinholt V, Schimerlik M, Dashwood R, Bailey G. Mechanisms of chlorophyllin anticarcinogenesis against aflatoxin B1: complex formation with the carcinogen. *Chem Res Toxicol* 1995; 8: 506–514.
64. Motohashi H, Yamamoto M. Nrf2-Keap1 defines a physiologically important stress response mechanism. *Trends Mol Med* 2004; 10: 549–557.

65. Kwak MK, Wakabayashi N, Itoh K, Motohashi H, Yamamoto M, Kensler TW. Modulation of gene expression by cancer chemopreventive dithiolethiones through the Keap1-Nrf2 pathway. Identification of novel gene clusters for cell survival. *J Biol Chem* 2003; 278: 8135–8145.
66. Sayer JM, Haruhiko Y, Wood AW, Conney AH, Jerina DM. Extremely facile reaction between the ultimate carcinogen benzo(a)pyrene-7,8-diol 9,10-epoxide and ellagic acid. *J Am Chem Soc* 1982; 104: 5562–5564.
67. Wood AW, Huang M-T, Chang RL, Newmark HL, Lehr RE, Yagi H, *et al.* Inhibition of the mutagenicity of bay-region diol epoxides of polycyclic aromatic hydrocarbons by naturally occurring plant phenols: exceptional activity of ellagic acid. *Proc Natl Acad Sci* 1982; 79: 5513–5517.
68. Chang RL, Huang M-T, Wood AW, Wong C-Q, Newmark HL, Yagi H, *et al.* Effect of ellagic acid and hydroxylated flavonoids on the tumorigenicity of benzo(a)pyrene and (+/−)-7β,8α-dihydroxy-9α,10α-epoxy-7,8,9,10-tetrahydrobenzo(a)pyrene on mouse skin and in the newborn mouse. *Carcinogenesis* 1986; 6: 1127–1133.
69. Polidori MC, Stahl W. Carotenoids and vitamin A. In *Chemoprevention of Cancer and DNA Damage by Dietary Factors*, Knasmuller DM, DeMarini D, Johnson I, Gerhauser C, eds., Chap. 23, Wiley-Blackwell, Weinheim, Federal Republic of Germany, 2007, pp. 371–383.
70. Carr A, Frei B. Does vitamin C act as a pro-oxidant under physiological conditions? *FASEB J* 1999; 13: 1007–1024.
71. Collins AR, Cadet J, Moller L, Poulsen HE, Vina J. Are we sure we know how to measure 8-oxo-7,8-dihydroguanine in DNA from human cells? *Arch Biochem Biophys* 2004; 423: 57–65.
72. Cadet J, Douki T, Gasparutto D, Ravanat JL. Oxidative damage to DNA: formation, measurement and biochemical features. *Mutat Res* 2003; 531: 5–23.
73. Swenberg JA, Lu K, Moeller BC, Gao L, Upton PB, Nakamura J, *et al.* Endogenous versus exogenous DNA adducts: their role in carcinogenesis, epidemiology, and risk assessment. *Toxicol Sci* 2011; 120: S130–S145.
74. Knasmuller S, Nersesyan A, Misik M, Gerner C, Mikulits W, Ehrlich V, *et al.* Use of conventional and -omics based methods for health claims of dietary antioxidants: a critical overview. *Br J Nutr* 2008; 99: ES3–ES52.
75. Nishida N, Arizumi T, Kitai S, Yada N, Hagiwara S, Inoue T, *et al.* Reactive oxygen species induce epigenetic instability through the formation of 8-hydroxydeoxyguanosine in human hepatocarcinogenesis. *Dig Dis* 2013; 31: 459–466.
76. The Alpha-Tocopherol Beta Carotene Cancer Prevention Study Group. The effect of vitamin E and beta carotene on the incidence of lung cancer and other cancers in male smokers. *N Engl J Med* 1994; 330: 1029–1035.

77. Omenn GS, Goodman GE, Thornquist MD, Balmes J, Cullen MR, Glass A, *et al.* Risk factors for lung cancer and for intervention effects in CARET, the Beta-Carotene and Retinol Efficacy Trial. *J Natl Cancer Inst* 1996; 88: 1550–1559.
78. VanHelden YGJ, Keijer J, Heil CP, Palou A, Oliver P, Munnia A, *et al.* Beta-carotene affects oxidative stress-related DNA damage in lung epithelial cells and in ferret lung. *Carcinogenesis* 2009; 30: 2070–2076.
79. Nagashima M, Kasai H, Yokota J, Nagamachi Y, Ichinose T, Sagai M. Formation of an oxidative DNA damage, 8-hydroxydeoxyguanosine, in mouse lung DNA after intratracheal instillation of diesel exhaust particles and effects of high dietary fat and beta-carotene on this process. *Carcinogenesis* 1995; 16: 1441–1445.
80. DiMascio P, Devasagayam TP, Kaiser S, Sies H. Carotenoids, tocopherols and thiols as biological singlet molecular oxygen quenchers. *Biochem Soc Transact* 1990; 18: 1054–1056.
81. Chen L, Stacewicz-Sapuntzakis M, Duncan C, Sharifi R, Ghosh L, van Breeman R, *et al.* Oxidative DNA damage in prostate cancer patients consuming tomato sauce-based entrees as a whole-food intervention. *J Natl Cancer Inst* 2001; 93: 1872–1879.
82. Shah GM, Goswami UC, Bhattacharya RK. Action of some retinol derivatives and their provitamins on microsome-catalyzed formation of benzo(a)pyrene-DNA adduct. *J Biochem Toxicol* 1992; 7: 177–181.
83. Bhattacharya RK, Firozi PF, Aboobaker VS. Factors modulating the formation of DNA adduct by aflatoxin B_1 *in vitro*. *Carcinogenesis* 1984; 5: 1359–1362.
84. Grosse Y, Chekir-Ghedira L, Huc A, Obrecht-Pflumio S, Dirheimer G, Bacha H, *et al.* Retinol, ascorbic acid and alpha-tocopherol prevent DNA adduct formation in mice treated with the mycotoxins ochratoxin A and zearalenone. *Cancer Lett* 1997; 114: 225–229.
85. Kramata P, Suh N. Selected vitamins, vitamin C. In *Chemoprevention of Cancer and DNA Damage by Dietary Factors*, Knassmuller S, DeMarini DM, Johnson I, Gerhauser C, eds., Chap. 24, Wiley-Blackwell, Weinheim, Federal Republic of Germany, 2009, pp. 385–393.
86. Duarte TL, Lunec J. Review: when is an antioxidant not an antioxidant? A review of novel actions and reactions of vitamin C. *Free Rad Res* 2005; 39: 671–686.
87. Javanbakht M, Keshvarz S, Mirshafiey A, Djalali M, Siassi F, Eshaghian M. The effects of vitamins E and D supplementation on erythrocyte superoxide dismutase and catalase in atopic dermatitis. *Iran J Public Health* 2010; 39: 57–63.

88. Samanta S, Chatterjee M, Ghosh B, Rajkumar M, Rana A, Chatterjee M. Vanadium and 1,25 (OH)2 vitamin D3 combination in inhibitions of 1,2-dimethylhydrazine-induced rat colon tumorigenesis. *Biochem Biophys Acta* 2008; 1780: 1106–1114.
89. Saha BK, Bishayee A, Kanjilal NB, Chatterjee M. 1 Alpha,25-dihydroxyvitamin D3 inhibits hepatic chromosomal aberrations, DNA strand breaks and specific DNA adducts during rat hepatocarcinogenesis. *Cell Mol Life Sci* 2001; 58: 1141–1149.
90. Wang X, Quinn PJ. Vitamin E and its function in membranes. *Prog Lipid Res* 1999; 38: 309–336.
91. Traber MG. Vitamin E regulatory mechanisms. *Annual Rev Nutr* 2007; 27: 347–362.
92. Hensley K, Benaksas EJ, Bolli R, Comp P, Grammas P, Hamdheydari L, *et al.* New perspectives on vitamin E: gamma-tocopherol and carboxyethylhydroxychroman metabolites in biology and medicine. *Free Rad Biol Med* 2004; 36: 1–159.
93. Jiang Q, Christen S, Shigenaga MK, Ames BN. Gamma-tocopherol, the major form of vitamin E in the US diet, deserves more attention. *Am J Clin Nutr* 2001; 74: 714–722.
94. Dietrich M, Traber MG, Jacques PF, Cross CE, Hu Y, Block G. Does gamma-tocopherol play a role in the primary prevention of heart disease and cancer? A review. *J Am Coll Nutr* 2006; 25: 292–299.
95. Grinberg-Funes RA, Singh VN, Perera FP, Bell DA, Young TL, Dickey C, *et al.* Polycyclic aromatic hydrocarbon-DNA adducts in smokers and their relationship to micronutrient levels and the glutathione-S-transferase M1 genotype. *Carcinogenesis* 1994; 15: 2449–2454.
96. Jacobson JS, Begg MD, Wang LW, Agarwal M, Norkus E, Singh VN, *et al.* Effects of 6-month vitamin intervention on DNA damage in heavy smokers. *Cancer Epidemiol Biomarkers Prev* 2000; 9: 1303–1311.
97. Kunnumakkara AB, Anand P, Harikumar KB, Aggarwal B. DNA damage and cancer chemoprevention by polyphenols. In *Chemoprevention of Cancer and DNA Damage by Dietary Factors*, Knassmuller S, DeMarini DM, Johnson I, Gerhauser C, eds., Chap. 27, Wiley-Blackwell, Weinheim, Federal Republic of Germany, 2009, pp. 455–482.

Chapter 11

Chemical Carcinogen Exposure, DNA Damage, and Epigenetic Alterations*

Kostiantyn Dreval[†] and Igor P. Pogribny[†]

[†]*Division of Biochemical Toxicology, National Center for Toxicological Research, Jefferson, Arkansas 72079, USA*

Abstract

The risk of the cancer development in humans is determined by the interaction of two main cancer-causing factors, genetics and carcinogen exposure. It is well-established that exposure to chemical carcinogens may trigger the carcinogenic process by affecting the genome, through DNA damage and mutations, and epigenome. Through examples of the accumulation of carcinogen-induced DNA damage and epigenetic alterations caused by genotoxic and nongenotoxic chemicals, this chapter summarizes current knowledge related to the role of DNA damage and epigenetic abnormalities as drivers of the carcinogenic process. It also highlights and emphasizes a tight link and mechanistic interconnectivity between DNA damage and epigenetic alterations in carcinogenesis.

*The views expressed in this manuscript do not necessarily represent those of the U.S. Food and Drug Administration.

1. Introduction

The risk of humans developing cancer is determined by two key etiological factors, genetics and environmental exposures, and their interactions [1]. The relative contribution of these two factors is highlighted by a recent report based on the analysis of 15,668 Western European monozygotic twins, which indicated that genetic factors might not be the principal cause of cancer development [2]. Furthermore, epidemiological evidence indicates that exposure to certain natural and man-made chemicals received during life is a major cause of cancers in humans [3]. Chemical carcinogens may be separated into two groups, a larger group consisting of genotoxic carcinogens, and a smaller group of nongenotoxic carcinogens. Genotoxic carcinogens damage DNA directly, or are metabolically activated to form DNA adducts [4], leading to permanent fixation of DNA mutations in critical genes required for growth control. Nongenotoxic carcinogens are agents that cause cancer, or accelerate the carcinogenic process, through a variety of mechanisms, which can include receptor-mediated pathways, altered gap junction and intercellular communication, cytochrome P450 (CYP450) overexpression, immunosuppression, increased cell proliferation, resistance to apoptotic cell death, and epigenetic dysregulation [5]. Many of these mechanisms lead to some form of DNA dysfunction, despite the lack of direct binding to DNA.

Recent evidence suggests that both genotoxic and nongenotoxic carcinogens can alter DNA by inducing major epigenetic abnormalities [6, 7], the consequences of which can be observed in the resulting tumors [8]. This chapter summarizes current knowledge related to the role and interconnectivity of DNA damage and epigenetic changes induced by some genotoxic carcinogens and some nongenotoxic chemical carcinogens, and emphasizes their importance in the carcinogenic process. The chemical carcinogens to be discussed below include the genotoxins benzo[*a*]pyrene (BP), 1,3-butadiene, benzene, 2-acetylaminofluorene (2-AAF), and tamoxifen (TAM), and the nongenotoxic carcinogens methapyrilene, furan, and Wyeth-14,643.

2. Brief Overview of Epigenetics

The term "epigenetics" defines the phenomena and mechanisms that cause chromosome-bound changes in gene expression, and which are unrelated to changes in the DNA sequence [9]. Cytosine methylation and posttranslational

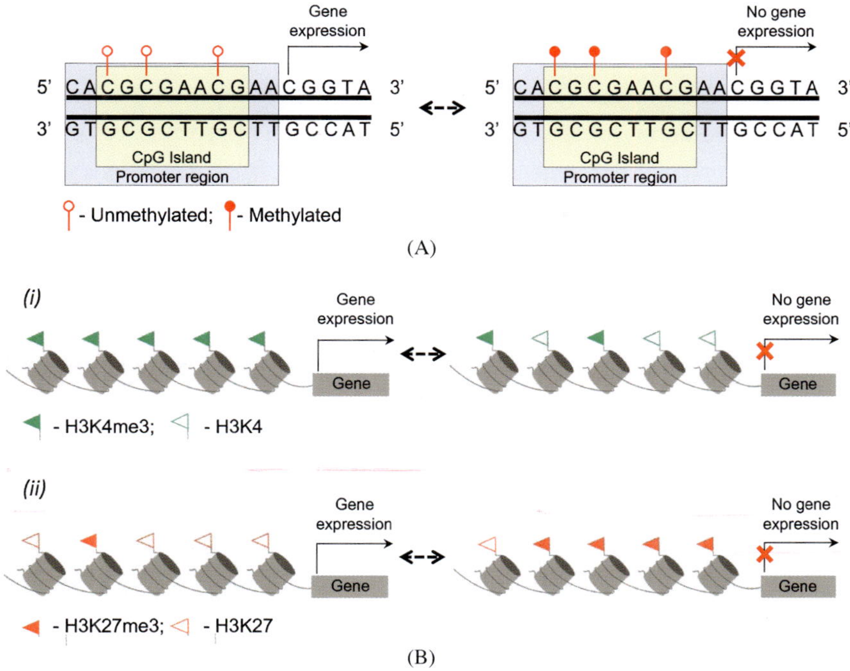

Figure 1. Association between DNA methylation (A), modifications of histone proteins (B), and gene transcription.

modifications of histone proteins are two major and much-studied epigenetic modifications (Figure 1). The addition of methyl groups from the universal methyl group donor *S*-adenosyl-l-methionine (SAM) to carbon five in cytosine residues results in the formation of 5-methylcytosine (5-meC) in DNA. This reaction is mediated by a family of DNA methyltransferases [10]. In somatic mammalian cells, methylation of DNA occurs almost exclusively at CpG dinucleotides, and approximately 70%–90% of them are methylated; however, the distribution of the methylated CpG sequences across the genome is uneven. In normal cells, most of the methylated CpG dinucleotides are located in repetitive sequences and exonic, intronic, and intergenic unique sequences, while the highly unmethylated high-density CpG sites are primarily found in gene promoters [11]. Methylation of DNA is a dynamic and well-balanced process between DNA methylation and DNA demethylation reactions. DNA methylation is initiated and established by the family of *de novo* DNA methyltransferases DNMT3 (DNMT3A and DNMT3B), and maintained during

DNA replication by DNMT1. DNA demethylation is achieved through two different mechanisms (1) a "passive" replication-dependent mechanism during the cell division and (2) an "active" replication-independent mechanism mediated by a family of ten-eleven translocation (TET) proteins, resulting in the formation of the cytosine-modified base 5-hydroxymethylcytosine (5-hmeC) [12]. An additional source of 5-hmeC formation is oxidative stress [13], which plays a fundamental role in the carcinogenic process.

A second major epigenetic mechanism, which is tightly and interdependently connected to DNA methylation, is posttranslational modification of histone proteins. There are several types of histone modifications, including methylation, acetylation, phosphorylation, sumoylation, and ubiquitination of specific amino acid residues on the N-terminal histone tails [14]. Histone modifications are established by "writers" (e.g., histone acetylases, methylases, and phosphorylases) that introduce chemical modification, interpreted by "readers," proteins that bind specifically to modified histone (e.g, bromodomains and chromodomains), and removed by "erasers" (e.g., histone deacetylases, demethylases, and phosphatases) that remove chemical modification from histone proteins [15, 16].

A primary function of 5-meC and 5-hmeC methylations, and histone modifications, is control of gene transcription. Figure 1A shows that DNA methylation at a gene promoter functions as a transcriptional "OFF–ON" switch, so that occurrence of DNA methylation at unmethylated CpG sites inhibits gene expression, and demethylation of methylated CpG sites activates transcription. In contrast, different types of histone modification, or some type modification at different amino acid residues, have differential effects on gene expression. For instance, trimethylation of histone 3 lysine 4 (H3K4me3) is associated with active gene transcription (Figure 1B, *i*), while trimethylation of histone 3 lysine 27 (H3K27me3) is linked to gene silencing (Figure 1B, *ii*). Well-balanced DNA methylation and histone modifications are essential for normal cell functioning, and their disruption can be associated with a number of pathological states, including carcinogenesis.

3. Genotoxic Carcinogens

3.1. *Benzo[a]pyrene (BP): genotoxicity and epigenetics*

BP (Figure 2), a representative polycyclic aromatic hydrocarbon, is one of the most well-studied genotoxic carcinogens, having been shown

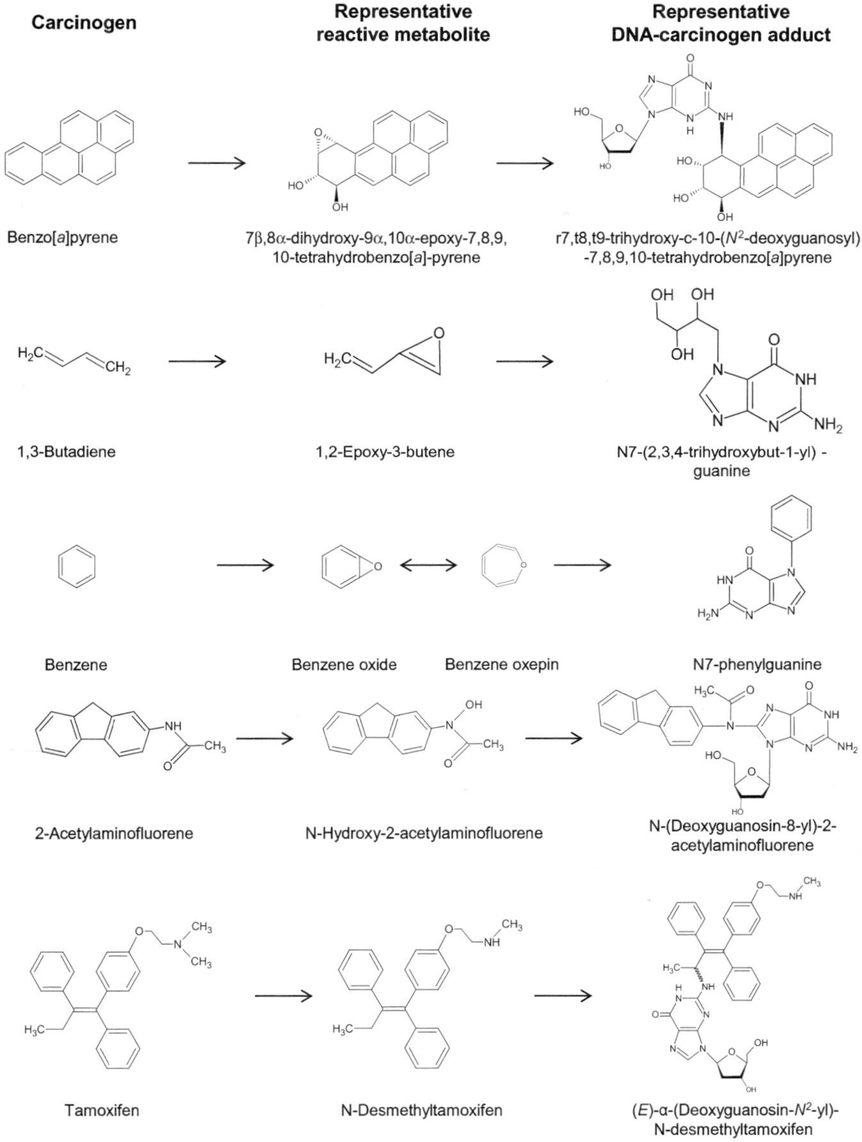

Figure 2. Representative genotoxic carcinogens, their reactive metabolites, and their carcinogen–DNA adducts.

carcinogenic in virtually all *in vivo* model systems evaluated, and recently having been declared a human carcinogen (Group 1) by the International Agency for Research on Cancer (IARC) [17]. This compound is metabolized by CYPs 1A1, 1A2, and 1B1 to several reactive metabolic diol epoxide intermediates (see chapter by Regina M. Santella). Arguably the most important metabolite is the 7β,8α-dihydroxy-9α,10α-epoxy-7,8,9,10-tetrahydrobenzo[*a*]-pyrene (BPDE), which binds directly to DNA [18] (Figure 2), forming a covalent adduct, the r7,t8,t9-trihydroxy-c-10-(N^2-deoxyguanosyl)-7,8,9,10-tetrahydrobenzo[*a*]pyrene (BPdG) at the N^2 position of guanine [18] (Figure 2). BPdG is the major stable DNA adduct of BP and is found in the liver, esophagus, forestomach, skin, and lungs of mice [17, 19], most of which are target organs for BP tumorigenicity. If not appropriately repaired, BPdG adducts lead to C:G → A:T transversion mutations, considered the major mutational signature of BP exposure. This is evidenced by a preferential formation of BPdG adducts at the most frequently mutated guanine nucleotides in the human tumor suppressor *TP53* gene [20, 21] and *KRAS* oncogene [22]. In addition to direct DNA binding, exposure to BP and its subsequent metabolism results in the generation of reactive oxygen species that may cause secondary damage to DNA [23].

The interaction between direct DNA adduct (BPdG) formation and epigenetic status is an important determinant of DNA damage disposition occurring as a result of BP exposure. For example, the formation of BPdG adducts depends greatly on the DNA sequence context and the status of 5-meC methylation of the neighboring bases [24]. Studies have shown that cytosine methylation enhances the binding of BPDE to CpG sites [24], including CpG sites in the human *TP53* and *KRAS* genes [22, 25]. Specifically, Glick *et al.* [26] demonstrated that CpG methylation at codons 157 and 158 of the human *TP53* gene, which are mutational "hot spots," enhances the formation of BPdG adducts by two- to four-fold, as compared to unmethylated sequences. Similar findings were reported by Hu *et al.* [22], who showed increased levels of BPdG adducts at the methylated mutational "hot spot" codon 14 in the human *KRAS* gene. Furthermore, Yoon *et al.* [27] reported a greater frequency of C:G → A:T transversion at methylated CpG sites, as compared to unmethylated CpG sites (42% *vs.* 23%).

Besides the CpG methylation–driven genotoxic effects of BP, the presence of BPdG adducts affects 5-meC levels in DNA by impairing the methylation efficiency of DNA methyltransferases [28, 29] in both global

DNA and specific regions. For example, long interspersed nuclear element 1 (LINE-1) repetitive elements and protein-coding genes exhibit hypomethylation after BP exposure [29–32]. Furthermore, BP exposure results in hypermethylation of the cyclin-dependent kinase inhibitor 2A (*CDKN2A*, $P16^{INK4A}$) [33], retinoic acid receptor β_2 (*RAR* β_2) [34], and hypermethylated in cancer 1 (*HIC1*) [35] genes. In addition to changes in 5-meC content, BP exposure alters histone proteins. Studies have demonstrated that BP and its metabolites induce phosphorylation of histone H2AX (γ-H2AX), and the extent of γ-H2AX formation was correlated with the number of BPdG adducts [36].

3.2. *1,3-Butadiene*

A gas monomer used in the production of synthetic rubber and plastics, 1,3-butadiene (Figure 2) is one of the most common representatives of conjugated genotoxic dienes. In addition to accumulated experimental evidence of 1,3-butadiene genotoxicity, epidemiological studies have demonstrated a positive association between exposure to 1,3-butadiene and risk of leukemias and other cancers in humans. Based on this evidence, the IARC has classified 1,3-butadiene as Group 1, or carcinogenic to humans [37].

Metabolism in the liver, by the CYP mixed-function oxidases, produces several genotoxic derivatives, including 1,2-epoxy-3-butene (Figure 2), 1,2,3,4-diepoxybutane, and 1,2-epoxy-3,4-butanediol [38]. These reactive metabolites interact with DNA at the N7 position of guanine, causing the formation of several DNA adducts, including N7-(2,3,4-trihydroxybut-1-yl)guanine (THB-Gua, Figure 2) and 1,4-bis-(guan-7-yl)-2,3-butanediol (bis-NG7-BD) cross-links [38, 39]. Additionally, 1,3-butadiene exposure causes the formation of DNA adducts with cytosine and adenine bases [40] and with proteins [41]. Based on this evidence, it has been postulated that genotoxicity plays a major role in the mechanism of 1,3-butadiene tumorigenicity.

In addition to genotoxicity, the accumulated experimental evidence indicates the importance of epigenetic mechanisms in tumorigenicity of 1,3-butadiene. Exposure of male C57BL/6J mice to 1,3-butadiene results in a decrease of whole genome and repetitive element (e.g., LINE-1) cytosine DNA methylation and hydroxymethylation [42]. Furthermore, there is reduced trimethylation of H3 histones at lysines 9 and 27 (H3K9me3 and H3K27me3), and H4 histone at lysine 20 (H4K20me3) [43]. In addition, in BP-exposed mice there is increased acetylation of histone H3 at lysine 27 (H3K27ac) in liver [42], and decreased acetylation of histone H3 at lysine 56 (H3K56) and H4 at lysine 16 (H4K16) in lung [42]. Both liver and lung are target organs for 1,3-butadiene-induced tumorigenesis.

3.3. *Benzene*

Benzene (Figure 2) is a natural constituent of crude oil and a high production volume industrial chemical. Wide usage of benzene in a variety of industries, including the manufacture of petrochemicals, as well as its presence as a common component of vehicle exhaust cause public health concern. Exposure to benzene has been linked to the development of lymphatic [44] and hematopoietic [45] cancers, in particular acute myeloid and chronic lymphocytic leukemia [46]. Bioactivation of benzene occurs primarily in the liver and is mainly mediated by CYP2E1 [47]. This results in the formation of several different reactive metabolites, including benzene oxide (Figure 2), phenol, (E,E)-muconaldehyde, catechol, and hydroquinone [48]. Benzene readily forms adducts with DNA (see N7-phenylguanine in Figure 2) and proteins [49, 50].

Several lines of evidence have demonstrated that the presence of benzene–DNA adducts affects cytosine methylation. Specifically, exposure to benzene at low concentrations was linked to reduced 5-meC levels at repetitive DNA elements (LINE-1 and AluI [51]) in humans, an epigenetic abnormality frequently found in acute myeloid leukemia patients [52]. Accumulation of DNA–benzene adducts, and subsequent genome instability, may be in part due to the compromised DNA repair caused by CpG island hypermethylation-associated silencing of DNA repair genes. These include the poly (ADP-ribose) polymerase 1 (*PARP-1*) [53] and the excision repair cross-complementation group 3 (*ERCC3*) [54]. Furthermore, inhibition of the essential DNA replication enzyme, DNA topoisomerase II (TOPOIIα), which is one of the major mechanisms in benzene hematotoxicity, is caused by a reduced level of histone transcription activating marks, including histone H3 and histone H4 acetylation, histone H3 lysine 4 dimethylation (Table 1), and an increased

Table 1. Carcinogens, DNA damage, and epigenetic changes

Carcinogen	Main metabolites	DNA damage	DNA methylation	Histone modifications
		Genotoxic carcinogens		
Benzo[*a*]pyrene	BPDE	BPdG	*Hypomethylation*:genomic, LINE-1 *Hypermethylation*: *CDKN2A*, *P16*, *RAR β2, HIC1*	*Increased:* gamma-H2AX *Histone adducts*: H2A, H2B, H4
1,3-Butadiene	1,2-epoxy-3-butene; 1,2,3,4-diepoxybutane; 1,2-epoxy-3,4-butanediol	THB-Gua bis-NG7-BD	*Hypomethylation*: genomic, LINE-1	*Increased*: H3K27ac *Reduced*: H3K9me3, H3K27me3, H4K20me3, H3K56ac, H4K16ac *Histone adducts*: H2B
Benzene	Benzene oxide; phenol; E,E-muconaldehyde; catechol; hydroquinone	N7-phenylguanine	*Hypomethylation*: LINE-1, AluI, *STAT3, IFNGR1*, *MPL* *Hypermethylation*: *PARP-1*, *ERCC3*, *PRKG1*, *PARD3*, *EPHA8*	*Increased*: H3K9me3, gamma-H2AX *Reduced*: H3ac, H4ac, H3K4me2

(*Continued*)

Table 1. *(Continued)*

Carcinogen	Main metabolites	DNA damage	DNA methylation	Histone modifications
2-AAF	N-hydroxy-2-acetylaminofluorene; N-sulphoxy-acetylaminofluorene	dG-8-AAF; dG-8-AF	*Hypermethylation*: *p16, Rassf1a, Socs1, Cdh1, Gjb2*	*Reduced*: H4K20me3
Tamoxifen	N-desmethyltamoxifen; endoxifen; 4-hydroxytamoxifen; α-hydroxytamoxifen	dG-TAM; dG-desMeTAM	*Hypomethylation*: genomic, LINE-1, *PAX2* *Hypermethylation*: *MGMT*	*Reduced*: H4K20me3
		Nongenotoxic carcinogens		
Methapyrilene	desmethylmethapyrilene; methapyrilene-N-oxide; methapyrilene O-glucuronide	8-oxodG	*Hypomethylation*: genomic *Hypermethylation*: *Prox1, Hnf1a, Ppara*	*Reduced*: H3K4me2, H4K20me3, H3K9ac, H3K56ac
Furan	BDA	8-oxodG	*Hypomethylation*: genomic, *p16, Rassf1a, Areg, Jag1* *Hypermethylation*: *Foxe1*	*Reduced*: H3K9ac, H3K56ac
Wy-14,643		8-oxodG	*Hypomethylation*: genomic, LINE-1, IAP, *Gstp, Gadd45b*	*Reduced*: H3K4me3, H4K20me3

level of the histone transcription repressing mark H3K9me3 in the promoter region of the *TOPOIIα* gene [55]. Additionally, as with BP, benzene and its metabolites cause phosphorylation of histone γ-H2AX [56].

Benzene exposure in humans has been associated with hypermethylation of cGMP-dependent protein kinase type I (*PRKG1*), par-3 family cell polarity regulator (*PARD3*), and EPH receptor A8 (*EPHA8*). Benzene-induced hypomethylation has been found in signal transducer and activator of transcription 3 (*STAT3*) and interferon gamma receptor 1 (*IFNGR1*) [57], genes that are related to the pathogenesis of acute myeloid leukemia development [58]. The results of *in vitro* studies have demonstrated that treatment of human cells with the benzene metabolite hydroquinone led to a decrease of 5-meC levels in LINE-1 repetitive elements [59], and to overexpression of the proto-oncogene, thrombopoietin receptor (*MPL*), as a consequence of hypomethylation [60].

3.4. *2-Acetylaminofluorene*

2-AAF (Figure 2), a representative aromatic amine, was an early experimental chemical carcinogen [61]. In animal studies, exposure to 2-AAF caused cancer in liver, intestine, mammary gland, and urinary bladder [62, 63]. Currently, 2-AAF is not manufactured for industrial use, but is still used in the experimental carcinogenesis studies. 2-AAF is metabolized in the liver by CYP mixed-function oxidases and, after bioactivation, forms N-(deoxyguanosin-8-yl)-2-acetylaminofluorene (dG-C8-AAF) (Figure 2) and N-(deoxyguanosin-8-yl)-2-aminofluorene (dG-C8-AF) DNA adducts [62–64]. The formation of dG-C8-AAF and dG-C8-AF adducts leads to base substitution and frameshift mutations in a critical cancer-related gene, the Harvey ras sarcoma virus oncogene (*Hras*) [65]. However, because formation of these DNA adducts is not always linear with tumor incidence [66], additional mechanisms may be required for tumorigenesis.

Accumulated experimental evidence in recent years has revealed an important role of epigenetic events in 2-AAF-induced tumor development. These include profound whole genome and region-specific epigenetic changes in the livers of rats fed 2-AAF [67, 68]. In particular, feeding male Sprague Dawley rats a 2-AAF-containing diet resulted in a time-dependent reduction of whole genome histone H4K20 trimethylation [67], a critically important histone mark for DNA damage repair, DNA replication, and chromatin organization [69]. A decreased level of H4K20me3 has been found in several forms of human cancers [70]. Furthermore, silencing of several critical cancer-related genes, including *p16^{INK4a}*, Ras association domain family member 1 (*Rassf1a*), suppressor of cytokine signaling 1 (*Socs1*), cadherin 1 (*Cdh1*), and gap junction protein beta 2 (*Gjb2*), has been detected in preneoplastic livers of rats exposed to 2-AAF [60].

3.5. *Tamoxifen*

TAM (Figure 2), a nonsteroidal selective estrogen receptor modulator, is widely used to treat and prevent occurrence and reoccurrence of breast cancer in women. The TAM mode of action in the breast is based on competitive binding to the estrogen receptor, causing downregulation of proliferative metabolic pathways. However, negative side effects include increased risk of thromboembolism and endometrial cancer [71, 72]. Based on epidemiological evidence, the IARC has classified TAM as Group 1, a known human carcinogen [73].

When still in the preclinical phase, TAM was shown to be a potent liver carcinogen in rats [74]. Subsequently, experimental data revealed TAM binding to DNA as a critical mechanistic component of TAM-induced rat liver carcinogenesis, with the major causative adducts considered to be the (*E*)-α-(deoxyguanosin-*N*^{2}-yl)-tamoxifen (dG-TAM) and (*E*)-α-(deoxyguanosin-*N*^{2}-yl)-N-desmethyltamoxifen (dG-desMeTAM) (Figure 2) [75]. The metabolic activation of TAM occurs *via* hepatic CYP450-dependent monooxygenases, through an α-hydroxylation pathway [76], with subsequent sulfation/esterification of the hydroxy function [76]. The occurrence of TAM-induced tumors in human endometrium may be partially due to TAM–DNA adduct formation, which has been reported in human uterus [77], but it is likely that other factors contribute. In addition to direct TAM–DNA damage, reactive oxygen species are also formed by TAM exposure, causing secondary oxidative damage to DNA and other macromolecules [76].

Advances in epigenetic research have demonstrated significant aberrations occurring in during TAM-induced hepatocarcinogenesis. In rat liver, TAM exposure led to a global loss of DNA methylation [78], decreased methylation of repetitive DNA elements such as LINE-1 [79], and a reduced level of H4K20me3. Interestingly, mouse studies showed no evidence of TAM-induced liver epigenetic changes, and TAM does not cause liver tumors in mice, despite robust formation of TAM–DNA adducts in mouse liver [80]. Furthermore, emerging evidence suggests that epigenetic alterations may contribute to TAM-associated endometrial carcinogenesis in humans. Wu *et al.* [81] and Nagy *et al.* [82] demonstrated substantial demethylation of the paired box 2 (*PAX2*) gene and hypermethylation in the promoter region of the O-6-methylguanine-DNA methyltransferase (*MGMT*) gene in TAM-induced human endometrial tumors.

4. Nongenotoxic Carcinogens

4.1. *Methapyrilene*

Methapyrilene is a histamine antagonist that was widely used in many nonprescription drugs until 1970, when it was demonstrated to cause liver tumors in rats [83]. Methapyrilene is metabolized in the liver by the CYP2C family, especially CYP2C11 [84], and forms several metabolites, including desmethyl-methapyrilene, methapyrilene-N-oxide, and methapyrilene O-glucuronide [85]. Because DNA adducts have not been detected in the livers of rats treated with methapyrilene [86], the hepatocarcinogenicity of methapyrilene is likely not associated with a direct genotoxic mechanism. However, several studies have demonstrated that the hepatocarcinogenicity of methapyrilene is associated with mitochondrial dysfunction, induction of oxidative stress, and subsequent secondary oxidative DNA damage [84, 87, 88]. These events comprise the major pathophysiological mechanisms associated with tumorigenesis, and particularly with liver tumorigenesis [89]. Specifically, treatment of rats with methapyrilene for 60 days resulted in a significant increase of 8-oxo-7-hydroxydeoxyguanosine (8-oxo-dG) in hepatic DNA. A major form of oxidative DNA damage, 8-oxo-dG formation in the livers resulted in altered expression of base-excision DNA repair genes, and increased levels of 4-hydroxynonenal (4-HNE)–protein adducts [88]. Both of these events are sensitive *in vivo* indicators of oxidative DNA damage [90, 91].

In addition to DNA damage, methapyrilene significantly affects the hepatic epigenetic profile by causing a dose-dependent loss of 5-meC and 5-hmeC in hepatic DNA, as well as demethylation of histones H3K4me2 and H4K20me3, and deacetylation of histones H3 lysine 9 (H3K9) and H3 lysine 56 (H3K56) in the liver [88]. Besides the global genome epigenetic changes, methapyrilene causes region-specific epigenetic changes, including gene-specific hypermethylation [92] and gene-specific histone modification changes [88]. In rats treated with methapyrilene, there was gene-specific histone H3K9 deacetylation and downregulation of critical cancer-related genes, including prospero homeobox 1 (*Prox1)*, HNF1 homebox A (*Hnf1a)*, and peroxisome proliferator–activated receptor alpha (*Ppara)*. These findings provide a mechanistic link between methapyrilene-induced epigenetic aberrations and liver carcinogenesis [88].

4.2. *Furan*

Furan is a significant food contaminant produced by thermal processing and found in a broad range of common foods. The presence of furan in foods has caused public health concern because of its hepatotoxicity and hepatocarcinogenicity in rodents [93, 94]. Based on evidence obtained in a 2-year rodent carcinogenicity bioassay, furan has been classified as Group 2B, a possible human carcinogen, by the IARC [95]. Until recently, mechanisms associated with furan hepatocarcinogenicity remained poorly understood, though both genotoxic and nongenotoxic mechanisms were suspected [96–100]. Studies conducted in the past few years point clearly to a nongenotoxic mode of carcinogenic action for furan. Specifically, Churchwell *et al.* [101] demonstrated that treatment of Fischer 344 rats with furan at 4.4 mg/kg body weight/day up to 360 days did not result in the formation of hepatic DNA adducts from *cis*-2-butene-1,4-dial, an electrophilic metabolite of furan. This finding is in good agreement with reports showing that furan exposure did not cause an increase in gene-specific mutation frequency [100, 102–104]. Furthermore, a report by Dong *et al.* [105] demonstrated that furan-induced liver tumorigenicity in male Fischer 344 rats may be associated with a nongenotoxic mechanism.

Several reports have shown that furan may cause liver toxicity and carcinogenicity through the induction of secondary oxidative DNA damage. Events associated with furan-induced liver tumroigenesis include oxidative stress-induced persistent cell proliferation [106, 107], marked transcriptomic changes, altered expression of protein coding [105], changes in the activity of microRNA and long noncoding RNA genes [99, 105, 108], and epigenetic alterations. It has been demonstrated that furan-induced hepatotoxicity is associated with global DNA hypomethylation [94] and promoter region hypomethylation in the tumor suppressor genes $p16^{IKN4a}$ and *Rassf1a*, accompanied by a decreased expression of *Rassf1a* [109]. Also, administration of furan caused a dose-dependent and sustained decrease in the acetylation of histone H3 at lysines 9 and 56, which was associated with the formation of heterochromatin and decreased gene expression [109, 110].

In further work, the development of cholangiofibrotic lesions in the livers of male Fisher 344 rats induced by 2-year exposure to 0.92 and 2 mg furan/kg body weight was accompanied by alterations in promoter-region DNA methylation and gene expression in a number of genes involved in key pathways associated with different aspects of liver tumorigenesis [111]. For example, decreased promoter methylation and increased expression were observed with the amphiregulin (*Areg*) and jagged 1 (*Jag1*) genes, whereas the opposite relationship was found with the forkhead box E1 (*Foxe1*) gene. Overexpression of *Areg* and *Jag1* and downregulation of *Foxe1* has been reported not only in full-fledged hepatocellular carcinomas but also in livers during earlier stages of carcinogenesis [111]. These results indicate that gene-specific changes in DNA methylation have functional consequences that may be important for the induction of cholangiofibrosis and subsequently cholangiocarcinoma.

4.3. *Wyeth-14,643*

Wyeth-14,643 (pirinixic acid), a well-known inducer of peroxisome proliferation, was developed in the pharmaceutical industry to decrease serum cholesterol levels [112]. However, it was never used as a drug due to the development of hepatocarcinoma in 100% of exposed rats and mice [113]. Wy-14,643, metabolized in the liver by CYP4A [114], does not bind to DNA and is not genotoxic [115, 116]. In contrast, it has been shown to suppress apoptosis, mediated by *Ppara* [117]. Furthermore, there is induction of oxidative stress [118], with increased levels of 8-oxo-dG [119, 120], accumulation of DNA single breaks [121], and changes in gene expression in the livers of exposed rats [122].

In addition to oxidative DNA damage, epigenetic alterations are prominent in Wy-14,643-related hepatocarcinogenesis. There was a decrease in 5-meC levels in hepatic DNA [121, 123], which corresponded to a loss of methylation at LINE-1 and IAP DNA repetitive elements [123], and site-specific hypomethylation of promoter regions of the placental-type glutathione-S-transferase (*Gstp*) [121] and growth arrest and DNA-damage-inducible 45 β (*Gadd45b*) [124] genes. In addition to the changes in DNA methylation status, Wy-14,643 also decreases global levels of H3K4me3 and H4K20me3 [123].

5. Mechanisms of Epigenetic Alterations Induced by Genotoxic and Nongenotoxic Carcinogens

DNA damage and epigenetic alterations are two of the key tightly linked characteristic events of the carcinogenic process caused by carcinogen exposure [8], and several possible mechanisms may be attributed to epigenetic alterations induced by genotoxic and nongenotoxic carcinogens (Figure 3). With respect to genotoxic carcinogens (Figure 3A), the formation of carcinogen–guanine adducts (as shown in Table 1) [4] may diminish the ability of DNA methyltransferases to methylate a target cytosine resulting in the demethylation of DNA [125]. Furthermore, miscoding properties of many DNA adducts may cause sequence variations at CpG base pairs, which could lead to the loss of 5-meC from CpG sites in DNA. Finally, activation of DNA repair and removal of carcinogen–DNA adducts could also lead to the demethylation of DNA.

A common form of DNA damage specific for both genotoxic and nongenotoxic carcinogens (Figure 3B) is the formation of 8-oxo-dG, a ubiquitous oxidative DNA lesion that impairs the enzymatic DNA methylation of adjacent cytosines [126, 127]. Two mechanisms may be involved. First, the presence of 8-oxo-dG in DNA inhibits the binding of DNA to the methyl-CpG binding protein 2 [128], and second, it diminishes the ability of DNA methyltransferases to methylate a target cytosine [129, 130]. Additionally, methylation of a cytosine adjacent and/or

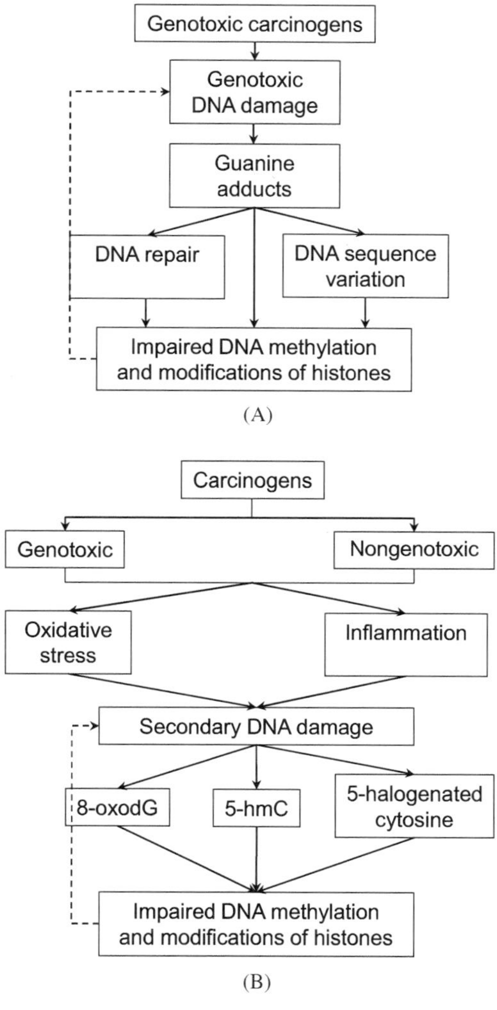

Figure 3. Association between carcinogen-induced DNA damage and impaired DNA methylation in genotoxic (A) and nongenotoxic (B) chemical carcinogenesis.

opposite to 8-oxo-dG decreases the ability of 8-oxoguanine-DNA glycosylase to repair that lesion in DNA [131]. A similar DNA demethylating mechanism is specific for 5-hmeC [127, 128, 132]. The inability of the DNA methyltrasferases to maintain a proper cytosine methylation status in carcinogen-damaged DNA could trigger compensatory overexpression of DNA methyltransferases and lead to random hypermethylation of normally unmethylated promoter CpG islands, with subsequent aberrant gene silencing [127].

Other potential mechanisms of hypermethylation of unmethylated regions in DNA may be attributed to the formation of 5-meC mediated by a free methyl radical mechanism [133] or inflammation-induced DNA damage [132]. The first mechanism is evidenced by the fact that genotoxic carcinogens, such as methylhydrazine and dimethylhydrazine, and products of carcinogen-induced oxidative stress, such as methionine sulfoxide, generate free methyl radicals that lead to the formation of 5-meC [133]. A second mechanism was revealed by Valinluck and Sowers [132], who showed that the formation of inflammation-induced 5-halogenated cytosine intermediates in DNA, including 5-chlorocytosine and 5-bromocytosine, induces abnormal cytosine hypermethylation at unmethylated CpG sites.

In addition to carcinogen–DNA damage-mediated changes in methylation of DNA, carcinogen exposure affects also histone proteins by causing histone protein adduct formation and aberrant histone modifications. In two separate *in vitro* studies, SooHoo *et al.* [134] reported that treatment of nuclei from human lymphoblast TK6 cells with anti-BPDE resulted in the formation of stable anti-BPDE–histone adducts with lysine residues, and Loecken *et al.* [135] demonstrated that 1,2,3,4-diepoxybutane caused the formation of DNA-histone H2B cross-links. Similar observation was also found in experiments *in vivo*. Specifically, Nunes *et al.* [136] have identified a cross-link between the glutathione conjugate of *cis*-2-butene-1,4-dial and lysine 107 of histone H2B in the livers of rats exposed to carcinogen furan. Additionally, extensive reports have shown substantial changes in the histone modifications, including the formation of γ-H2AX and changes in histone H3K9, H3K27, and H4K20 methylation; and H3K9, H3K56, and H4K16 acetylation (Table 1); modifications that are associated with changes in chromatin structure, DNA damage response; and DNA repair [137].

The interconnection between DNA damage and epigenetics can be illustrated further by the fact that several epigenetic marks may serve as the indicators of DNA damage. In addition to the well-established role of γ-H2AX as a marker of DNA damage [138], Kafer *et al.* [139] reported that 5-hmeC marks sites of DNA damage, and growing evidence indicates a key role of histone modifications, such as H4K20 methylation [69] and H3K56 acetylation [140], in DNA repair. Therefore, carcinogen-induced DNA damage can severely impair epigenetic mechanisms and alter the status of the cellular epigenome, while changes in the epigenome can also impact the extent of DNA damage induced by the carcinogenic exposure. For instance, genetic aberrations in the one-carbon metabolism of genes and/or chromatin-modifying genes [141, 142], and metabolic alterations [143] can enhance carcinogen-induced DNA damage. Furthermore, the severity of DNA damage and repair efficiency may be associated with the epigenetically mediated interindividual differences in the DNA damage response to carcinogen exposure [144].

6. Conclusions

The development of cancer is a complex multifactorial process. Major contributions come from continuous exposure to an increasing number of natural and man-made chemicals in the form of agricultural, industrial, pharmaceutical, and dietary exposures [3]. Multiple mechanisms may be involved in the carcinogenic process. Recently, Smith *et al.* [8] proposed that all carcinogenic mechanisms can be grouped into 10 key characteristic categories. However, each of these 10 key mechanisms is not necessarily specific to each carcinogen. Furthermore, each mechanism may not contribute equally to tumorigenesis. As highlighted in this chapter, a wealth of data obtained from exposed humans, human cells *in vitro*, and experimental animal studies convincingly indicates that induction of DNA damage and epigenetic alterations are specific to all carcinogenic agents regardless of their known mode of action (Figure 3). It has become increasingly clear that genotoxic events and epigenetic changes are interdependent and inextricably connected in the tumorigenic process, and both types of changes must occur simultaneously in the target organ. Therefore, simultaneous evaluation of both carcinogen-induced DNA damage and epigenetic aberrations will facilitate the identification and classification of newly introduced chemical agents.

Abbreviations

2-AAF	2-acetylaminofluorene
4-HNE	4-hydroxynonenal
5-hmeC	5-hydroxymethylcytosine
5-meC	5-methylcytosine
8-oxo-dG	8-oxo-7-hydroxydeoxyguanosine
ac	acetylation
Bis-NG7-BD	1,4-bis-(guan-7-yl)-2,3-butanediol
BP	benzo[*a*]pyrene
BPDE	7β,8α-dihydroxy-9α,10α-epoxy-7,8,9,10-tetrahydrobenzo[*a*]-pyrene
BPdG	r7,t8,t9-trihydroxy-c-10-(N^2-deoxyguanosyl)-7,8,9,10-tetrahydrobenzo[*a*]pyrene
CYPs	cytochrome P450 mixed function oxidases
dG-C8-AAF	N-(deoxyguanosin-8-yl)-2-acetylaminofluorene
dG-C8-AF	N-(deoxyguanosin-8-yl)-2-aminofluorene
dG-desMeTAM	(*E*)-α-(deoxyguanosin-N^2-yl)-N-desmethyltamoxifen
dG-TAM	(*E*)-α-(deoxyguanosin-N^2-yl)-tamoxifen
DNMT	DNA methyltransferase
H2, H3, H4	histones 2, 3, 4
IARC	International Agency for Research on Cancer
K	lysine
LINE-1	long interspersed nuclear element 1
me	methyl
SAM	*S*-adenosyl-l-methionine
TAM	tamoxifen
THB-Gua	N7-(2,3,4-trihydroxybut-1-yl)guanine
γH2AX	phosphorylated form of histone H2AX
TET	ten-eleven translocation

References

1. Walker CL, Ho SM. Developmental reprogramming of cancer susceptibility. *Nat Rev Cancer* 2012; 12: 479–486.

2. Rappaport SM. Genetic factors are not the major causes of chronic diseases. *PLOS ONE* 2016; 11: e0154387.
3. Stewart BW, Wild CP. *World Cancer Report 2014*. International Agency for Research on Cancer, France, 2014.
4. Poirier MC. Linking DNA adduct formation and human cancer risk in chemical carcinogenesis. *Environ Mol Mutagen* 2016; 57: 499–507.
5. Jacobs MN, Colacci A, Louekari K, Luijten M, Hakkert BC, Paparella M, *et al*. International regulatory needs for development of an IATA for non-genotoxic carcinogenic chemical substances. *ALTEX* 2016; 33: 359–392.
6. Pogribny IP, Rusyn I. Environmental toxicants, epigenetics, and cancer. *Adv Exp Med Biol* 2013; 754: 215–232.
7. Pogribny IP, Beland FA. DNA methylome alterations in chemical carcinogenesis. *Cancer Lett* 2013; 334: 39–45.
8. Smith MT, Guyton KZ, Gibbons CF, Fritz JM, Portier CJ, Rusyn I, *et al*. Key characteristics of carcinogens as a basis for organizing data on mechanisms of carcinogenesis. *Environ Health Perspect* 2016; 124: 713–721.
9. Deans C, Maggert KA. What do you mean, "epigenetic"?. *Genetics* 2015; 199: 887–896.
10. Ooi SK, O'Donnell AH, Bestor TH. Mammalian cytosine methylation at a glance. *J Cell Sci* 2009; 122: 2787–2791.
11. Edwards JR, O'Donnell AH, Rollins RA, Peckham HE, Lee C, Milekic MH, *et al*. Chromatin and sequence features that define the fine and gross structure of genomic methylation patterns. *Genome Res* 2010; 20: 972–980.
12. Song J, Pfeifer GP. Are there specific readers of oxidized 5-methylcytosine bases? *Bioessays* 2016; 38: 1038–1047.
13. Madugundu GS, Cadet J, Wagner JR. Hydroxyl-radical-induced oxidation of 5-methylcytosine in isolated and cellular DNA. *Nucleic Acids Res* 2014; 42: 7450–7460.
14. Kouzarides T. Chromatin modifications and their function. *Cell* 2007; 128: 693–705.
15. Torres IO, Fujimori DG. Functional coupling between writers, erasers and readers of histone and DNA methylation. *Curr Opin Struct Biol* 2015; 35: 68–75.
16. Paluch BE, Naqash AR, Brumberger Z, Nemeth MJ, Griffiths EA. Epigenetics: a primer for clinicians. *Blood Rev* 2016; 30: 285–295.
17. World Health Organization. IARC Monographs on the Evaluations of Carcinogenic Risks to Humans. A review of human carcinogens. Part F: Chemical agents and related occupations. 2012; 100: 111–144.
18. Beland FA, Churchwell MI, Von Tungeln LS, Chen S, Fu PP, Culp SJ, *et al*. High-performance liquid chromatography electrospray ionization

tandem mass spectrometry for the detection and quantitation of benzo[a] pyrene-DNA adducts. *Chem Res Toxicol* 2005; 18: 1306–1315.
19. Zuo J, Brewer DS, Arlt VM, Cooper CS, Phillips DH. Benzo[a]pyrene-induced DNA adducts and gene expression profiles in target and non-target organs for carcinogenesis in mice. *BMC Genomics* 2014; 15: 880.
20. Kucab JE, van Steeg H, Luijten M, Schmeiser HH, White PA, Phillips DH, *et al.* TP53 mutations induced by BPDE in Xpa-WT and Xpa-Null human TP53 knock-in (Hupki) mouse embryo fibroblasts. *Mutat Res* 2015; 773: 48–62.
21. Menzies GE, Reed SH, Brancale A, Lewis PD. Base damage, local sequence context and TP53 mutation hotspots: a molecular dynamics study of benzo[a]pyrene induced DNA distortion and mutability. *Nucleic Acids Res* 2015; 43: 9133–9146.
22. Hu W, Feng Z, Tang MS. Preferential carcinogen-DNA adduct formation at codons 12 and 14 in the human K-ras gene and their possible mechanisms. *Biochemistry* 2003; 42: 10012–10023.
23. Briedé JJ, Godschalk RW, Emans MT, De Kok TM, Van Agen EB, Van Maanen JM, *et al. In vitro* and *in vivo* studies on oxygen free radical and DNA adduct formation in rat lung and liver during benzo[a]pyrene metabolism. *Free Radic Res* 2004; 38: 995–1002.
24. Wang HT, Weng MW, Chen WC, Yobin M, Pan J, Chung FL, *et al.* Effect of CpG methylation at different sequence context on acrolein- and BPDE-DNA binding and mutagenesis. *Carcinogenesis* 2013; 34: 220–227.
25. Tretyakova N, Guza R, Matter B. Endogenous cytosine methylation and the formation of carcinogen carcinogen–DNA adducts. *Nucleic Acids Symp Ser* 2008; 52: 49–50.
26. Glick J, Xiong W, Lin Y, Noronha AM, Wilds CJ, Vouros P. The influence of cytosine methylation on the chemoselectivity of benzo[a]pyrene diol epoxide-oligonucleotide adducts determined using nanoLC/MS/MS. *J Mass Spectrom* 2009; 44: 1241–1248.
27. Yoon JH, Smith LE, Feng Z, Tang MS, Lee CS, Pfeifer GP. Methylated CpG dinucleotides are the preferential targets for G-to-T transversion mutations induced by benzo[a]pyrene diol epoxide in mammalian cells similarities with the p53 mutation spectrum in smoking-associated lung cancers. *Cancer Res* 2001; 61: 7110–7117.
28. Pfeifer GP, Grunberger D, Drahovsky D. Impaired enzymatic methylation of BPDE-modified DNA. *Carcinogenesis* 1984; 5: 931–935.
29. Teneng I, Montoya-Durango DE, Quertermous JL, Lacy ME, Ramos KS. Reactivation of L1 retrotransposon by benzo (a) pyrene involves complex genetic and epigenetic regulation. *Epigenetics* 2011; 6: 355–367.

30. Pavanello S, Bollati V, Pesatori AC, Kapka L, Bolognesi C, Bertazzi PA, *et al.* Global and gene-specific promoter methylation changes are related to *anti*-B[a]PDE-DNA adduct levels and influence micronuclei levels in polycyclic aromatic hydrocarbon-exposed individuals. *Int J Cancer* 2009; 125: 1692–1697.
31. Ouyang B, Baxter CS, Lam HM, Yeramaneni S, Levin L, *et al.* Hypomethylation of dual specificity phosphatase 22 promoter correlates with duration of service in firefighters and is inducible by low-dose benzo[a]pyrene. *J Occup Environ Med* 2012; 54: 774–780.
32. Herbstman JB, Tang D, Zhu D, Qu L, Sjödin A, Li Z, *et al.* Prenatal exposure to polycyclic aromatic hydrocarbons, benzo[a]pyrene-DNA adducts, and genomic DNA methylation in cord blood. *Environ Health Perspect* 2012; 120: 733–738.
33. Yang P, Ma J, Zhang B, Duan H, He Z, Zeng J, *et al.* CpG site–specific hypermethylation of p16INK4α in peripheral blood lymphocytes of PAH-exposed workers. *Cancer Epidemiol Biomarkers Prev* 2012; 21: 182–190.
34. Ye F, Xu XC. Benzo[a]pyrene diol epoxide suppresses retinoic acid receptor-β 2 expression by recruiting DNA (cytosine-5-)-methyltransferase 3A. *Mol Cancer* 2010; 9: 93.
35. Peluso ME, Munnia A, Bollati V, Srivatanakul P, Jedpiyawongse A, Sangrajrang S, *et al.* Aberrant methylation of hypermethylated-in-cancer-1 and exocyclic DNA adducts in tobacco smokers. *Toxicol Sci* 2014; 137: 47–54.
36. Mattsson Å, Jernström B, Cotgreave IA, Bajak E. H2AX phosphorylation in A549 cells induced by the bulky and stable DNA adducts of benzo [*a*] pyrene and dibenzo [*a, l*] pyrene diol epoxides. *Chem Biol Interact* 2009; 177: 40–47.
37. IARC Monographs on the Evaluations of Carcinogenic Risks to Humans. A review of human carcinogens. Part F: Chemical Agents and Related Occupations. 2012; 100: 309–338.
38. Koc H, Tretyakova NY, Walker VE, Henderson RF, Swenberg JA. Molecular dosimetry of N-7 guanine adduct formation in mice and rats exposed to 1,3-butadiene. *Chem Res Toxicol* 1999; 12: 566–574.
39. Park S, Tretyakova N. Structural characterization of the major DNA-DNA cross-link of 1,2,3,4-diepoxybutane. *Chem Res Toxicol* 2004; 17: 129–136.
40. Selzer RR, Elfarra AA. Chemical modification of deoxycytidine at different sites yields adducts of different stabilities: characterization of N^3- and O^2-deoxycytidine and N^3-deoxyuridine adducts of butadiene monoxide. *Arch Biochem Biophys* 1997; 343: 63–72.

41. Boysen G, Georgieva NI, Bordeerat NK, Šram RJ, Vacek P, Albertini RJ, *et al.* Formation of 1,2:3,4-diepoxybutane-specific hemoglobin adducts in 1,3-butadiene exposed workers. *Toxicol Sci* 2012; 125: 30–40.
42. Chappell G, Kobets T, O'Brien B, Tretyakova N, Sangaraju D, Kosyk O, *et al.* Epigenetic events determine tissue-specific toxicity of inhalational exposure to the genotoxic chemical 1,3-butadiene in male C57BL/6J mice. *Toxicol Sci* 2014; 142: 375–384.
43. Koturbash I, Scherhag A, Sorrentino J, Sexton K, Bodnar W, Tryndyak V, *et al.* Epigenetic alterations in liver of C57BL/6J mice after short-term inhalational exposure to 1,3-butadiene. *Environ Health Perspect* 2011; 119: 635–640.
44. Collins JJ, Anteau SE, Swaen GM, Bodner KM, Bodnar CM. Lymphatic and hematopoietic cancers among benzene-exposed workers. *J Occup Environ Med* 2015; 57: 159–163.
45. Pyatt D. Benzene and hematopoietic malignancies. *Clin Occup Environ Med* 2004; 4: 529–525.
46. Stenehjem JS, Kjærheim K, Bråtveit M, Samuelsen SO, Barone-Adesi F, Rothman N, *et al.* Benzene exposure and risk of lymphohaematopoietic cancers in 25000 offshore oil industry workers. *Br J Cancer* 2015; 112: 1603–1612.
47. Seaton MJ, Schlosser PM, Bond JA, Medinsky MA. Benzene metabolism by human liver microsomes in relation to cytochrome P450 2E1 activity. *Carcinogenesis* 1994; 15: 1799–1806.
48. Snyder R, Hedli CC. An overview of benzene metabolism. *Environ Health Perspect* 1996; 104: 1165–1171.
49. Arfellini G, Grilli S, Colacci A, Mazzullo M, Prodi G. *In vivo* and *in vitro* binding of benzene to nucleic acids and proteins of various rat and mouse organs. *Cancer Lett* 1985; 28: 159–168.
50. Mazzullo M, Bartoli S, Bonora B, Colacci A, Grilli S, Lattanzi G, *et al.* Benzene adducts with rat nucleic acids and proteins: dose-response relationship after treatment *in vivo*. *Environ Health Perspect* 1989; 82: 259–266.
51. Bollati V, Baccarelli A, Hou L, Bonzini M, Fustinoni S, Cavallo D, *et al.* Changes in DNA methylation patterns in subjects exposed to low-dose benzene. *Cancer Res* 2007; 67: 876–880.
52. Melki JR, Clark SJ. DNA methylation changes in leukaemia. *Semin Cancer Biol* 2002; 12: 347–357.

53. Gao A, Zuo X, Liu Q, Lu X, Guo W, Tian L. Methylation of PARP-1 promoter involved in the regulation of benzene-induced decrease of PARP-1 mRNA expression. *Toxicol Lett* 2010; 195: 114–118.
54. Xing C, Chen Q, Li G, Zhang L, Zheng M, Zou Z, *et al.* Microsomal epoxide hydrolase (EPHX1) polymorphisms are associated with aberrant promoter methylation of ERCC3 and hematotoxicity in benzene-exposed workers. *Environ Mol Mutagen* 2013; 54: 397–405.
55. Yu K, Shi YF, Yang KY, Zhuang Y, Zhu RH, Xu X, *et al.* Decreased topoisomerase IIα expression and altered histone and regulatory factors of topoisomerase IIα promoter in patients with chronic benzene poisoning. *Toxicol Lett* 2011; 203: 111–117.
56. Ishihama M, Toyooka T, Ibuki Y. Generation of phosphorylated histone H2AX by benzene metabolites. *Toxicol In Vitro* 2008; 22: 1861–1868.
57. Yang J, Bai W, Niu P, Tian L, Gao A. Aberrant hypomethylated STAT3 was identified as a biomarker of chronic benzene poisoning through integrating DNA methylation and mRNA expression data. *Exp Mol Pathol* 2014; 96: 346–353.
58. Bruserud Ø, Nepstad I, Hauge M, Hatfield KJ, Reikvam H. STAT3 as a possible therapeutic target in human malignancies: lessons from acute myeloid leukemia. *Expert Rev Hematol* 2015; 8: 29–41.
59. Coulter JB, O'Driscoll CM, Bressler JP. Hydroquinone increases 5-hydroxymethylcytosine formation through ten eleven translocation 1 (TET1) 5-methylcytosine dioxygenase. *J Biol Chem* 2013; 288: 28792–28800.
60. Liu L, Ling X, Liang H, Gao Y, Yang H, Shao J, *et al.* Hypomethylation mediated by decreased DNMTs involves in the activation of proto-oncogene MPL in TK6 cells treated with hydroquinone. *Toxicol Lett* 2012; 209: 239–245.
61. Cox AJ, Wilson RH, DeEds F. The carcinogenic activity of 2-acetaminofluorene characteristics of the lesions in albino rats. *Cancer Res* 1947; 7: 647–657.
62. Beland FA, Kadlubar FF. Formation and persistence of arylamine DNA adducts *in vivo. Environ Health Perspect* 1985; 62: 19–30.
63. Beland FA, Kadlubar FF. Metabolic activation and DNA adducts of aromatic amines and nitroaromatic hydrocarbons. In *Handbook of Experimental Pharmacology*, Cooper CS, Grover PL, eds., Vol. 94, Springer-Verlag Berlin, Heidelberg, 1990, pp. 267–325.
64. Poirier MC, Hunt JM, True BA, Laishes BA, Young JF, Beland FA. DNA adduct formation, removal and persistence in rat liver during one month of feeding 2-acetylaminofluorene. *Carcinogenesis* 1984; 5: 1591–1596.

65. Wiseman RW, Stowers SJ, Miller EC, Anderson MW, Miller JA. Activating mutations of the c-Ha-ras protooncogene in chemically induced hepatomas of the male B6C3 F_1 mouse. *Proc Natl Acad Sci USA* 1986; 83: 5825–5829.
66. Poirier MC, Fullerton NF, Kinouchi T, Smith BA, Beland FA. Comparison between DNA adduct formation and tumorigenesis in livers and bladders of mice chronically fed 2-acetylaminofluorene. *Carcinogenesis* 1991; 12: 895–900.
67. Bagnyukova TV, Tryndyak VP, Montgomery B, Churchwell MI, Karpf AR, James SR, *et al.* Genetic and epigenetic changes in rat preneoplastic liver tissue induced by 2-acetylaminofluorene. *Carcinogenesis* 2008; 29: 638–646.
68. Pogribny IP, Muskhelishvili L, Tryndyak VP, Beland FA. The role of epigenetic events in genotoxic hepatocarcinogenesis induced by 2-acetylaminofluorene. *Mutat Res Genet Toxicol Environ Mutagen* 2011; 722:106–113.
69. Jørgensen S, Schotta G, Sørensen CS. Histone H4 lysine 20 methylation: key player in epigenetic regulation of genomic integrity. *Nucleic Acids Res* 2013; 41: 2797–2806.
70. Fraga MF, Ballestar E, Villar-Garea A, Boix-Chornet M, Espada J, Schotta G, *et al.* Loss of acetylation at Lys16 and trimethylation at Lys20 of histone H4 is a common hallmark of human cancer. *Nat Genet* 2005; 37: 391–400.
71. Cuzick J, Powles T, Veronesi U, Forbes J, Edwards R, Ashley S, *et al.* Overview of the main outcomes in breast-cancer prevention trials. *Lancet* 2003; 361: 296–300.
72. Jones ME, van Leeuwen FE, Hoogendoorn WE, Mourits MJE, Hollema H, van Boven H, Press MF, *et al.* Endometrial cancer survival after breast cancer in relation to tamoxifen treatment: pooled results from three countries. *Breast Cancer Res* 2012; 14: R91.
73. IARC Monographs on the Evaluations of Carcinogenic Risks to Humans. A review of human carcinogens. Part A: Pharmaceuticals. 2011; 100: 131–162.
74. Dragan YP, Fahey S, Nuwaysir E, Sattler C, Babcock K, Vaughan J, *et al.* The effect of tamoxifen and two of its non-isomerizable fixed-ring analogs on multistage rat hepatocarcinogenesis. *Carcinogenesis* 1996; 17: 585–594.
75. da Costa GG, McDaniel-Hamilton LP, Heflich RH, Marques MM, Beland FA. DNA adduct formation and mutant induction in Sprague–Dawley rats treated with tamoxifen and its derivatives. *Carcinogenesis* 2001; 22: 1307–1315.

76. Phillips DH. Understanding the genotoxicity of tamoxifen? *Carcinogenesis* 2001; 22: 839–849.
77. Hernandez-Ramon EE, Sandoval NA, John K, Cline JM, Wood CE, Woodward RA, *et al.* Tamoxifen–DNA adduct formation in monkey and human reproductive organs. *Carcinogenesis* 2014; 35: 1172–1176.
78. Tryndyak VP, Muskhelishvili L, Kovalchuk O, Rodriguez-Juarez R, Montgomery B, Churchwell MI, *et al.* Effect of long-term tamoxifen exposure on genotoxic and epigenetic changes in rat liver: implications for tamoxifen-induced hepatocarcinogenesis. *Carcinogenesis* 2006; 27: 1713–1720.
79. Tryndyak VP, Kovalchuk O, Muskhelishvili L, Montgomery B, Rodriguez-Juarez R, Melnyk S, *et al.* Epigenetic reprogramming of liver cells in tamoxifen-induced rat hepatocarcinogenesis. *Mol Carcinog* 2007; 46: 187–197.
80. de Conti A, Tryndyak V, Churchwell MI, Melnyk S, Latendresse JR, Muskhelishvili L, *et al.* Genotoxic, epigenetic, and transcriptomic effects of tamoxifen in mouse liver. *Toxicology* 2014; 325: 12–20.
81. Wu H, Chen Y, Liang J, Shi B, Wu G, Zhang Y, *et al.* Hypomethylation-linked activation of PAX2 mediates tamoxifen-stimulated endometrial carcinogenesis. *Nature* 2005; 438: 981–987.
82. Nagy E, Gajjar KB, Patel II, Taylor S, Martin-Hirsch PL, Stringfellow HF, *et al.* MGMT promoter hypermethylation and K-RAS, PTEN and TP53 mutations in tamoxifen-exposed and non-exposed endometrial cancer cases. *Br J Cancer* 2014; 110: 2874–2880.
83. Lijinsky W, Reuber MD, Blackwell BN. Liver tumors induced in rats by oral administration of the antihistaminic methapyrilene hydrochloride. *Science* 1980; 209: 817–819.
84. Ratra GS, Cottrell S, Powell CJ. Effects of induction and inhibition of cytochromes P450 on the hepatotoxicity of methapyrilene. *Toxicol Sci* 1998; 46: 185–196.
85. Kelly DW, Holder CL, Korfmacher WA, Getek TA, Lay JO, Casciano DA, *et al.* Metabolism of methapyrilene by Fischer-344 rat and $B_6C_3F_1$ mouse hepatocytes. *Xenobiotica* 1992; 22: 1367–1381.
86. Casciano DA, Shaddock JG, Talaska G. The potent hepatocarcinogen methapyrilene does not form DNA adducts in livers of Fischer 344 rats. *Mutat Res* 1988; 208: 129–135.
87. Craig A, Sidaway J, Holmes E, Orton T, Jackson D, Rowlinson R, *et al.* Systems toxicology: integrated genomic, proteomic and metabonomic analysis of methapyrilene induced hepatotoxicity in the rat. *J Proteome Res* 2006; 5: 1586–1601.

88. Shpyleva S, Dreval K, de Conti A, Kindrat I, Melnyk S, Chen T, *et al.* Organ-specific epigenetic changes induced by non-genotoxic liver carcinogen methapyrilene in Fischer 344 rats. *Toxicol Sci* 2017; 156: 190–198.
89. Marquardt JU, Andersen JB, Thorgeirsson SS. Functional and genetic deconstruction of the cellular origin in liver cancer. *Nat Rev Cancer* 2015; 15: 653–667.
90. Powell CL, Swenberg JA, Rusyn I. Expression of base excision DNA repair genes as a biomarker of oxidative DNA damage. *Cancer Lett* 2005; 229: 1–11.
91. Uchida K. 4-Hydroxy-2-nonenal: a product and mediator of oxidative stress. *Prog Lipid Res* 2003; 42: 318–343.
92. Ozden S, Kara NT, Sezerman OU, Durasi İM, Chen T, Demirel G, *et al.* Assessment of global and gene-specific DNA methylation in rat liver and kidney in response to non-genotoxic carcinogen exposure. *Toxicol Appl Pharmacol* 2015; 289: 203–212.
93. NTP (National Toxicology Program). Toxicology and carcinogenesis studies of furan (CAS No. 110-00-9) in F344/N rats and B6C3F1 mice (gavage studies). National Toxicology Program Technical Report Series, Vol. 402, 1993, pp. 1–77.
94. Moser GJ, Foley J, Burnett M, Goldsworthy TL, Maronpot R. Furan-induced dose–response relationships for liver cytotoxicity, cell proliferation, and tumorigenicity (furan-induced liver tumorigenicity). *Exp Toxicol Pathol* 2009; 61: 101–111.
95. IARC. *IARC Monographs on the Evaluations of Carcinogenic Risks to Humans. Dry Cleaning, Some Chlorinated Solvents and Other Industrial Chemicals,* Vol. 63, IARC, France, 1995, pp. 393–407.
96. Johansson E, Reynolds S, Anderson M, Maronpot R. Frequency of Ha-ras-1 gene mutations inversely correlated with furan dose in mouse liver tumors. *Mol Carcinog* 1997; 18: 199–205.
97. Peterson LA. Electrophilic intermediates produced by bioactivation of furan. *Drug Metab Rev* 2006; 38: 615–626.
98. Neuwirth C, Mosesso P, Pepe G, Fiore M, Malfatti M, Turteltaub K, *et al.* Furan carcinogenicity: DNA binding and genotoxicity of furan in rats *in vivo*. *Mol Nutr Food Res* 2012; 56: 1363–1374.
99. Chen T, Williams TD, Mally A, Hamberger C, Mirbahai L, Hickling K, *et al.* Gene expression and epigenetic changes by furan in rat liver. *Toxicology* 2012; 292: 63–70.

100. McDaniel LP, Ding W, Dobrovolsky VN, Shaddock JG, Mittelstaedt RA, Doerge DR, *et al*. Genotoxicity of furan in Big Blue rats. *Mutat Res* 2012; 742: 72–78.
101. Churchwell MI, Scheri RC, Von Tungeln LS, da Costa GG, Beland FA, Doerge DR. Evaluation of serum and liver toxicokinetics for furan and liver DNA adduct formation in male Fischer 344 rats. *Food Chem Toxicol* 2015; 86: 1–8.
102. Ding W, Petibone DM, Latendresse JR, Pearce MG, Muskhelishvili L, White GA, *et al*. *In vivo* genotoxicity of furan in F344 rats at cancer bioassay doses. *Toxicol Appl Pharmacol* 2012; 261: 164–171.
103. Banda M, Recio L, Parsons BL. ACB-PCR measurement of spontaneous and furan-induced H-ras codon 61 CAA to CTA and CAA to AAA mutation in B6C3F1 mouse liver. *Environ Mol Mutagen* 2013; 54: 659–667.
104. Terrell AN, Huynh M, Grill AE, Kovi RC, O'Sullivan MG, Guttenplan JB, *et al*. Mutagenicity of furan in female Big Blue B6C3F1 mice. *Mutat Res Genet Toxicol Environ Mutagen* 2014; 770: 46–54.
105. Dong H, Gill S, Curran IH, Williams A, Kuo B, Wade MG, *et al*. Toxicogenomic assessment of liver responses following subchronic exposure to furan in Fischer F344 rats. *Arch Toxicol* 2016; 90: 1351–1367.
106. Hickling KC, Hitchcock JM, Oreffo V, Mally A, Hammond TG, Evans JG, *et al*. Evidence of oxidative stress and associated DNA damage, increased proliferative drive, and altered gene expression in rat liver produced by the cholangiocarcinogenic agent furan. *Toxicol Pathol* 2010; 38: 230–243.
107. Mally A, Graff C, Schmal O, Moro S, Hamberger C, Schauer UM, *et al*. Functional and proliferative effects of repeated low-dose oral administration of furan in rat liver. *Mol Nutr Food Res* 2010; 54: 1556–1567.
108. Recio L, Phillips SL, Maynor T, Waters M, Jackson AF, Yauk CL. Differential expression of long noncoding RNAs in the livers of female B6C3F1 mice exposed to the carcinogen furan. *Toxicol Sci* 2013; 135: 369–379.
109. de Conti A, Kobets T, Escudero-Lourdes C, Montgomery B, Tryndyak V, Beland FA, *et al*. Dose-and time-dependent epigenetic changes in the livers of Fisher 344 rats exposed to furan. *Toxicol Sci* 2014; 139: 371–380.
110. de Conti A, Kobets T, Tryndyak V, Burnett SD, Han T, Fuscoe JC, *et al*. Persistence of furan-induced epigenetic aberrations in the livers of F344 rats. *Toxicol Sci* 2015; 144: 217–226.
111. Tryndyak V, de Conti A, Doerge DR, Olson GR, Beland FA, Pogribny IP. Furan-induced transcriptomic and gene-specific DNA methylation changes

in the livers of Fischer 344 rats in a 2-year carcinogenicity study. *Arch Toxicol* 2016; 7: 1–1.
112. Santilli AA, Scotese AC, Tomarelli RM. A potent antihypercholesterolemic agent: [4-chloro-6-(2, 3-xylidino)-2-pyrimidinylthio] acetic acid (Wy-14643). *Cell Mol Life Sci* 1974; 30: 1110–1111.
113. Reddy JK, Rao MS, Azarnoff DL, Sell S. Mitogenic and carcinogenic effects of a hypolipidemic peroxisome proliferator, [4-chloro-6-(2, 3-xylidino)-2-pyrimidinylthio] acetic acid (Wy-14,643), in rat and mouse liver. *Cancer Res* 1979; 39: 152–161.
114. Cowart LA, Wei S, Hsu MH, Johnson EF, Krishna MU, Falck JR, *et al.* The CYP4A isoforms hydroxylate epoxyeicosatrienoic acids to form high affinity peroxisome proliferator-activated receptor ligands. *J Biol Chem* 2002; 277: 35105–35112.
115. Gupta RC, Goel SK, Earley K, Singh B, Reddy JK. ^{32}P-Postlabeling analysis of peroxisome proliferator-DNA adduct formation in rat liver *in vivo* and hepatocytes *in vitro*. *Carcinogenesis* 1985; 6: 933–936.
116. Lefevre PA, Tinwell H, Galloway SM, Hill R, Mackay JM, Elcombe CR, *et al.* Evaluation of the genetic toxicity of the peroxisome proliferator and carcinogen methyl clofenapate, including assays using Muta Mouse and Big Blue transgenic mice. *Hum Exp Toxicol* 1994; 13: 764–775.
117. Green S, Tugwood JD, Issemann I. The molecular mechanism of peroxisome proliferator action: a model for species differences and mechanistic risk assessment. *Toxicol Lett* 1992; 64: 131–139.
118. Clayson DB, Mehta R, Iverson F. Oxidative DNA damage — the effects of certain genotoxic and operationally non-genotoxic carcinogens. *Mutat Res* 1994; 317: 25–42.
119. Rusyn I, Asakura S, Pachkowski B, Bradford BU, Denissenko MF, Peters JM, *et al.* Expression of base excision DNA repair genes is a sensitive biomarker for *in vivo* detection of chemical-induced chronic oxidative stress identification of the molecular source of radicals responsible for DNA damage by peroxisome proliferators. *Cancer Res* 2004; 64: 1050–1057.
120. Reddy JK, Rao MS. Oxidative DNA damage caused by persistent peroxisome proliferation: its role in hepatocarcinogenesis. *Mutat Res* 1989; 214: 63–68.
121. Pogribny IP, Tryndyak VP, Boureiko A, Melnyk S, Bagnyukova TV, Montgomery B, *et al.* Mechanisms of peroxisome proliferator-induced DNA hypomethylation in rat liver. *Mutat Res* 2008; 644: 17–23.

122. Ellinger-Ziegelbauer H, Stuart B, Wahle B, Bomann W, Ahr HJ. Comparison of the expression profiles induced by genotoxic and nongenotoxic carcinogens in rat liver. *Mutat Res* 2005; 575: 61–84.
123. Pogribny IP, Tryndyak VP, Woods CG, Witt SE, Rusyn I. Epigenetic effects of the continuous exposure to peroxisome proliferator WY-14,643 in mouse liver are dependent upon peroxisome proliferator activated receptor α. *Mutat Res* 2007; 625: 62–71.
124. Kim JH, Wahyudi LD, Kim KK, Gonzalez FJ. PPARα activation drives demethylation of the CpG islands of the *Gadd45b* promoter in the mouse liver. *Biochem Biophys Res Commun* 2016; 476: 293–298.
125. Valinluck V, Sowers LC. Inflammation-mediated cytosine damage: a mechanistic link between inflammation and the epigenetic alterations in human cancers. *Cancer Res* 2007; 67: 5583–5586.
126. Weitzman SA, Turk PW, Milkowski DH, Kozlowski K. Free radical adducts induce alterations in DNA cytosine methylation. *Proc Natl Acad Sci USA* 1994; 91: 1261–1264.
127. Wu Q, Ni X. ROS-mediated DNA methylation pattern alterations in carcinogenesis. *Curr Drug Targets* 2015; 16: 13–19.
128. Valinluck V, Tsai HH, Rogstad DK, Burdzy A, Bird A, Sowers LC. Oxidative damage to methyl-CpG sequences inhibits the binding of the methyl-CpG binding domain (MBD) of methyl-CpG binding protein 2 (MeCP2). *Nucleic Acids Res* 2004; 32: 4100–4108.
129. Turk PW, Laayoun A, Smith SS, Weitzman SA. DNA adduct 8-hydroxyl-2′-deoxyguanosine (8-hydroxyguanine) affects function of human DNA methyltransferase. *Carcinogenesis* 1995; 16: 1253–1255.
130. Maltseva DV, Baykov AA, Jeltsch A, Gromova ES. Impact of 7, 8-dihydro-8-oxoguanine on methylation of the CpG site by Dnmt3a. *Biochemistry* 2009; 48: 1361–1368.
131. Kasymov RD, Grin IR, Endutkin AV, Smirnov SL, Ishchenko AA, Saparbaev MK, *et al.* Excision of 8-oxoguanine from methylated CpG dinucleotides by human 8-oxoguanine DNA glycosylase. *FEBS Lett* 2013; 587: 3129–3134.
132. Valinluck V, Sowers LC. Endogenous cytosine damage products alter the site selectivity of human DNA maintenance methyltransferase DNMT1. *Cancer Res* 2007; 67: 946–950.
133. Kasai H. What causes human cancer? Approaches from the chemistry of DNA damage. *Genes Environ* 2016; 38: 19.
134. SooHoo CK, Singh K, Skipper PL, Tannenbaum SR, Dasari RR. Characterization of benzo[*a*]pyrene anti-diol epoxide adducts to human histones. *Chem Res Toxicol* 1994; 7: 134–138.

135. Loecken EM, Dasari S, Hill S, Tabb DL, Guengerich FP. The *bis*-electrophile diepoxybutane cross-links DNA to human histones but does not result in enhanced mutagenesis in recombinant systems. *Chem Res Toxicol* 2009; 22: 1069–1076.
136. Nunes J, Martins IL, Charneira C, Pogribny IP, de Conti A, Beland FA, *et al.* New insights into the molecular mechanism of chemical carcinogenesis: *in vivo* adduction of histone H2B by a reactive metabolite of the chemical carcinogen furan. *Toxicol Lett* 2016; 264: 106–113.
137. Cao LL, Shen C, Zhu WG. Histone modifications in DNA damage response. *Sci China Life Sci* 2016; 59: 257–270.
138. Mah LJ, El-Osta A, Karagiannis TC. γH2AX: a sensitive molecular marker of DNA damage and repair. *Leukemia* 2010; 24: 679–686.
139. Kafer GR, Li X, Horii T, Suetake I, Tajima S, Hatada I, *et al.* 5-Hydroxymethylcytosine marks sites of DNA damage and promotes genome stability. *Cell Rep* 2016; 14: 1283–1292.
140. Wurtele H, Kaiser GS, Bacal J, St-Hilaire E, Lee EH, Tsao S, *et al.* Histone H3 lysine 56 acetylation and the response to DNA replication fork damage. *Mol Cell Biol* 2012; 32: 154–172.
141. Zijno A, Andreoli C, Leopardi P, Marcon F, Rossi S, Caiola S, *et al.* Folate status, metabolic genotype, and biomarkers of genotoxicity in healthy subjects. *Carcinogenesis* 2003; 24: 1097–1103.
142. Kaji K, Factor VM, Andersen JB, Durkin ME, Tomokuni A, Marquardt JU, *et al.* DNMT1 is a required genomic regulator for murine liver histogenesis and regeneration. *Hepatology* 2016; 64: 582–598.
143. Tice RR, Yager JW, Andrews P, Crecelius E. Effect of hepatic methyl donor status on urinary excretion and DNA damage in B6C3F1 mice treated with sodium arsenite. *Mutat Res* 1997; 386: 315–334.
144. Koturbash I, Scherhag A, SorSrentino J, Sexton K, Bodnar W, Swenberg JA, *et al.* Epigenetic mechanisms of mouse inter-strain variability in genotoxicity of the environmental toxicant 1,3-butadiene. *Toxicol Sci* 2011; 122: 448–456.

Chapter 12

Adductomics

David H. Phillips

Department of Analytical, Environmental and Forensic Sciences, School of Public Health and Environmental Sciences, Faculty of Life Sciences and Medicine; MRC-PHE Centre for Environmental Health; and NIHR Health Protection Research Unit in Health Impact of Environmental Hazards at King's College London in partnership with Public Health England

King's College London, Franklin-Wilkins Building, 150 Stamford Street, London SE1 9NH, United Kingdom

Abstract

DNA and protein adducts are validated biomarkers of human exposure to carcinogens and, in some instances, of risk of developing cancer. As derivatives of reactive chemicals, adducts may also provide unique mechanistic insights. Until recently studies have adopted a "bottom-up" approach, whereby exposure to known or suspected agents is monitored. The advent of the "exposome (see Glossary)" concept, which seeks to define the totality of the influence of the environment on disease susceptibility, has led to a more untargeted top-down "adductomics" approach to identifying multiple modifications to DNA or protein (hemoglobin and albumin), attempting to capture entire populations of adducts. Although still in its infancy, the approach offers the prospect, particularly when combined with other omic approaches, such as proteomics, transcriptomics and metabolomics, all of which have potential for high-throughput screening of populations, of capturing snapshots of human exposure to genotoxicants and the resultant biological consequences at multiple stages throughout life. This should shed significant light on the causes and courses of chronic diseases afflicting humans and provide new opportunities for early interventions for disease prevention.

1. The Exposome

It is generally accepted that the majority of important chronic diseases are likely to result from a combination of environmental exposures to chemical and physical stressors and of human genetics. There is growing realization that it is environmental factors, rather than inherited characteristics that play a major role in disease causation [1]; but the information on environmental factors is often incomplete or inaccurate, so estimation of overall risks associated with these factors is severely hindered and important associations can go undetected. This limitation has recently been framed within the context of the *exposome* (see Glossary), the environmental counterpart of the genome. The concept of the exposome refers to the totality of environmental exposures from conception onwards, including its external (exogenous environmental agents) and internal components (endogenous cellular processes that give rise to altered stasis or function) [2–5] (Figure 1). Unravelling the causes of chronic human diseases and accurately assessing the risks are essential to the goal of effective disease prevention and require a broad perspective that encompasses many health, environmental, educational, socioeconomic, and political factors [6–9].

Electrophiles have long been suspected in the causality of cancer and other chronic diseases. Because they are reactive they can be measured indirectly through the *adducts* they form with protein and DNA. The exposome concept encompasses a "top-down" approach to identifying environmental factors that determine susceptibility to disease throughout the entire lifespan. In parallel, a "bottom-up" approach can investigate biomarkers specific for certain environmental exposures, based on the knowledge of environmental carcinogens and their pathways of metabolic activation involved in the formation of DNA or protein adducts. As part of this

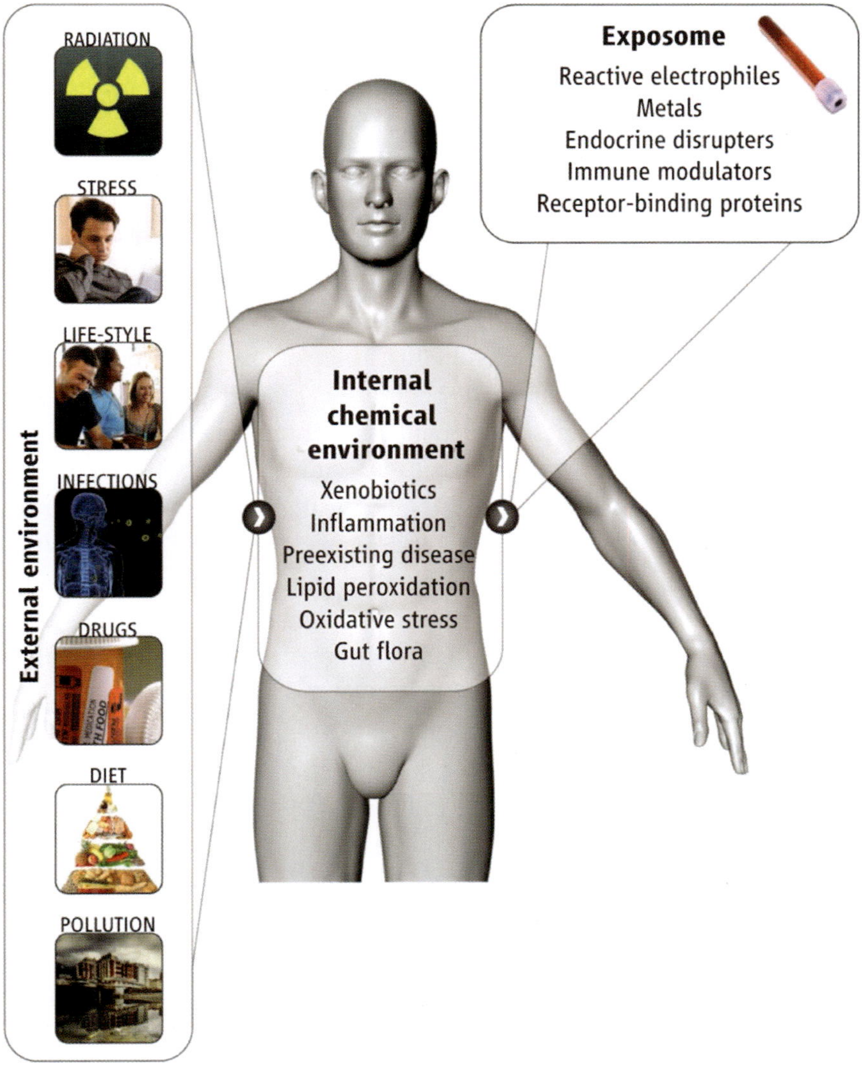

Figure 1. Characterizing the exposome. The exposome represents the combined exposures from all sources that reach the internal chemical environment. Toxicologically important classes of exposome chemicals are shown. Signatures and biomarkers can detect these agents in blood or serum.

Reproduced from Rappaport and Smith (2010) [2] with permission.

approach, *adductomics* constitutes the untargeted investigation of modification to endogenous macromolecules by endogenous or exogenous agents.

Application of the exposome concept to investigating environmental impacts on human health has already been adopted in two Europe-wide collaborative projects. These studies aim to assess environmental exposure at personal and population levels within existing short- and long-term population studies. The *EXPOsOMICS* project has focused on the measurement and impact of air and water pollution, studied in a number of adult and child populations [10]. The *HELIX* project has concentrated on early-life events, examining exposure to a range of chemicals and physical agents in existing European birth cohorts [11]. Both projects combine exposure monitoring, using mobile and static monitors, smartphone, and satellite data, with omics techniques to determine biomarkers associated with exposures. The methodologies include *metabolome*, *proteome*, *transcriptome,* and *epigenome* profiles and, in the case of *EXPOsOMICS*, protein adductomics [10]. While many of the results of these interrelated analyses have yet to emerge, it is anticipated that new insights into the importance of environmental factors in the etiology of human diseases will ensue and that these studies will point the way to improved strategies for monitoring human exposures and their health consequences.

Broader ecological issues may also be encompassed within the exposome concept. The adverse outcome pathway (AOP) concept seeks to define the initial molecular events that culminate in adverse (toxicological) endpoints [12], as a means to refine hazard identification and risk assessment. It has been proposed that considering the relative contributions of components of the exposome in relation to complex mixtures combined with mechanistic understanding of the induced adverse effects may improve the integrated risk assessment for both human and environmental health [13].

2. DNA Adductomics

In the years following the Second World War, the commercial availability of radioisotopes made it possible for chemicals of interest to be radiolabeled, providing a sensitive means of following the fate of those chemicals in biological systems. This led to the detection of covalent binding of chemical carcinogens to cellular macromolecules in experimental models, and from these studies it became established that the majority of carcinogens are genotoxic, exerting their biological activity through *modification of DNA* and leading to mutations when the DNA replicates. Mutations are a feature of tumors, whose growth is driven by the altered function of critical genes for cellular stasis.

The reliance on radiolabeled chemicals limited the early studies to experimental systems and precluded the investigation of the formation of carcinogen-DNA adducts in human tissues. However, since the early 1980s a number of techniques have been refined that are capable of detecting and identifying carcinogen-DNA adducts in human tissues without the need for radiolabeled test compounds. These included antibody-based methods, whereby monoclonal or rabbit antibodies to synthesized DNA adducts or modified DNA were employed in immunohistochemistry, immunoassay, or immunoaffinity procedures to allow detection of DNA adducts in human DNA [14, 15]. Similar antibody-based methods have also been used with modified proteins [16, 17]. Additional approaches comprised physicochemical methods, including mass spectrometry and fluorescence spectroscopy [18]. All of these methods require the use of DNA or protein adduct standards and can be considered to be targeted, "bottom-up" approaches to adduct analysis, that is, the approach involves the identification of an adduct of interest with which to form the basis of a method for its detection. A more "top-down" approach that does not require prior knowledge of the structure of a DNA adduct in order to detect it, is the ^{32}P-postlabeling assay [19]. Using this approach, DNA adducts are detected by enzymatic addition of a phosphate group, which contains the radioactive ^{32}P isotope. However, while the method is very sensitive and suitable for the detection of multiple adducts arising from exposures to complex mixtures, limited structural information can be obtained without synthetic standards, and identification is largely dependent in the demonstration of cochromatography rather than more definitive or unequivocal characterization.

Mass spectrometry (see Table 1) is considered the "gold standard" for DNA adduct analysis [20], as it is the only method capable of delivering unequivocal chemical characterization of a specific adduct. It is the method of choice for the newer adductome approach to "top down" detection of DNA damage in human tissues, and its sensitivity has

Table 1. Summary of mass spectrometry approaches to adductomics.

Material	Method	Abbreviation	References
DNA	Electrospray ionization tandem mass spectrometry	LC/ESI-MS/MS	[22, 23]
DNA	Electrospray ionization tandem mass spectrometry with selected reaction monitoring	LC/ESI-MS/MS SRM	[24]
DNA	Isotope-dilution ultra-performance liquid chromatography tandem mass spectrometry	UPLC-MS/MS	[25]
DNA	Nanoelectrospray coupled with Orbitrap high resolution mass spectrometry	LC-HRMS	[26]
DNA	HPLC hybrid linear ion trap triple quadrupole mass spectrometry	HPLC Q-TRAP MS	[27]
Protein — hemoglobin	Liquid chromatography tandem mass spectrometry with electrospray ionization	LC-MS/MS	[30–33]
Protein — albumin	Micro liquid chromatography tandem mass spectrometry	Micro-LC-MS/MS	[35]
Protein — albumin	Fixed step selected reaction monitoring with triple quadrupole mass spectrometry	FS-SRM with TQ-MS	[36, 37, 39]
Protein — albumin	Liquid chromatography with high resolution Orbitrap mass spectrometry	LC-HRMS	[38, 41]

become comparable to that of ^{32}P-postlabeling in some situations [21]. Methodological improvements and advances in instrumentation will drive limits of detection lower still.

One of the earliest ventures into this genre was the use of liquid chromatography with electrospray ionization tandem mass spectrometry (LC/ESI-MS/MS) to analyze for multiple adducts in human lung DNA [22]. The strategy was to detect the neutral loss of 2′-deoxyribose (dR) (with a molecular ion mass of 116 amu) from positively ionized 2′-deoxynucleoside adducts (see Figure 2) and it resulted in a total of 374 transitions in the mass range m/z 228.8 to 602.8. The analysis required 100 mg

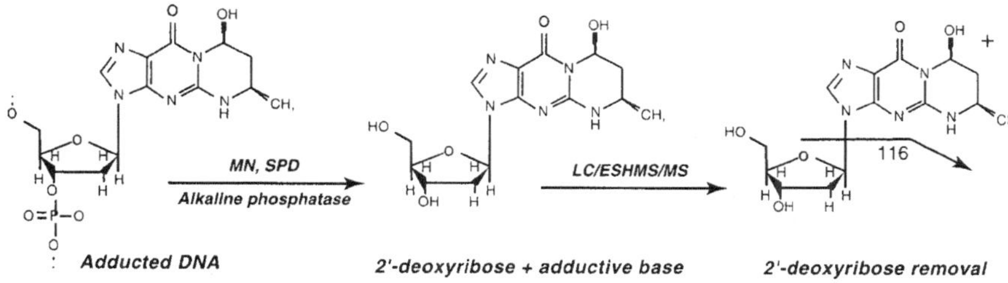

Figure 2. Depiction of the general process for the untargeted detection of multiple DNA adducts by LC/ESI-MS/MS. DNA is digested to nucleosides. Mass spectrometry screens for $[M + H]^+ \rightarrow [M + H - 116]^+$ transitions. In this scheme, 1,N^2-PdG_1 is used as an example of a DNA adduct.

Reproduced from Kanaly *et al.* (2006) [22] with permission.

of DNA and involved 12 injections of material into the mass spectrometer, each with a total run time of 28 minutes. DNA from a smoker's lung was compared with that of a nonsmoker; of the 20 most abundant adducts, 11 were common to both samples. Quantitatively, these 20 accounted for well over half of the total adduct levels, comprising 778 adducts in the case of the nonsmoker and 677 in the case of the smoker (see Figure 3). Considering the 200 most abundant adducts, 142 of these were common to both samples. Some adducts were identified through stable isotope dilution and comparison with authentic standards: 1,N^2-PdG_1 (α-*S*-methyl-γ-hydroxy-1,N^2-propano-2′-deoxyguanosine), 1,N^2-PdG_2 (α-*R*-methyl-γ-hydroxy-1,N^2-propano-2′-deoxyguanosine), N^2-ethyl-dG (N^2-ethyl-2′-deoxyguanosine), and εdA (1,N^6-etheno-2′-deoxyadenosine).

In a subsequent study [23], the same methodology was applied to a comparison of the putative adducts in two lung tissue samples and one esophageal sample from a single human individual. Considering the 50 most abundant adducts (out of more than 1000 adducts detected overall), the two lung samples shared 90% similarity, while the esophagus sample was 84% similar to the peripheral lung sample and 80% similar to the central lung sample. In addition to the four adducts identified previously [22], a further three were characterized in this study, all products of reactive oxygen species (ROS): 8-OH-PdG [3-(2′-deoxyribosyl)-5,6,7,8,-tetrahydro-8-hydroxy-pyrimido[1,2-*a*]purine-(3*H*)-one] and two isomers

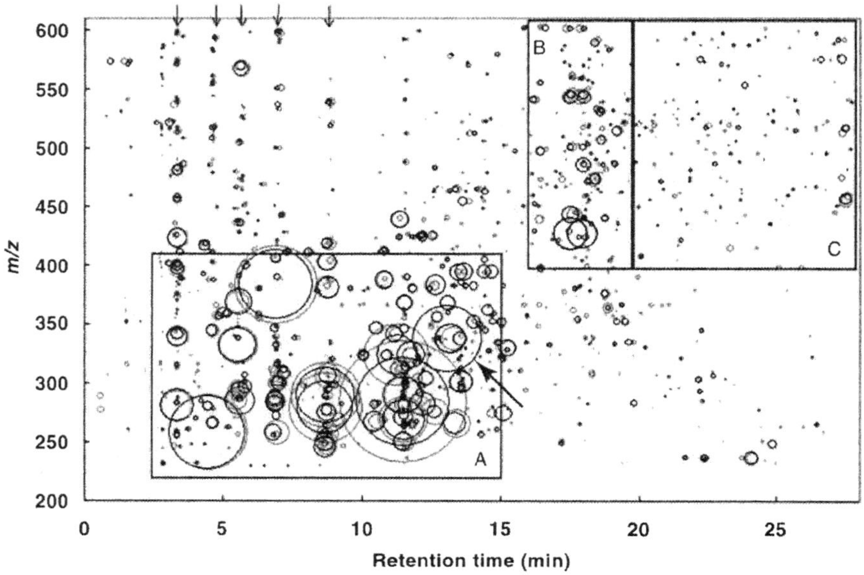

Figure 3. Adductome map of putative DNA adducts detected in human lung tissue from a nonsmoker (light circles) and a smoker (dark circles). The size of the circles reflects the magnitude of the signal. In all, there are 374 transitions in the mass range from *m/z* 228.8 to *m/z* 602.8. The arrow indicates a putative adduct + that was detected in the smoker at 4.8 times the level in the nonsmoker.

Reproduced from Kanaly *et al.* (2006) [22] with permission.

of 6-OH-PdG [3-(2′-deoxyribosyl)-5,6,7,8,-tetrahydro-6-hydroxy-pyrimido[1,2-*a*]purine-(3*H*)-one].

Thus, the key characteristic of adductome analysis is the objective to detect *multiple* adducts in samples, as this is a truer reflection of the reality of human exposure. These early untargeted explorations of the adductome demonstrate the breadth and complexity of the endogenous and exogenous DNA damage residing in human tissues, but they required rather large amounts of human material and prolonged instrument time, both of which put limitations on the application of the method to population studies.

In a method developed for *targeted* adductomic detection of multiple DNA adducts formed by polycyclic aromatic hydrocarbons (PAHs), Singh *et al.* [24] used LC/ECI-MS/MS with selected reaction monitoring (SRM) to detect the neutral loss of dR (116 amu) from a series of PAH-modified DNA samples. DNA adducts derived from the *anti*-diol-epoxides of benzo[*a*]pyrene, benzo[*b*]fluoranthene, dibenzo[*a,l*]pyrene, and dibenz[*a,h*] anthracene displayed the expected fragmentation pattern, although one of the adducts deriving from dibenzo[*a,l*]pyrene, that formed at the N^6-position of adenine, showed only the loss of the entire PAH moiety. The study demonstrates the feasibility of mass spectrometric methods for simultaneously detecting different adducts when present together in a DNA sample.

Another approach to the simultaneous detection of multiple adducts used isotope-dilution ultra-performance liquid chromatography tandem MS/MS (UPLC-MS/MS) [25]. In this case, the instrumentation was a linear tandem quadrupole mass spectrometer with an electrospray interface operated in the positive ion mode. The procedure is based on detection and quantification of 16 different DNA adducts formed by environmental carcinogens, including benzo[a]pyrene, 1-methylpyrene, 4-aminobiphenyl, 2-amino-1-methyl-6-phenylimidazo[4,5-*b*]pyridine (PhIP), malondialdehyde, and others. When the method was applied to 10 lung tumor samples, which included both smokers and nonsmokers, most of these adducts were not detectable. However, two that were found were those formed by furfuryl alcohol and methyleugenol. Also detected were some adducts of endogenous origin formed from lipid peroxidation products, namely 3,N^4-etheno-2′-deoxycytidine (εdC) and 1,N^6-etheno-2′-deoxyadenosine (εdA).

A third study that has explored the simultaneous detection of multiple adducts has used a high-resolution mass spectrometry approach, with a nanoelectrospray coupled to an Orbitrap mass spectrometer [26]. The method was optimized for a mixture of 18 different DNA adduct standards, including some of those already mentioned, and it was also used to analyze DNA from liver tissue of mice exposed to 4-(methylnitrosamino)-1-(3-pyridyl)-1-butanone (NNK), a tobacco-specific nitrosamine. This experiment resulted in the detection of several adducts in the mouse tissue, some expected and some unexpected, which illustrates the potential of the untargeted "top-down" adductomic approach to identify new adducts, not simply those for which a preconceived expectation has been raised.

Yao *et al.* [27] have used a nontargeted approach to investigate the formation of DNA adducts in cultured mouse ovarian follicles exposed to benzo[*a*]pyrene and to cigarette smoke condensate. Using high pressure liquid chromatography (HPLC) and hybrid linear ion trap triple quadrupole mass spectrometry (Q-Trap) to monitor for the neutral loss of dR from DNA adducts, they detected the dose-dependent formation of three adducts in the benzo[*a*]pyrene-treated cells, *r*-7,*t*-8,*t*-9-trihydroxy-*c*-10-(N^2-deoxyguanosyl)-7,8,9,10-tetrahydrobenzo[*a*]pyrene (BPdG), benzo[*a*]pyrene-7,8-quinone-deoxyguanosine (BPQ-dG) and phenathrene-1,2-quinone-deoxyguanosine (PheQ-dG). When cultures treated with cigarette smoke condensate were analyzed, these same three adducts were detected, plus a fourth, identified as 4-aminobiphenyl-dG. It was postulated that the phenanthrene-derived adducts formed in benzo[*a*]pyrene-treated cultures were the result of benzo[*a*]pyrene biodegradation, but those formed following exposure to cigarette smoke condensate were the result of metabolic activation of phenanthrene itself [27].

As a research field, DNA adductomics is new and relatively undeveloped [28]. Its emergence and growth is being driven largely by significant increases in sensitivity and specificity of mass spectrometry. While triple quadrupole instrumentation offers high sensitivity, simplicity, and relatively low cost, high-resolution instrumentation, such as Orbitrap, provides accurate mass measurement, including the accurate mass monitoring of the neutral loss of the dR moiety (116.0474 amu), which increases the specificity of identification of putative adducts and eliminates artefacts. Most studies thus far have been designed to develop methods and provide proof-of-principle using semi-targeted approaches rather than a true "top-down" untargeted approach. Those studies that have taken this latter approach, by Kanaly *et al.* for example [22, 23], reveal a highly complex pattern of adducts in human tissues. The challenge will be to identify the unknown adducts and to determine their biological significance in the etiology of chronic diseases [29]. A database of DNA adducts, characterized by mass spectrometry, will need to be established to aid this endeavor.

3. Protein Adductomics

Electrophiles that can react with DNA are also capable of reaction with proteins, which also have multiple nucleophilic sites. Reactivity may not be identical however and, in addition, other toxic processes besides mutagenesis and carcinogenesis may be associated with protein adducts, for example, allergy, ageing, and other degenerative conditions. In addition to leading to the identification of exogenous electrophiles as biomarkers of exposure or disease state, there is the opportunity to investigate perturbations of endogenous processes that produce electrophiles in the genesis of noncommunicative chronic diseases.

Advantages of monitoring proteins for adducts include the abundance of material that can be obtained from tissue banks; for example, red blood cells are an abundant source of hemoglobin and blood plasma or serum is an abundant source of albumin. These two proteins are thus the main focus of adductomic approaches with proteins. The lifespan of the proteins, approximately 120 days in the case of hemoglobin and 22–25 days in the case of albumin, means that there is a substantial "capture period" for monitoring exposure to genotoxicants. Also, unlike DNA adducts, protein adducts are not subject to loss or removal through repair processes.

3.1. *Hemoglobin*

Studies on interactions between toxicants and hemoglobin have, to date, been more common than studies with albumin. Analysis of hemoglobin adducts involves observing modifications to the N-terminal valine (Val) of the protein [30]. This methodology is based on the FIRE procedure, which is a modified Edman degradation reaction where adducts to N-terminals in hemoglobin are selectively detached by the reagent fluorescein isothiocyanate (FITC) and analyzed as fluorescein thiohydantoin (FTH) derivatives by LC-MS/MS with positive ionization. It is applicable to a wide range of electrophiles, such as ethylene oxide (an epoxide) and acrylamide (an α,β-unsaturated carbonyl compound), but it does not work for all types of adduct, for example, ring-closed adducts such as the *N,N*-(2,3-dihydroxy-1,4-butadiyl) Val adduct formed by diepoxybutane.

In the first adductomic application of this approach, samples from smokers and nonsmokers were analyzed and compared [31]. In all samples, seven adducts at the *N*-terminal Val were identified; these were the addition of methyl and ethyl groups, and adducts formed by ethylene oxide, acrylonitrile, methyl vinyl ketone, acrylamide, and glycidamide; in addition, a further 19 unknown adducts were detected in all samples. Subsequently, one of these

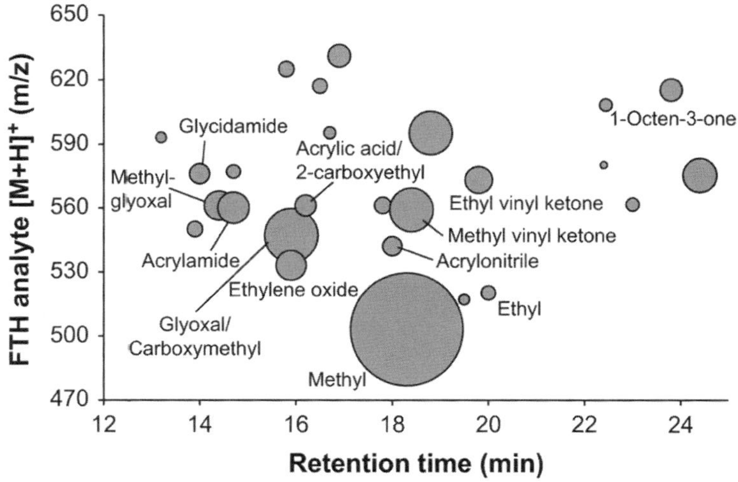

Figure 4. Adductome map showing the relative average levels of 26 detected adducts in 12 human blood samples. The estimated levels range from approx. 5–1100 pmol/g Hb, with the size of each circle corresponding to the relative level of the corresponding adduct. The 12 adducts or the corresponding precursors identified to date are written out in the figure. The retention times (min.) in the LC-MS analysis are shown on the x-axis, and the m/z of the adduct derivatives (precursor ions, $[M + H]^+$) on the y-axis.

Reproduced from Carlsson and Tornqvist (2016) [30] with permission.

unknown adducts has been identified as derived from ethyl methyl ketone [32] and a further four have been attributed to the precursor electrophiles glyoxal, methylglyoxal, acrylic acid, and 1-octen-3-one [33] (see Figure 4).

These studies demonstrate the potential of the adductomic (untargeted) approach to both validate the detectability of known electrophilic species and to discover new ones that, with application of a strategy for their characterization, can lead to the identification of novel exogenous or endogenous agents whose electrophilicity leads to protein damage and possibly DNA damage as well.

3.2. *Albumin*

Many carcinogens and toxicants have been demonstrated to form covalent adducts with albumin [34]. Albumin is the most abundant protein in human serum, with a mean concentration of 43 mg/mL. It contains

35 cysteine (Cys) residues, 34 of which are involved in disulfide bonds between pairs of the amino acid. The one remaining unpaired Cys is at position 34 and is thus a unique site for covalent modification by electrophiles. Eighty per cent of the covalent modification of albumin occurs at Cys34. Digestion of albumin with trypsin generates a 21-amino acid peptide (T3; ALVLIAFAQYLQQCPFEDHVK), which contains adducts of interest. To isolate albumin from serum, unwanted proteins (e.g., immunoglobulins) are salted out and low-molecular-weight solutes (e.g., metabolites) are removed *via* buffer exchange. The yield of albumin from 50 μL of serum is typically 2 mg.

An early study that demonstrated the utility of monitoring albumin for alkylated Cys involved exposure of human blood to ^{14}C-labeled sulfur mustard (the chemical warfare agent — mustard gas) [35]. Isolation and tryptic digestion of albumin produced the 21-amino acid fragment containing a sulfur mustard-Cys adduct, detected by micro-LC-MS/MS. A modified tripeptide (Cys-Pro-Phe) was also detected and characterized following pronase digestion. The method was used to analyze samples of blood from nine Iranians exposed to sulfur mustard during the Iran–Iraq war (1986). In all nine cases, the sulfur mustard-adducted tripeptide was detected.

In 2011, a protocol for human serum albumin (HSA) adductomics using fixed-step selected reaction monitoring (FS-SRM), a triple quadrupole mass spectrometry (TQ-MS) method, was published [36]. The authors detected modifications within a short sequence of amino acid residues in HSA (residues 27–41) *via* shifts in a sequence tag of the third-heaviest tryptic peptide ("T3"). Such a shift may be characterized by the mass of a putative "R" group (presumably attached to the sulfur atom of Cys-34). By devoting more time to fewer measurements (cf. conventional scanning), stepped methods should benefit from enhanced sensitivity, accuracy, and precision. These benefits are gained at the expense of resolution, meaning that the usefulness of FS-SRM comes not from an ability to identify particular adducts, but rather from the characterization of their distribution as a whole. Application of this method to analyses of archived plasma proteins that had been pooled according to subjects' ethnicities and tobacco smoking habits demonstrated differences between pools [36] and suggested that FS-SRM might be able to detect statistically significant differences between groups of individual samples that had not been

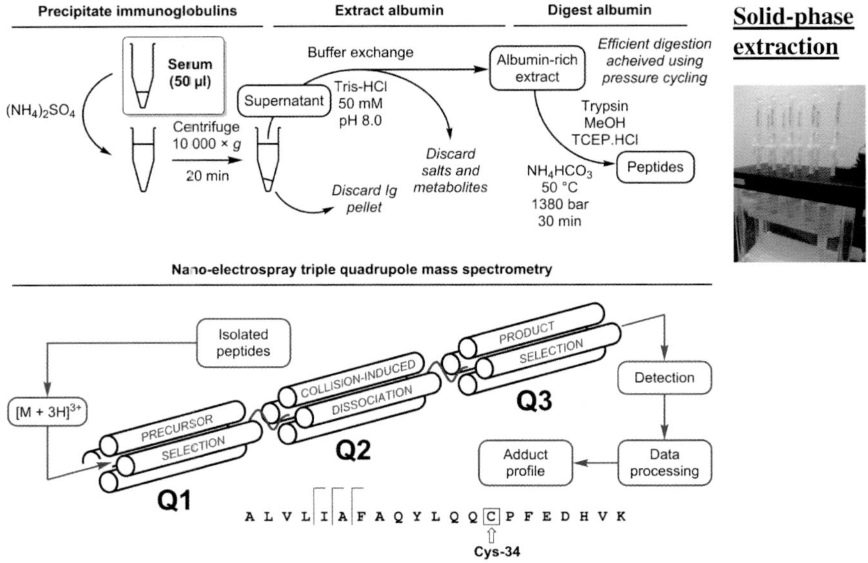

Figure 5. Workflow for analysis of adducts in a 21-amino acid peptide of albumin, adapted from Preston *et al.* (2017) [37] from the method originally described by Li *et al.* (2011) [36].

pooled. More recently, the method has been developed and validated for larger scale processing of samples, to make it suitable for molecular epidemiology studies, with quality control and assessment of reproducibility of a batch workflow of samples [37] (see Figure 5). Challenges posed by technical and biological variation were addressed in the processing and interpretation of the data. A pilot study of 20 smokers and 20 never-smokers provided evidence of an effect of smoking on levels of putative serum albumin adducts. Differences between smokers and never-smokers were most apparent in putative adducts with net gains in mass between 105 Da and 114 Da (relative to unmodified albumin). This provides the first example of FS-SRM being used to investigate associations between environmental factors and levels of putative HSA adducts in a quantitative fashion [37].

The FS-SRM approach has the scope for screening populations for difference in adduct spectra that may indicate difference in exposure or health status, but it does not provide accurate masses or facilitate annotation readily. In order to identify the structures and putative origins of adducts that might be detected by the FS-SRM approach, high-resolution mass spectrometry coupled with liquid chromatography as has been reported recently to be advantageous [38]. Further investigations of the effects of tobacco smoking revealed around 43 adduct features, some of which are positively associated with smoking and but also some that are negatively associated. The former result from genotoxic constituents of tobacco smoke, such as ethylene oxide and acrylonitrile, while the latter, which include Cys34 oxidation products and disulfides, may reflect alterations in the serum redox state of smokers, resulting in reduced adduct levels [38].

Another study that utilized FS-SRM approach focused on interactions between the Cys34 and ROS [39]. Chronic exposure to ROS is linked to many chronic diseases and in this study, a number of adducts originating from ROS were detected in human serum: Cys-SO_2H (sulfinic acid), Cys-SO_3H (sulfonic acid), and a proposed sulfinamide structure (a mono-oxygenated moiety also with the loss of two hydrogen atoms).

Antibody enrichment may pave the way to a more sensitive assay. A polyclonal antibody, raised against the T3 peptide but with cross-reactivity to the peptide containing adducts has been developed [40]. Using this approach, ten modified T3 peptides were detected in human plasma samples; eight of them were characterized and they included Cys34 oxidation products, modification involving loss of water or lysine, cysteinylation, and transpeptidation of arginine.

In a study of women from the Xuanwei and Fuyuan counties in China, where extensive use of smoky coal for heating and cooking has resulted in very high rates of lung cancer among nonsmokers, the adducts at the Cys34 locus of albumin were compared in 29 females who used smoky coal and 10 controls using other energy sources [41]. Fifty different modified T3 peptides were identified, including oxidation products, mixed disulfides, rearrangements, and truncations. Two peptides that were detected at significantly *lower* levels in the smoky coal group were adducts of *S*-glutathione (*S*-GSH) and *S*-γ-glutamylcysteine (*S*-γ-GluCys). The results are interpreted as showing evidence that exposure to the indoor combustion products results in depletion of GSH, an essential antioxidant, as well as its precursor γ-GluCys [41].

4. Prospects for the Future

A key consideration in the applicability of the new approaches described here is whether there is a readily available source of human material to investigate. In the case of DNA adductomics, the amount of DNA (from white blood cells) that is normally available from biobanks is limited; for the untargeted (but not so informative) technique of ^{32}P-postlabeling analysis, this amount of material can be sufficient, but for some of the mass spectrometry techniques, it may be inadequate at present. Nevertheless, with improvements in sensitivity, this constraint may yet be overcome. Another potential source of material is archival tissue, frequently in the form of formalin-fixed paraffin-embedded (FFPE) tissue blocks. While many such tissue samples have been previously found to be unsuitable, recent studies have shown that DNA suitable for mass spectrometric adduct analysis can be recovered from paraffin blocks [42, 43]. This was demonstrated in the detection of aristolochic acid-DNA adducts in FFPE mouse liver and kidney and in human kidney tissues [42], and then in a second study in fixed rodent tissues, adducts were detected from four known or putative human carcinogens representing four chemical classes — PAHs, aromatic amines, heterocyclic amines, and N-nitroso compounds [43]. It is probable that fixation for longer periods will progressively degrade the DNA and render it unrecoverable or unanalyzable, and thus only FFPE preserved tissues that have been processed promptly, with a minimum period of formalin fixation, will be usable [44].

For protein adductomics, the abundance of albumin in human serum and plasma, commonly obtained and stored in biobanks in milliliter quantities, makes this an attractive material for human biomonitoring, and its suitability is demonstrated in the studies described in this article. For hemoglobin, the requirement for white blood cells (as for DNA adductomics) means that the material in biobanks can be a sufficient source for protein adductomics studies. Alternatively dried blood spots can be a suitable sources of hemoglobin for investigation [45]. If obtained from neonatal blood spots (Guthrie spots), the material then provides a valuable opportunity for investigating exposures *in utero*. A single blood spot of about 50 mL is estimated to contain about 9.6 mg of protein, of which about 7.7 mg will be hemoglobin and 1.2 mg HSA [45].

There is also the potential to broaden the scope of adductomics by investigating novel modifiable loci in blood proteins. Cys-34 of HSA is undoubtedly important (see above), as is the N-terminus of hemoglobin, but the literature hints at wider reactivity within the blood proteome. Consequently, a hitherto-untapped source of analytes for protein adductomics can be envisaged. Mapping the loci at which adducts can form will be beneficial and for this purpose, new chemical tools will be required.

Adductomics is an emerging science. It is a component of the exposome concept still in its infancy that is attempting to capture the totality of all biologically relevant exposures, including electrophiles of endogenous as well as exogenous origin. Rapid advances in mass spectrometry instrumentation, with increasing sensitivity and improving resolution, are the driving forces for both DNA and protein adductomics. When coupled with other omics approaches, such as proteomics, transcriptomics, and metabolomics, all of which have potential for high-throughput screening of populations, a future can be envisaged in which it will be possible to capture snapshots of human exposure to genotoxicants and the resultant biological consequences at multiple stages throughout life. Such a comprehensive picture will, it is reasonable to expect, shed significant light on the causes and courses of chronic diseases afflicting humans. With such knowledge will surely come new opportunities for early intervention to prevent the diseases before they occur and for improving the quality of life of millions of humans.

Acknowledgments

Research in the author's laboratory is funded by Cancer Research UK, the Wellcome Trust, the European Commission and the National Institute for Health Research. The views expressed are those of the author and not necessarily those of the National Health Service, the NIHR, the Department of Health or Public Health England.

Abbreviations

amu	atomic mass unit
AOP	adverse outcome pathway
dR	2′-deoxyribose
FFPE	formalin-fixed paraffin-embedded
FITC	fluorescein isothiocyanate
FTH	fluorescein thiohydantoin
GSH	glutathione
HPLC	high pressure (or performance) liquid chromato graphy
Hb	hemoglobin
HSA	human serum albumin
LC/ESI-MS/MS	liquid chromatography with electrospray ionization tandem mass spectrometry
NNK	4-(methylnitrosamino)-1-(3-pyridyl)-1-butanone
PhIP	2-amino-1-methyl-6-phenylimidazo[4,5-*b*]pyridine
PAH(s)	polycyclic aromatic hydrocarbon(s)
ROS	reactive oxygen species
SRM	selective reaction monitoring TQ-MS — triple quadrupole mass spectrometry
UPLC-MS/MS	ultra-performance liquid chromatography tandem mass spectrometry

DNA Adduct Abbreviations

1,N^2-PdG_1	α-*S*-methyl-γ-hydroxy-1,N^2-propano-2′-deoxyguanosine
1,N^2-PdG_2	α-*R*-methyl-γ-hydroxy-1,N^2-propano-2′-deoxyguanosine
N^2-ethyl-dG	N^2-ethyl-2′-deoxyguanosine
εdA	1,N^6-etheno-2′-deoxyadenosine
8-OH-PdG	3-(2′-deoxyribosyl)-5,6,7,8,-tetrahydro-8-hydroxy-pyrimido [1,2-*a*]purine-(3*H*)-one
6-OH-PdG	3-(2′-deoxyribosyl)-5,6,7,8,-tetrahydro-6-hydroxy-pyrimido [1,2-*a*]purine-(3*H*)-one

εdC	3,N^4-etheno-2′-deoxycytidine
BPdG	*r*-7,*t*-8,*t*-9-trihydroxy-*c*-10-(N^2-deoxyguanosyl)-7,8,9,10-tetrahydrobenzo[*a*]pyrene
BPQ-dG	benzo[*a*]pyrene-7,8-quinone-deoxyguanosine
PheQ-dG	phenathrene-1,2-quinone-deoxyguanosine

References

1. Rappaport SM. Genetic factors are not the major causes of chronic diseases. *PLoS One* 2016; 11: e0154387.
2. Rappaport SM, Smith MT. Epidemiology. Environment and disease risks. *Science* 2010; 330: 460–461.
3. Wild CP. Complementing the genome with an "exposome": the outstanding challenge of environmental exposure measurement in molecular epidemiology. *Cancer Epidemiol Biomarkers Prev* 2005; 14: 1847–1850.
4. Wild CP. The exposome: from concept to utility. *Int J Epidemiol* 2012; 41: 24–32.
5. Rappaport SM, Barupal DK, Wishart D, Vineis P, Scalbert A. The blood exposome and its role in discovering causes of disease. *Environ Health Perspect* 2014; 122: 769–774.
6. Rappaport SM. Biomarkers intersect with the exposome. *Biomarkers* 2012; 17: 483–489.
7. Dennis KK, Marder E, Balshaw DM, Cui Y, Lynes MA, Patti GJ, *et al.* Biomonitoring in the era of the exposome. *Environ Health Perspect* 2017; 125: 502–510.
8. Stewart BW, Bray F, Forman D, Ohgaki H, Straif K, Ullrich A, *et al.* Cancer prevention as part of precision medicine: 'plenty to be done'. *Carcinogenesis* 2016; 37: 2–9.
9. Wild CP, Scalbert A, Herceg Z. Measuring the exposome: a powerful basis for evaluating environmental exposures and cancer risk. *Environ Mol Mutagen* 2013; 54: 480–499.
10. Vineis P, Chadeau-Hyam M, Gmuender H, Gulliver J, Herceg Z, Kleinjans J, *et al.* (2016) The exposome in practice: design of the EXPOsOMICS project. *Int J Hyg Environ Health* 2016.
11. Vrijheid M, Slama R, Robinson O, Chatzi L, Coen M, van den Hazel P, *et al.* The human early-life exposome (HELIX): project rationale and design. *Environ Health Perspect* 2014; 122: 535–544.

12. Ankley GT, Bennett RS, Erickson RJ, Hoff DJ, Hornung MW, Johnson RD, *et al.* Adverse outcome pathways: a conceptual framework to support ecotoxicology research and risk assessment. *Environ Toxicol Chem* 2010; 29: 730–741.
13. Escher BI, Hackermüller J, Polte T, Scholz S, Aigner A, Altenburger R, *et al.* From the exposome to mechanistic understanding of chemical-induced adverse effects. *Environ Int*: 97–106.
14. Poirier MC. Linking DNA adduct formation and human cancer risk in chemical carcinogenesis. *Environ Mol Mutagen* 2016; 57: 499–507.
15. Santella RM. Immunological methods for detection of carcinogen-DNA damage in humans. *Cancer Epidemiol Biomarkers Prev* 1999; 8: 733–739.
16. Santella RM, Lin CD, Dharmaraja N. Monoclonal antibodies to a benzo[a] pyrene diolepoxide modified protein. *Carcinogenesis* 1986; 7: 441–444.
17. Chen CJ, Wang LY, Lu SN, Wu MH, You SL, Zhang YJ, *et al.* Elevated aflatoxin exposure and increased risk of hepatocellular carcinoma. *Hepatology* 1996; 24: 38–42.
18. Poirier MC, Santella RM, Weston A. Carcinogen macromolecular adducts and their measurement. *Carcinogenesis* 2000; 21: 353–359.
19. Phillips DH, Arlt VM. The ^{32}P-postlabeling assay for DNA adducts. *Nat Protoc* 2007; 2: 2772–2781.
20. Farmer PB, Brown K, Tompkins E, Emms VL, Jones DJ, Singh R, *et al.* DNA adducts: mass spectrometry methods and future prospects. *Toxicol Appl Pharmacol* 2005; 207: 293–301.
21. Klaene JJ, Sharma VK, Glick J, Vouros P. The analysis of DNA adducts: the transition from (32)P-postlabeling to mass spectrometry. *Cancer Lett* 2013; 334: 10–19.
22. Kanaly RA, Hanaoka T, Sugimura H, Toda H, Matsui S, Matsuda T. Development of the adductome approach to detect DNA damage in humans. *Antioxid Redox Signal* 2006; 8: 993–1001.
23. Kanaly RA, Matsui S, Hanaoka T, Matsuda T. Application of the adductome approach to assess intertissue DNA damage variations in human lung and esophagus. *Mutat Res* 2007; 625: 83–93.
24. Singh R, Teichert F, Seidel A, Roach J, Cordell R, Cheng MK, *et al.* Development of a targeted adductomic method for the determination of polycyclic aromatic hydrocarbon DNA adducts using online column-switching liquid chromatography/tandem mass spectrometry. *Rapid Commun Mass Spectrom* 2010; 24: 2329–2340.
25. Monien BH, Schumacher F, Herrmann K, Glatt H, Turesky RJ, Chesne C. Simultaneous detection of multiple DNA adducts in human lung samples by isotope-dilution UPLC-MS/MS. *Anal Chem* 2015; 87: 641–648.

26. Balbo S, Hecht SS, Upadhyaya P, Villalta PW. Application of a high-resolution mass-spectrometry-based DNA adductomics approach for identification of DNA adducts in complex mixtures. *Anal Chem* 2014; 86: 1744–1752.
27. Yao C, Foster WG, Sadeu JC, Siddique S, Zhu J, Feng YL. Screening for DNA adducts in ovarian follicles exposed to benzo[a]pyrene and cigarette smoke condensate using liquid chromatography-tandem mass spectrometry. *Science Total Environ* 2017; 575: 742–749.
28. Balbo S, Turesky RJ, Villalta PW. DNA adductomics. *Chem Res Toxicol* 2014; 27: 356–366.
29. Villalta PW, Balbo S. The future of DNA adductomic analysis. *Int J Mol Sci* 2017; 18.
30. Carlsson H, Tornqvist M. An adductomic approach to identify electrophiles in vivo. *Basic Clin Pharmacol Toxicol* 2016; 121: 44–54. doi:10.1111/bcpt.12715
31. Carlsson H, von Stedingk H, Nilsson U, Tornqvist M. LC-MS/MS screening strategy for unknown adducts to N-terminal valine in hemoglobin applied to smokers and nonsmokers. *Chem Res Toxicol* 2014; 27: 2062–2070.
32. Carlsson H, Motwani HV, Osterman Golkar S, Tornqvist M. Characterization of a hemoglobin adduct from ethyl vinyl ketone detected in human blood samples. *Chem Res Toxicol* 2015; 28: 2120–2129.
33. Carlsson H, Tornqvist M. Strategy for identifying unknown hemoglobin adducts using adductome LC-MS/MS data: Identification of adducts corresponding to acrylic acid, glyoxal, methylglyoxal, and 1-octen-3-one. *Food Chem Toxicol* 2016; 92: 94–103.
34. Sabbioni G, Turesky RJ. Biomonitoring human albumin adducts: the past, the present, and the future. *Chem Res Toxicol* 2017; 30: 332–366.
35. Noort D, Hulst AG, de Jong LP, Benschop HP. Alkylation of human serum albumin by sulfur mustard in vitro and in vivo: mass spectrometric analysis of a cysteine adduct as a sensitive biomarker of exposure. *Chem Res Toxicol* 1999; 12: 715–721.
36. Li H, Grigoryan H, Funk WE, Lu SS, Rose S, Williams ER, *et al.* Profiling Cys34 adducts of human serum albumin by fixed-step selected reaction monitoring. *Mol Cell Proteomics* 2011; 10: M110 004606.
37. Preston GW, Plusquin M, Sozeri O, van Veldhoven K, Bastian L, Nawrot TS, *et al.* Refinement of a methodology for untargeted detection of serum albumin adducts in human populations. *Chem Res Toxicol* 2017; 30: 2120–2129.
38. Grigoryan H, Edmands W, Lu SS, Yano Y, Regazzoni L, Iavarone AT, *et al.* Adductomics pipeline for untargeted analysis of modifications to Cys34 of human serum albumin. *Anal Chem* 2016; 88: 10504–10512.

39. Grigoryan H, Li H, Iavarone AT, Williams ER, Rappaport SM. Cys34 adducts of reactive oxygen species in human serum albumin. *Chem Res Toxicol* 2012; 25: 1633–1642.
40. Chung MK, Grigoryan H, Iavarone AT, Rappaport SM. Antibody enrichment and mass spectrometry of albumin-Cys34 adducts. *Chem Res Toxicol* 2014; 27: 400–407.
41. Lu SS, Grigoryan H, Edmands WM, Hu W, Iavarone AT, Hubbard A, *et al.* Profiling the serum albumin Cys34 adductome of solid fuel users in Xuanwei and Fuyuan, China. *Environ Sci Technol* 2017; 51: 46–57.
42. Yun BH, Rosenquist TA, Nikolic J, Dragicevic D, Tomic K, Jelakovic B, *et al.* Human formalin-fixed paraffin-embedded tissues: an untapped specimen for biomonitoring of carcinogen DNA adducts by mass spectrometry. *Anal Chem* 2013; 85: 4251–4258.
43. Guo J, Yun BH, Upadhyaya P, Yao L, Krishnamachari S, Rosenquist TA, *et al.* Multiclass carcinogenic DNA adduct quantification in formalin-fixed paraffin-embedded tissues by ultraperformance liquid chromatography-tandem mass spectrometry. *Anal Chem* 2016; 88: 4780–4787.
44. Hewer A, Phillips DH. Effect of tissue fixation on recovery of DNA adducts in the ^{32}P-postlabelling assay. In *Postlabelling Methods for Detection of DNA Adducts,* Phillips DH, Castegnaro M, Bartsch H, eds., IARC, Lyon, 1993, pp. 211–214.
45. Funk WE, Waidyanatha S, Chaing SH, Rappaport SM. Hemoglobin adducts of benzene oxide in neonatal and adult dried blood spots. *Cancer Epidemiol Biomarkers Prev* 2008; 17: 1896–1901.

Glossary

Exposome: The environmental counterpart of the genome. The concept of the exposome refers to the totality of environmental exposures from conception onwards, including its external (exogenous environmental agents) and internal components (endogenous cellular processes that give rise to altered stasis or function).

Electrophile: A reagent attracted to electrons. Electrophiles are positively charged or neutral species having vacant orbitals that are attracted to an electron rich center. It participates in a chemical reaction by accepting an electron pair in order to bond to a nucleophile. For adduct formation, cellular macromolecules such as DNA, RNA, and protein have multiple sites that are nucleophilic, and opposites attract each other.

Adduct: An addition product. Generally refers to a chemical modification to a cellular macromolecule (DNA, RNA, protein), where a chemical or a chemical group is covalently attached (bound) to a base (in a nucleic acid) or an amino acid (in a protein).

Adductomics: The untargeted analysis of all the chemically modified DNA, RNA, or protein components in a cell, tissue or organism.

EXPOsOMICS: A collaborative project funded by the European Union with a multinational team that aims to predict individual disease risk related to the environment, by developing a new approach to environmental exposure and health assessment during critical periods of life. http://www.exposomicsproject.eu/

HELIX: A collaborative project funded by the European Union. Its "early-life exposome" approach involves linking the combined environmental hazards to which mothers and children are exposed, to the health, growth, and development of children. http://www.projecthelix.eu/

Metabolome: The total number of metabolites present within an organism, cell, or tissue

Proteome: The entire complement of proteins that is or can be expressed by a cell, tissue, or organism

Transcriptome: The sum total of mRNA transcripts expressed by an organism

Epigenome: The sum total of heritable alterations in a cell that are not due to modifications of the DNA sequence. It can include DNA methylation and histone modification, which alter DNA accessibility and chromatin structure, thereby regulating or altering patterns of gene expression.

Modification of DNA: A change to the structure or integrity of DNA, caused by formation of an adduct, loss of a base or a break in the sugar-phosphate backbone. Such lesions may be premutagenic, that is, if unrepaired, they may lead to errors in DNA replication whereby mutation (an altered base sequence, a loss of a nucleotide or nucleotides from the sequence, or a larger scale rearrangement of the DNA sequence) may occur.

Index